DEPARTMENT OF THE INTERIOR

BULLETIN

OF THE

UNITED STATES

GEOLOGICAL SURVEY

No. 182

SERIES A, ECONOMIC GEOLOGY, 12

WASHINGTON
GOVERNMENT PRINTING OFFICE
1901

UNITED STATES GEOLOGICAL SURVEY

CHARLES D. WALCOTT, DIRECTOR

A REPORT

ON THE

ECONOMIC GEOLOGY

OF THE

SILVERTON QUADRANGLE, COLORADO

BY

FREDERICK LESLIE RANSOME

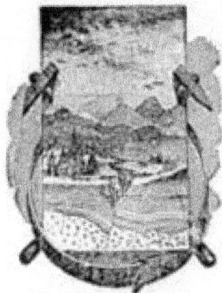

WASHINGTON

GOVERNMENT PRINTING OFFICE

1901

CONTENTS.

5

CONTENTS.

ILLUSTRATIONS.

11

BAKERS PARK AND THE TOWN OF SILVERTON, FROM THE SOUTH

On the left is Anvil Mountain, between Mineral and Cement creeks; in the background is Boulder Mountain, between Cement Creek and the Animas River. The three streams unite at Silverton and enter the gorge of the Animas near the lower right-hand corner of the picture. Photograph by Whitman Cross.

ECONOMIC GEOLOGY OF THE SILVERTON QUADRANGLE, COLORADO.

By F. L. RANSOME.

PART I. GENERAL DESCRIPTION AND DISCUSSION OF THE ORE DEPOSITS.

INTRODUCTION AND ACKNOWLEDGMENTS.

The field work upon which this report is based occupied the season of 1899 from the beginning of July to the middle of October. During that summer nearly all of the area was examined and most of the mines and prospects were visited.

The work was resumed in the autumn of 1900, with Mr. Alfred M. Rock as assistant, a period of six weeks in August, September, and October being devoted to those portions of the quadrangle that had not been thoroughly studied in the preceding year and to an investigation of such mining developments as had been effected during the winter.

It was at one time hoped that the report of Mr. Whitman Cross, of this Survey, on the general geology of the Silverton quadrangle, forming the Silverton folio of the Geologic Atlas of the United States, might be published in advance of this investigation of the economic resources of the district, but owing to various causes his field work is not yet finished. I am therefore unable to use or to cite his results in their entirety, or to reproduce his geological map as an adjunct to this report. This is to be regretted, as both a thoroughly satisfactory treatment and a proper understanding of the ore deposits require a knowledge of the geological relationships of the rocks in which they lie. This want has been partly supplied by a brief preliminary sketch of the geology, kindly furnished by Mr. Cross. This is necessarily an imperfect outline, to be later replaced by his final results. To Mr. Cross I owe, also, the privilege of publishing a preliminary geological map (Pls. IV, V) of a portion of the area. This map is based upon field work done by himself and his assistants. The geological

13

boundaries shown are not final and will undoubtedly be somewhat changed when the field work in this region is completed.

To Mr. S. F. Emmons, geologist in charge of the investigation of metalliferous deposits, my thanks are due for various helpful suggestions and criticisms and for access to his unpublished notes on the Yankee Girl and Guston mines.

The descriptions of the abandoned mines of the Red Mountain district would be far more imperfect than they are were it not for much information courteously supplied by Mr. Bedford McNeill, of London, liquidator of the New Guston Company; Mr. T. E. Schwarz, of Denver, and Mr. Otto J. Schulz, of St. Louis, trustee of the Genesee-Vanderbilt mines.

To the mining men of the region I am indebted for much cordial assistance and for courtesies which have many times been recalled with pleasure.

GEOGRAPHICAL POSITION.

The Silverton quadrangle, embracing one-sixteenth of a square degree of the earth's surface, lies between the meridians 107° 30′ and 107° 45′ west longitude and the parallels 37° 45′ and 38° 00′ north latitude. On the west it adjoins the Telluride quadrangle and on the south the Needle Mountain quadrangle. Its area is approximately 235.3 square miles. It lies in southwestern Colorado, in the highest and most rugged portion of the San Juan Mountains, including a part of the continental divide. There are comprised within its boundaries the headwaters of the Animas River and portions of the headwaters of the Rio Grande, Gunnison, Uncompahgre, and San Miguel rivers. The principal settlement is Silverton, a town of about 900 inhabitants. The position of the Silverton quadrangle with reference to the adjacent map units of the Geological Survey is shown in the index map, Pl. II.

LITERATURE.

The published literature relating to that portion of the San Juan Mountains included within the Silverton quadrangle is not voluminous. It is chiefly comprised in the official reports of the members of the Hayden survey, who visited the region in 1874, 1875, and 1876; in descriptive notes of various travelers through the San Juan; in scattered descriptions of mines and minerals occurring in the district, and in references to the history of mining development found in the more general works on Colorado and its resources. The following bibliography includes most of the important publications concerning the Silverton region. The data contained in them have been freely used in the preparation of the following historical sketch, for the purpose of supplementing information obtained at first hand from those who actually participated in the development of the district.

INDEX MAP SHOWING POSITION OF THE SILVERTON QUADRANGLE

Scale of miles

0   10   20   30   40   50   60   70   80   90   100

A. HOEN & CO. BALTIMORE

BIBLIOGRAPHY.

RHODA, F.   Report on the topography of the San Juan country: Hayden Survey Report for 1874, p. 451.

Contains a few notes on early history of the region.

ENDLICH, F. M.   Report on the San Juan region: Hayden Survey Report for 1874, pp. 185–240.

Contains brief notes on the mines as developed at that time. He divides the veins into two systems—one striking nearly northwest and southeast, and the other nearly northeast and southwest. The veins were formed at the close of the Cretaceous or beginning of the Tertiary.

ENDLICH, F. M.   Report on the San Juan region: Hayden Survey, Annual Report for 1875, pp. 176–191.

General notes on geology.

ENDLICH, F. M.   Report on the San Juan region: Hayden Survey, Annual Report for 1876, pp. 120–121.

Brief notes on mines and ores.

FOSSETT, F.   Colorado.   Denver, 1876.   2d ed., New York, 1880.

Contains much useful data, chiefly of a historical character, in regard to the early settlement and development of the San Juan region.

KOENIG, G. A.   On alaskaite, a new member from the series of bismuth sulphosalts: Proc. Am. Philos. Soc., Vol. XIX, 1881, pp. 472–477.

Describes alaskaite from Alaska mine, Poughkeepsie Gulch.

COMSTOCK, T. B.   Notes on the geology and mineralogy of San Juan County, Colorado: Trans. Am. Inst. Min. Eng., Vol. XI, 1882–83, pp. 165–191.

Discusses chiefly the character and distribution of the veins. Accompanied by a sketch map.

CROSS, WHITMAN.   A list of specially noteworthy minerals of Colorado: Proc. Colo. Sci. Soc., Vol. I, 1883–84, pp. 134–144.

Refers to occurrence of various minerals in the San Juan region.

SCHWARZ, T. E.   Remarks on occurrence of ore in mines near Silverton: Proc. Colo. Sci. Soc., Vol. I, 1883–84, p. 132.

Describes principal deposits of the Red Mountain district.

HILLEBRAND, W. F.   Mineralogical notes: Proc. Colo. Sci. Soc., Vol. I, 1883–84, pp. 121–123.

Describes zinkenite from Brobdignag mine.

HILLEBRAND, W. F.   On zunyite and guitermanite, two new minerals from Colorado: Proc. Colo. Sci. Soc., Vol. I, 1883–84, pp. 124–131.

HILLS, R. C.   Kaolinite from Red Mountain, Colorado: Am. Jour. Sci., 3d ser., Vol. XXVII, 1884, p. 472.

Describes kaolinite from National (American) Belle mine.

HILLS, R. C.   Extinct glaciers of the San Juan Mountains, Colorado: Am. Jour. Sci., 3d ser., Vol. XXVII, 1884, pp. 391–396. Also Proc. Colo. Sci. Soc., Vol. I, 1883–84, pp. 39–46.

Notes on extent and location of former glaciers of San Juan region.

COMSTOCK, T. B.   The distribution of San Juan County ores: Eng. and Min. Jour., Vols. XXXVIII and XXXIX, 1884–85.
A series of brief papers.

HILLEBRAND, W. F.   Miscellaneous mineral notes: Bull. U. S. Geol. Survey No. 20, 1885, pp. 89–99.
Describes kaolinite from National Belle mine and zinkenite from Brobdignag mine, on Red Mountain Range, near Chattanooga.

HILLEBRAND, W. F.   New mineral species from Colorado: Bull. U. S. Geol. Survey No. 20, 1885, pp. 100–107.
Describes zunyite and guitermanite from the Zuñi mine, Anvil Mountain, near Silverton.

EMMONS, S. F.   Geological sketch of the Rocky Mountain division: Tenth Census, Vol. XIII (Precious Metals), 1885, pp. 82–85.
A general sketch of the San Juan region by counties.   A table of mines, giving country rock, dip, and strike of vein and ore minerals.

EMMONS, S. F.   Notes on some Colorado ore deposits: Proc. Colo. Sci. Soc., Vol. II, 1885–1887, pp. 85–105.
San Juan region an excellent field for the study of fissure deposits.   Points out that so-called "geyser mounds" of the Red Mountain district are merely masses of country rock bleached and altered by solfataric action.

EMMONS, S. F.   On the origin of fissure veins: Proc. Colo. Sci. Soc., Vol. II, 1885–1887, p. 207.
Ascribes shape of Yankee Girl ore body to the formation of a shattered zone at intersection of three or more fissure planes.

COMSTOCK, T. B.   The metallurgy of San Juan County ores: Eng. and Min. Jour., Vol. XXXIX, 1885, pp. 69–71; also Vols. XLI and XLII, 1886.
A series of brief papers dealing chiefly with causes of failure of various smelters.

IHLSENG, M. C.   Review of mining interests of the San Juan region: Colo. State School of Mines, annual report of field work and analyses, 1886, pp. 19–63.
General notes on the mines.

COMSTOCK, T. B.   The geology and vein structure of southwestern Colorado: Trans. Am. Inst. Min. Eng., Vol. XV, 1886–87, pp. 218–265.
Advances speculative views as to the origin of the ore deposits and divides the veins into groups.   The Red Mountain deposits are supposed to occupy the throats of extinct geysers and the bowls of nonsintering hot springs.

LAKES, A.   Geology of the Colorado ore deposits: Ann. Rept. State School of Mines, 1887, pp. i–clix.
Contains scattered and unimportant references to veins of the San Juan Mountains.

REUSCH, H.   Krystallizirter Kaolin von Denver in Colorado: Neües Jahrbuch für Mineralogie, etc., 1887, II, pp. 70–72.
Description and chemical analysis of kaolin from National Belle mine. Red Mountain.

KEDZIE, G. E. The bedded ore deposits of Red Mountain mining district, Ouray County, Colo.: Trans. Am. Inst. Min. Eng., Vol. XVI, 1887–88, pp. 570–581.

Classifies deposits in northern San Juan County as fissure veins, "chimneys," and "bedded" or "contact" deposits. Describes the deposit of the Saratoga mine as of the latter class.

SCHWARZ, T. E. Notes on the ore occurrence of the Red Mountain district: Proc. Colo. Sci. Soc., Vol. III, 1888, pp. 77–85.

Ascribes Red Mountain deposits to hot spring action. Describes changes in ore bodies, character of ores, and origin of ore caves.

EMMONS, S. F. Structural relations of ore deposits: Trans. Am. Inst. Min. Eng., Vol. XVI, 1888, pp. 804–839.

Describes ore body of Yankee Girl mine and criticises the speculative statements of T. B. Comstock. Discusses fractures and their causes. Refers to zones of sheeting in the rocks of the San Juan. Suggests that complicated systems of fractures there found seem to require torsional strains for their formation.

COMSTOCK, T. B. Hot spring formations in the Red Mountain district, Colorado: Trans. Am. Inst. Min. Eng., Vol. XVII, 1888–89, pp. 261–264.

A reply to criticism by Mr. S. F. Emmons.

HILLS, R. C. Orographic and structural features of Rocky Mountain geology: Proc. Colo. Sci. Soc., Vol. III, 1888–1890, pp. 362–458.

A few references of general character to the geology of the San Juan region.

SCHWARZ, T. E. The ore deposits of Red Mountain, Ouray County, Colo.: Trans. Am. Inst. Min. Eng., Vol. XVIII, 1889–90, pp. 139–145.

A careful general description of the peculiar deposits characteristic of the Red Mountain district. Some hypotheses are advanced, but the value of the paper lies in the facts presented relative to the main ore bodies, which are no longer accessible.

KELLER, H. F. Ueber Kobellit von Ouray, Colorado, und über die chemische Zusammensetzung dieser Species: Zeitschrift für Krystallographie, Vol. XVII, 1890, pp. 67–72.

Describes kobellite from the Silver Belle mine.

ARGALL, PHILIP. Discussion of paper by P. H. Van Diest: Proc. Colo. Sci. Soc., Vol. IV, 1891–1893, p. 353.

States that in the San Juan region many of the mineral veins in the eruptive rocks occupy fissures of contraction and do not extend into the underlying rocks.

BANCROFT, H. H. History of Nevada, Colorado, and Wyoming; San Francisco, 1890, pp. 495–503.

Sketch of early history of San Juan region.

HILLS, R. C. Types of past eruptions in the Rocky Mountains: Proc. Colo. Sci. Soc., Vol. IV, 1891–1893, p. 25.

San Juan Mountains said to correspond to Tushar type, of Powell, carved from thick volcanic masses in nearly horizontal attitudes.

HALL, FRANK. History of Colorado, Vols. II–IV; Chicago, 1890–1895.

Early history of prospecting in the San Juan.

HURLBURT, E. B. On alunite from Red Mountain, Ouray County,
Colo.: Am. Jour. Sci., 3d ser., Vol. XLVIII, 1894, pp. 130–131.

RICKARD, T. A. The development of Colorado's mining industry:
Trans. Am. Inst. Min. Eng., Vol. XXVI, 1896, pp. 834–898.
Contains brief historical notes on the San Juan region.

SCHWARZ, T. E. Discussion of paper by T. A. Rickard: Trans. Am.
Inst. Min. Eng., Vol. XXVI, 1896, pp. 1056–1060.
Describes "ore break" of Genesee-Vanderbilt mine.

PURINGTON, C. W. Preliminary report on the mining industries of
the Telluride quadrangle, Colorado: Eighteenth Ann. Rept. U. S.
Geol. Survey, Part III, 1898, pp. 745–850.
A valuable study of the ore deposits lying immediately west of the Silverton
quadrangle and including also a few mines, such as the Tomboy, within the
former area. The general similarity of conditions obtaining within the Silverton and Telluride quadrangles renders the material of this report specially important, and it will frequently be referred to in the following pages.

CROSS AND PURINGTON. The Telluride folio: Geologic Atlas U. S.,
folio 57; U. S. Geol. Survey, 1899.
Maps and general geology of the quadrangle adjoining the Silverton quadrangle
on the west. In this folio are defined and described many of the geological formations which extend into the Silverton area.

ENGINEERING AND MINING JOURNAL, Vols. XXIII (1877) to the
present.
Numerous brief reports from correspondents in the San Juan region. Their
value at the present time is in their historical data.

RAYMOND, R. W. Report on the Mineral Resources of the States
and Territories West of the Rocky Mountains, Vols. I to VIII,
comprising the years 1869–1876.
Contains many notes on the mines of the San Juan.

BURCHARD, H. C. Report of the Director of the Mint on the Production of Gold and Silver in the United States, 1880–1884.
Contains numerous notes on the development and production of the mines in
the Silverton quadrangle.

KIMBALL, JAMES P. Report of the Director of the Mint on the Production of Gold and Silver in the United States, 1885–1888.
Brief notes on the San Juan region, chiefly in regard to production.

LEECH, E. O. Report of the Director of the Mint on the Production
of Gold and Silver in the United States, 1889–1892.
Brief notes on mines, chiefly in regard to production.

ROBERTS, GEORGE E. Report of the Director of the Mint on the Production of Gold and Silver in the United States, 1897.
Brief notes on mines and production of region, by counties.

LEE, H. A. Report of the State Bureau of Mines for 1897; Denver,
pp. 109–112.
Gives production for year by counties and some historical data on San Juan County.

VAN HORN, FRANK R. Andesitic rocks near Silverton, Colorado:
Bull. Geol. Soc. America, Vol. XII, 1900, pp. 4–9.
Petrographical descriptions and chemical analyses of various more or less altered
specimens from the southeastern quarter of the Silverton quadrangle.

HISTORICAL SKETCH.

Deeply covered with snow for a great part of the year, and standing remote from lines of transportation and points of supply, it is not surprising that the rugged San Juan Mountains should, until a comparatively late period, have discouraged the advance of prospectors into their lonely ravines.   Prior to 1860 the area now included in the Silverton quadrangle had been visited by but few white men, and at a time when every gulch of the Sierra Nevada was a scene of picturésque activity the Indian and the mountain sheep were as yet undisturbed in their possession of the San Juan.   But no natural obstacles have ever long withstood the restlessness and indomitable perseverance of the seekers after precious metals.   In 1860 (according to some, 1861) a large party of miners, under the leadership of John Baker,[1] penetrated to the little mountain-rimmed "park" where the town of Silverton now stands.   They had hoped to find profitable gulch mining, but, overtaken by the heavy winter snows and harassed by the Ute Indians, many of the party perished miserably and the remnant escaped over the mountains only after suffering great hardships.   For several years the memory of this unfortunate expedition seems to have discouraged further attempts at prospecting in the neighborhood of Bakers Park.   It was not until the early seventies that reports of mineral wealth again began to draw the more adventurous miners into the San Juan region.   Some gold was early obtained by washing in Arrastra Gulch, and this led, in 1870, to the discovery by a party of prospectors sent out by Governor Pile, of New Mexico, of the first mine which was successfully operated, the Little Giant, on the north side of Arrastra Gulch.[2]   This produced a gold ore, of which some 27 tons were treated in arrastres, yielding $150 a ton.   The first shipment of ore from the district is said to have been from this mine.   In 1872 troops were sent into the region to keep out the miners, as their presence constituted a violation of the treaty of 1868, by which the Utes were secured in sole possession.   In the same year a commission was appointed by Congress to negotiate a new treaty with the Indians to reduce the extent of their reservation.   The Little Giant Company was organized in Chicago in 1872, and in 1873 the arrastres were replaced by an amalgamating mill equipped with a Dodge crusher, a ball pulverizer, and five stamps.   Power was furnished by a 12-horsepower engine.   The mill was built 1,000 feet below the mine and the ore was brought down on the first wire-rope tramway built in the region.   This year the mine produced about $12,000 out of a total of about $15,000 for the entire region.   The pay shoot, however, began

---

[1] Accounts differ as to the Christian name of this pioneer. Thus Bancroft (History of Nevada, Colorado, and Wyoming, San Francisco, 1890, p. 495) refers to him as *John* Baker; T. A. Rickard, in his paper on The Development of Colorado's Mining Industry (Trans. Am. Inst. Min. Eng., Vol. XXVI, 1896), gives his name as *Jim* Baker; while Frank Hall, in his History of Colorado, makes him *Charles* Baker.

[2] According to Rickard, opened by Miles T. Johnson in 1871.

to diminish, and after the milling of a few hundred tons of ore mine and mill were abandoned. Several lode. had by this time been opened in the region and some small amounts of rich ore had been taken out, but it was not until 1874 that the main rush to the country began. In September of the previous year a treaty, known as the Brunot treaty, had been drawn up with the Utes, whereby the San Juan Mountains were thrown open to settlement. The ratification of this treaty by the Senate in April, 1874, was followed by a sudden influx of miners, chiefly from the northern camps of Colorado, but including also a few from the south, and some even from the far West. It is estimated that about 2,000 men came into the district during the summer of 1874, and Endlich[1] reports that more than this number of lodes were then staked out.[2] At that time La Plata, Hinsdale, and Rio Grande were the only counties into which the former reservation had been divided. The chief settlement and the county seat of La Plata County was Howardsville; but in the autumn of 1874 the county seat was moved to Silverton, then a growing town of some dozen houses, admirably situated in Bakers Park. The nearest post-office at this time was Del Norte, about 125 miles distant. In 1876 San Juan County was formed from a portion of La Plata County, with Silverton as the county seat. At this time the town is said to have had a population of about 500 voters. Ouray, San Miguel, and Dolores counties were subsequently formed by legislative enactment from the territory originally included in La Plata County.

In 1874 real mining began, principally on Hazelton Mountain, and several hundred tons of gray copper and galena ore were taken out from the Aspen, Prospector, Susquehanna, and neighboring claims during this and the immediately succeeding years. This ore was treated chiefly in Greene & Co's. smelter, which was erected just north of Silverton in 1874, but which was not successfully blown in until the following year. The machinery was brought in on burros from Colorado Springs, then the terminus of the Denver and Rio Grande Railroad. The product of the entire quadrangle for 1875 was about $35,000, and an estimate made in 1877 places the total product from the beginning of mining to the close of 1876 at a little over $1,000,000. The Greene smelter was in intermittent operation until 1879, and was the first successful water-jacket furnace in the State. Its daily capacity was about 12 tons, and it is said to have smelted nearly $400,000 worth of silver-lead bullion. The bullion was shipped by pack train and wagon to Pueblo. The cost of transporting it to the railway terminus was $60 per ton in 1876, $56 per ton in 1877, and $40 per ton in 1878. The average price for treatment was not far from $100 per ton. During the seventies the chief route into the Animas mining district was by the trail from Del Norte on the Rio

[1] U. S. Geol. and Geog. Surv. Terr., Report for 1876, pp. 120-121.
[2] According to Bancroft, "more than 1,000 lodes claimed." Loc. cit., p. 501.

Grande, by way of Antelope Park and Cunningham Gulch.   Over this route the first ore sold from the Pride of the West mine, in Cunningham Gulch, was taken out in 1874.   It was not until 1879 that the wagon road from Antelope Park was completed by way of Stony Gulch, and ore could be hauled out to Del Norte by teams at $30 a ton.

The founding of Lake City, about the year 1875, and the establishment there by Crooke & Co. of a smelting plant, afforded a market for the ores of the northeastern portion of the quadrangle.   The first ore shipped out from this part of the district was from the Mountain Queen mine, at the head of California Gulch, in 1877.   It amounted to 370 tons, and contained 64 per cent of lead and 30 ounces of silver per ton.   It was carried by pack animals to the end of the road at Rose's cabin, at a cost of $3 per ton.   Crooke & Co., of Lake City, and Mather & Geist, of Pueblo, both had ore-buying agencies in Silverton in 1879.   During this year about 500 tons of ore, worth about $60,000, were sent to the Lake City smelter, and about 185 tons went to Pueblo. The value of the latter was probably about $25,000.

In 1879 a road was completed from Silverton up Cement Creek to the head of Poughkeepsie Gulch, where prospecting and mining was going on with great activity on the Old Lout, Alabama, Poughkeepsie, Red Roger, Saxon, Alaska, Bonanza, and other claims.   Chlorination and lixiviation works were erected at Gladstone about this time, to treat these ores by the Augustin process.   Their capacity was about 6 tons per day.

During the seventies the eastern and northeastern portions of the quadrangle were actively prospected, and nearly every lode which has subsequently proved valuable was then located.   In some cases paying ore was taken out in large quantities, as from the North Star mine on Sultan Mountain and others already referred to.   But this activity was in great part feverish and unwholesome.   The success of a few encouraged extravagance in the incompetent, and opened a rich field to unscrupulous and dishonest promoters.   Smelting plants and mills were erected before the presence of ore was ascertained.   Reduction processes were installed without any pains having been taken to ascertain their applicability to the particular ores to be treated.   Thus in 1876 Animas Forks was a lively town of some 30 houses and 2 mills, and in 1883 boasted of a population of 450.   But there was never any real justification for its existence.   Built upon hopes never realized, its decline was almost as rapid as its rise, and the town is now ruined and desolate.   Its principal mill was put up in 1875 or 1876 to treat ore from the Red Cloud mine, but was never successful. The Eclipse smelter, erected by James Cherry as late as 1880, at the mouth of Grouse Gulch, ostensibly to run on lead ores from the Mountain Queen and other claims, was also a costly failure.   The Bonanza tunnel, a mile and a half west of the town, was run 1,000 feet at the extravagant cost of $300,000 or $400,000, and then abandoned. Around Mineral Point probably $2,000,000 or $3,000,000 were squandered in mining operations which resulted in no permanent improvements or actual development. Numerous similar cases might be cited from this region, many of them unfortunately of much more recent date. Capital thus invested serves only to build monuments of failure, folly, and dishonesty, which may operate to delay for years such development and improvement as the mineral resources of a district really warrant.

In 1881 the remarkable deposits between Red Mountain and Ironton were discovered, and in 1882 and 1883 prospectors swarmed into this new field. The Yankee Girl ore body was struck in 1882, and, with the Guston, shipped large quantities of high-grade silver ore for over fourteen years. These two mines alone have probably produced at least $6,000,000 or $7,000,000, but a very large amount of capital has been vainly expended in attempts to find other ore bodies in the vicinity equally large and rich. Nowhere else in the quadrangle has mining been carried on so extensively. In 1888 the railroad was extended from Silverton to Ironton, greatly facilitating the marketing of the ore. Red Mountain and Ironton were formerly thriving mining towns of 400 or 500 inhabitants, and the activity steadily increased almost up to 1893, when the fall in the value of silver and the exhaustion of the phenomenally rich portions of the ore bodies caused the boom to collapse. The Yankee Girl and Guston mines continued working until 1896, but the low price of silver and the increased expense involved in deep workings and in handling the troublesome corrosive waters of these mines, together with the lower grade of the ores in depth, finally compelled them too to shut down. From that time to the present Red Mountain and Ironton have remained practically dead camps.

Previous to the advent of the railroad in Silverton, ores running less than $100 per ton could seldom be handled with profit, but with the completion of the Silverton branch of the Denver and Rio Grande narrow-gauge railroad in July, 1882, the rate of transportation on low-grade ores was much reduced, and many mines hitherto unavailable became productive. Freight charges, at first $16 per ton to Denver or Pueblo, were soon dropped to $12, at which high figure they stood for some time. Over 6,000 tons of ore were shipped from Silverton during the first six months after the advent of the railroad. The Greene smelter had some years previously (about 1880) come into the possession of the New York and San Juan Smelting Company, which in 1881 moved the Silverton plant to Durango and in 1882 started the present smelter in that town. In September, 1887, the name was changed to the Durango Smelting Company, which operated until April 1, 1888. From that date until May 1, 1895, business was carried on under the name of the San Juan Smelting and Mining Company, a corporation organized through the consolidation of the Durango Smelting Company of Durango, and the Hazelton Mountain Mining Company, of Silverton, owners of the Aspen group of mines. The Martha Rose smelter, with a capacity of about 20 tons, began operations in Silverton in 1882, but after smelting about 11 tons of bullion shut down and was never successfully reopened.

The year 1883 was a busy one in Silverton, and the population of the town rose to over 1,500 inhabitants. Sampling works had previously been erected by E. T. Sweet and T. B. Comstock & Co. Late in the season a third plant was opened by Stoiber Brothers. The North Star on Sultan, the Belcher, Aspen, Gray Eagle, North Star on Solomon, the Green Mountain, as well as the Red Mountain mines, were all actively producing, while great strikes were announced in the Ben Franklin and Sampson mines. The Silver Lake mine also came into prominence and shipped that year 72 tons of ore to Sweet's sampling works.

It was not until about 1890 that any real attempt was made to concentrate low-grade ores. The credit of thus initiating a procedure upon which largely depends the future of the whole district must be divided between J. H. Terry, of the Sunnyside mine, and E. G. Stoiber, of the Silver Lake mine. Both men have been successful, and Mr. Stoiber in particular has shown how low-grade veins may be worked successfully on a large scale with a modern plant. About this time the North Star mine, on Sultan Mountain, put up the present mill, run by water power, and in 1894 Thomas Walsh and others erected the matte smelter just west of Silverton. Walsh treated by the Austin process the lower-grade Guston ore, and bought siliceous ores wherever he could obtain them. This smelter ran pretty steadily for three years and finally shut down. Its capacity was about 100 tons of ore a day for ten months in the year. In all, about 100,000 tons of low-grade siliceous and pyritiferous ores were treated. There was no further attempt made to smelt ores in Silverton until the construction in 1900 of the pyritic smelter, near the mouth of Cement Creek. The smelter at Durango, which was leased in 1895 by the Omaha and Grant Company, and which on May 1, 1899, became the property of the American Smelting and Refining Company, has continued to handle the bulk of the ore and concentrates from the Silverton region.

With a few notable exceptions the mines of the Silverton quadrangle produce ores in which silver and lead are the predominant metals. Naturally the rapid decline in the value of silver in 1892 and succeeding years resulted in the closing of many mines hitherto productive, and in a general decrease of mining activity. At the present time, however, there are signs of a favorable reaction and a marked increase of activity. The success of Messrs. Stoiber and Terry in handling low-grade ores has demonstrated that when wasteful and inadequate methods are replaced by modern appliances and shrewd management, mines carrying abundant low-grade ore may be made profitable, even with the present low price of silver. Probably, as in the past, periods of depression will sometimes retard mining improvements. Yet it is clear that the resources of the district have been far too often superficially, ignorantly, or wastefully exploited. Although examples of extravagance are still not hard to find, there is now a more general realization of the distinction between such wasteful expenditure and the legitimate and often large outlay that makes possible the economical working of known low-grade ore bodies. It is very probable that the future will see a great and permanent increase in the productive development of such large and persistent ore bodies of low average grade. The decline in the price of silver, and also in that of lead and copper, which are important constituents of most of the ores, has greatly reduced the margin of profit and rendered economical working imperative, as is illustrated in the following table:

Table showing the average annual prices of silver, lead, and copper, from 1880 to 1899.

Year.	Silver, per fine ounce.	Lead, per 100 pounds.	Copper, per pound.a	Year.	Silver, per fine ounce.	Lead, per 100 pounds.	Copper, per pound.a
1880	$1.145	$5.03	$0.216	1890	$1.050	$4.35	$0.156
1881	1.138	4.81	.188	1891	.988	4.33	.128
1882	1.136	4.91	.185	1892	.874	4.07	.115
1883	1.110	4.32	.159	1893	.779	4.10	.108
1884	1.113	3.74	.139	1894	.634	3.28	.094
1885	1.065	3.94	.110	1895	.658	3.24	.107
1886	.995	4.61	.110	1896	.674	2.97	.109
1887	.978	4.50	.113	1897	.603	3.60	.112
1888	.940	4.41	.168	1898	.588	3.75	.119
1889	.936	3.81	.134	1899	.601	4.44	.177

a Lake Superior ingot.

The prices of silver are taken from the annual reports of the Director of the Mint. The lead and copper values are compiled from tables showing monthly prices and monthly range in the reports on mineral resources issued as parts of the annual reports of the Geological Survey.

Although the first mine to be worked in the district was a gold producer, yet it is an interesting fact that for many years prospecting was practically restricted to a search for silver and lead ores. It was apparently owing to this adherence to an established routine that the Una and Gertrude claims in Imogene Basin, worked twenty years ago for silver and lead, were subsequently abandoned, with no knowledge of the remarkable gold ore which lay close alongside the argentiferous streak, and which was thrown out as waste. Masses of this rich ore were discovered by Thomas F. Walsh on the Camp Bird claim, and subsequently on the dump of the old workings of the Una and Gertrude, and he purchased the latter in 1896 for $10,000. It is to-day one of the greatest producing mines in the quadrangle, and certainly excels all others in the average richness and regularity of its ore bodies.

Placer mining has never been extensively practiced within the Silverton quadrangle. In former years a little washing was done on the east side of California Mountain, in Picayune Gulch, and in Arrastra Gulch, but there are no extensive deposits of auriferous gravels in the district, and the total output from placer mining is probably insignificant.

PRODUCTION.

Accurate figures from which the total production of the Silverton quadrangle might be computed are not available. The area comprises portions of several counties, and has shipped ore to various smelters. An attempt to ascertain and combine the products of the individual mines within the quadrangle has been only partially successful. In the case of some abandoned mines no records can be found, and in two instances requests to the owners of active mines for confidential statements of total output have met with no satisfactory response; but by combining actual figures with individual estimates it appears that the total production of the Silverton quadrangle from the beginning of mining activity to the close of 1900 has been at least $35,000,000. The greater part of this has undoubtedly been in silver, but during recent years, largely owing to the activity of the Camp Bird, Tomboy, and Gold King mines, and to the lower price of silver, the value of the gold output has predominated.

CLIMATE.

The climate of the Silverton quadrangle is, in general, somewhat rigorous. Meteorological records are lacking for this elevated region, and no accurate data can be given in regard to temperature and precipitation. The winters are long, the snowfall is heavy, and the temperature sometimes falls to 20° to 30° below zero. The first heavy snow usually arrives late in October, but heavy falls are not unknown in September. This early storm may be followed by fine weather, lasting well into December. In midwinter snowshoes afford the only means of communication over most of the higher roads and trails. Snowslides are frequent and dangerous, and a winter seldom passes without loss of life and property from this cause. It is often late in June before the snow has disappeared from the higher trails, and on northern slopes, sheltered from the sun, patches frequently linger until long after midsummer. The period extending from the beginning of July to the middle of August is referred to by the inhabitants as the "rainy season." During this period in normal years showers are of almost daily occurrence, usually coming up early in the afternoon and lasting for an hour or more. It is during this season that the wild flowers in the upland basins, notably many species of gentians and a magnificent lilac-colored columbine (*Aquilegia cœrulea*) attain the height of their luxuriant beauty. The temperature at this time is changeable. The transformation from balmy air, genial sunshine, and humming insects to cold winds, dark clouds, and chilling rains is by no means uncommon, and may take place in an hour. The month of September is usually delightful. The rainy season has passed, and toward the end of this month the aspens assume their yellow and orange autumnal coloring, enlivening and softening for a brief period a landscape in which bleak grandeur is the rather too dominant element.

VEGETATION.

Most of the slopes, up to a timber line varying from 11,000 to 11,500 feet, are covered with firs, spruces, and aspens, the last flourishing particularly on the southern exposures and on the talus slopes. Although unsuitable for good sawn lumber, the firs and spruces furnish round timber of sufficient size and strength for ordinary timbering in the mines. The basins, being mostly above 11,500 feet, are treeless, but are covered in summer with a luxuriant growth of grass and flowers.

TOPOGRAPHY.

For more complete accounts of the topography of the region the reader is referred to the general geological text of the forthcoming Silverton folio. The present sketch is confined to such considerations of topographic form as are necessary to enable the reader to fully understand the modes of occurrence of the ore bodies and the conditions governing their exploitation.

The lowest point within the quadrangle is found on its northern edge in the canyon of the Uncompahgre, where the contours (Pl. IV) show an elevation of somewhat over 8,100 feet. The highest point is Handies Peak, on the extreme eastern edge of the area, with an altitude of 14,008 feet. Between these extremes the region exhibits such abrupt and frequent variations in relief as to constitute a topography of the most rugged character. This assemblage of sharp peaks, jagged ridges, steep-walled basins, and deep canyons is the result of vigorous erosion, acting for the most part on nearly horizontal volcanic rocks. The present drainage system is apparently irregular or autogenous— that is, uncontrolled to any recognizable degree by regular structure in the rocks—and was undoubtedly established in essentially its present form prior to the epoch of glaciation. Of the character of the initial land surface, upon which the present streams began their work, we know very little. We may assume, with some degree of probability, that it was the surface of a broad, dome-like elevation extending beyond the bounds of the quadrangle, which, after its original structural doming, had been reduced by erosion to an irregular plateau near the end of the Cretaceous. This was buried to a depth of several

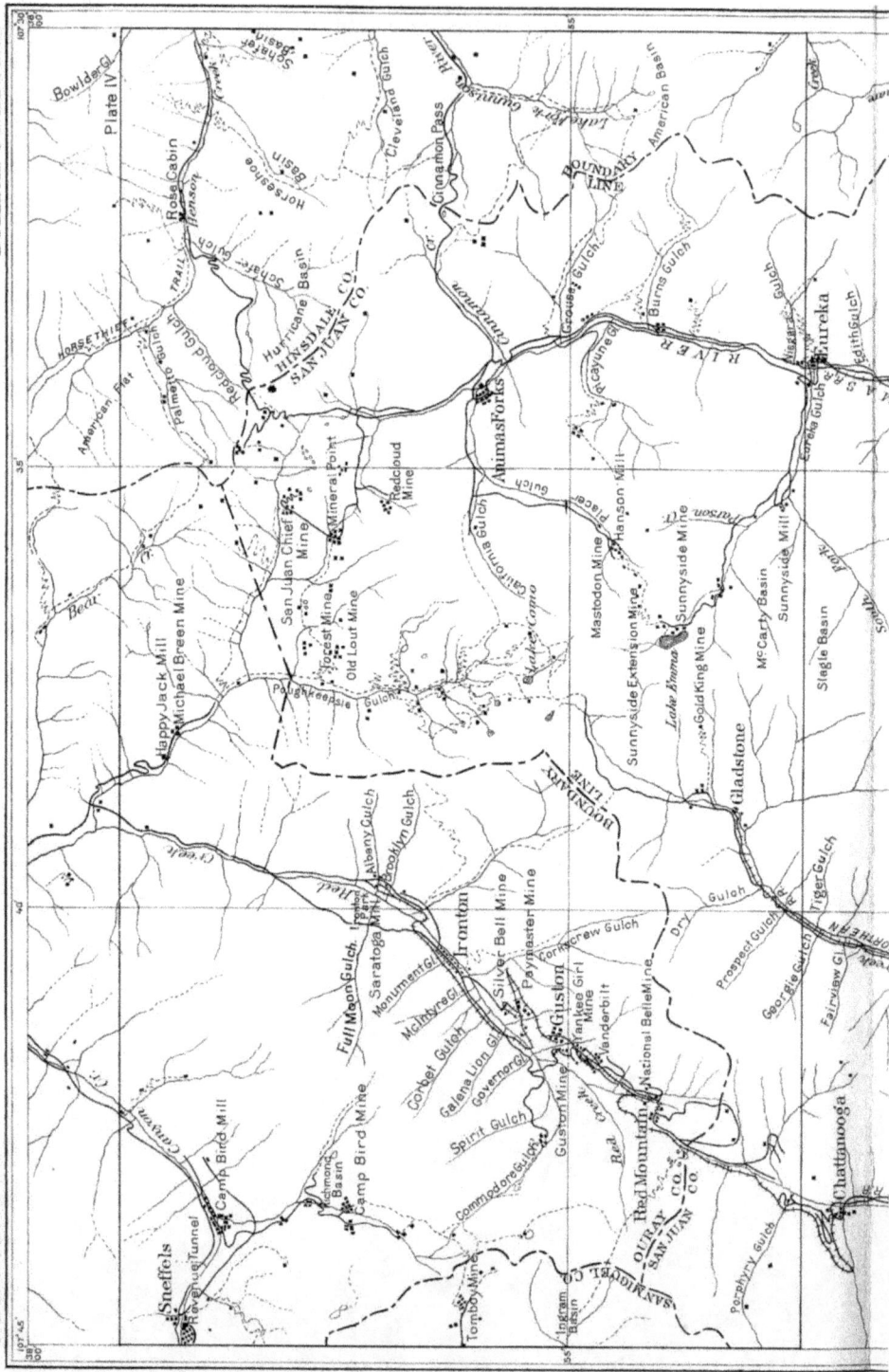

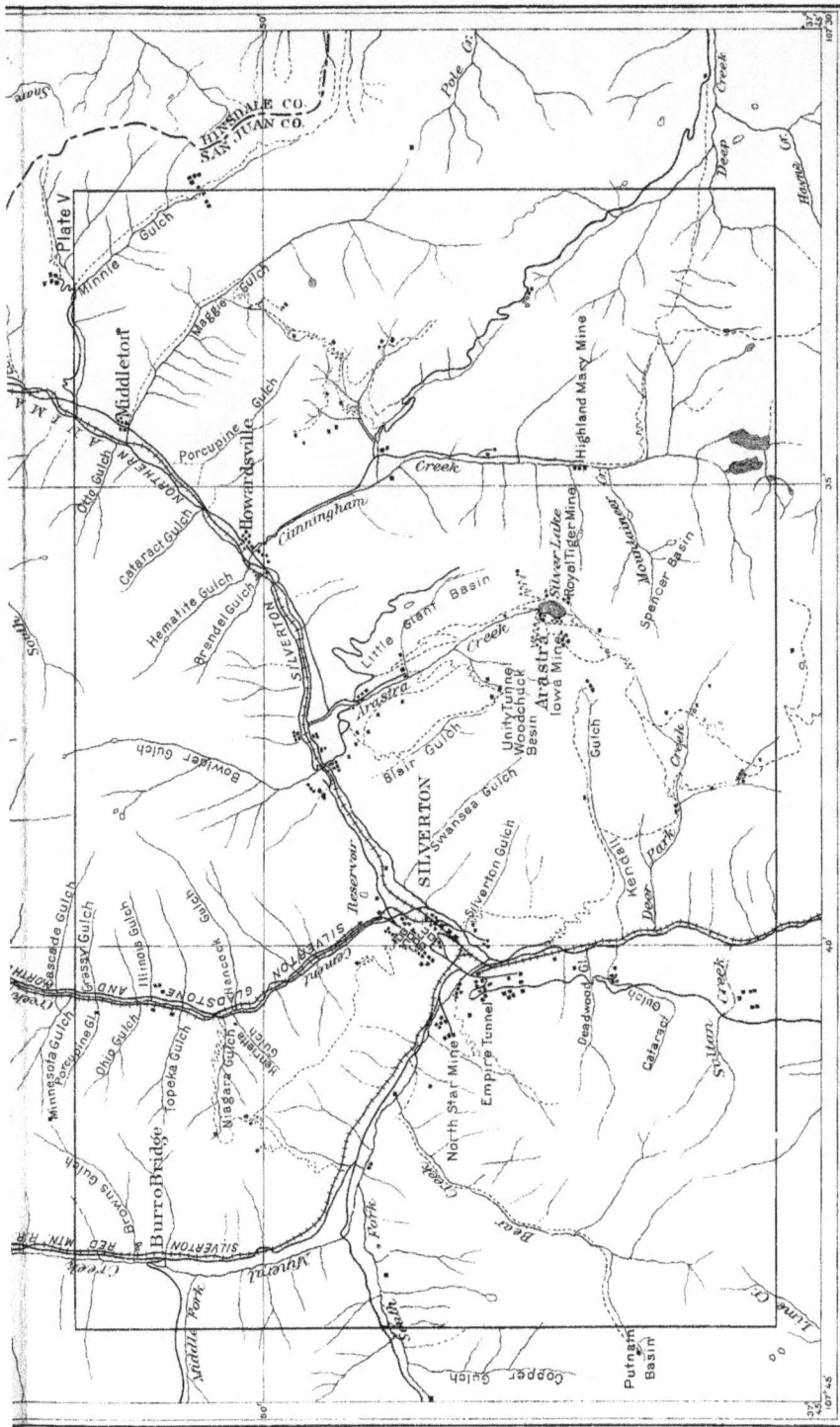

OUTLINE MAP OF THE SILVERTON QUADRANGLE, COLORADO

Showing areas covered by Plates IV and V

Scale

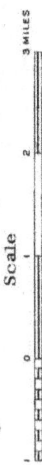

3 MILES

thousand feet by volcanic accumulations in Tertiary times, and has been complicated by faulting and minor oscillations. No recognized trace of this surface remains. The topography, as we know it to-day, is the product of stream erosion, modified by glaciation and the powerful disintegrating action of frost at these high altitudes. It is essential in all discussions relative to the lodes to keep in mind the fact that the entire topography is due to erosion, and that the highest peaks are but residuals which owe their relative height to the resistant materials of which they are composed, or to their distance from the main streams, but in no sense to direct and local elevations.

With the exception of Bakers Park, Ironton Park, and perhaps the gently sloping bit of upland at the head of Henson Creek, commonly known as American Flats, no considerable areas of approximately level land occur within the quadrangle. The mountains rise in steep slopes, or in inaccessible cliffs, often from 3,000 to 4,000 feet in height from the bottoms of the main canyons. The smaller streams descend these precipitous declivities in successions of waterfalls, or occupy small ravines or gulches of high gradient, such as Niagara Gulch near Eureka, Porcupine Gulch, or the gulches northwest of Howardsville. The larger tributaries have frequently excavated long canyons of moderate gradient, such as Cunningham and Poughkeepsie gulches, but in these there is nearly always a point near the headwaters where a moderate gradient is succeeded by a much steeper one. Most of the streams, large and small, head in cirques, or "basins" as they are locally termed, which often contain one or more lakelets. These basins form a very characteristic feature of the topography, and their general shape and character are admirably illustrated by the amphitheaters just north of Sultan and Bear peaks (Pl. V).

The rock floors of these basins, often of somewhat hummocky character, generally slope upward on three sides and pass beneath the talus or "slide rock," with which they are always partly filled. Above the talus and inclosing the basin on three sides are usually more or less precipitous cliffs, from which have fallen the fragments making up the talus at their base. As may be seen from the maps, these cliffs are merely the sides of narrow ridges, surmounted by peaks and intersected by cols or saddles which separate adjacent basins. The talus frequently extends up to a saddle, and it is then possible to pass from one basin to another over the divide. At their lower ends the basins are usually terminated by a precipitous descent to a gulch or to a second basin at the lower level. This relation is well shown in the case of Silver Lake Basin and Arrastra Gulch. When the connecting gulch is very short and steep, as Niagara Gulch, the basins have the form of hanging valleys. The talus that always lies against the cliffs and slopes rimming the basins represents in some cases the gradual accumulation of comparatively small fragments, such as may frequently be heard rattling down the cliffs during the spring and summer. But in many instances it is plainly the result of one or more rock slides of some magnitude, by which large masses of the cliff have fallen and slid out onto the floor of the basin.

The lakelets lying in these basins or amphitheaters are usually rock-rimmed and often of considerable depth. The existence of the lakes depends merely upon the characteristic form of the main basins or cirques in which they lie. This form is immediately due to the eroding action of local glaciers which once occupied the basins. The outlet of Silver Lake flows over a sheet of hard, massive andesite, whose surface has been rounded and scored by the ice. The lake basin itself has been excavated, however, in softer, more readily erodible volcanic breccias. Ordinary erosion by running water would have been powerless to excavate such a basin below the level of its present outlet. Ice is the only eroding agent that could produce such a result.

The evolution of the topography has undoubtedly been influenced to some extent by the fissures and veins that are so abundant in this region. But it is difficult or impossible to reduce their effect to the form of a simple and general statement. Mineralized fissures are frequently lines of comparatively easy oxidation, rapid decomposition, and ready erosion. Thus very many of the lodes determine the position of the cols or saddles in the ridges, and can be distinguished from a distance by the yellowish color of their surface detritus. Yet this is by no means universally true. It often happens that in the present stage of erosion the croppings of a vein traverse an even slope or cross the summit of a high peak. As an example of the last may be cited the important Titusville vein where it crosses Kendall Mountain. It is plain that in some cases the fact that erosion has resulted in cliffs rather than steep slopes is due to the presence of nearly vertical fissures. A system of such fissures running nearly northeast and southwest is responsible for the cliffs which separate Silver Lake Basin from Arrastra Gulch.

In this elevated region the disintegrating action of frost and the active erosion far outstrip the chemical processes of decay which in mild and moist climates are efficient in reducing the rocks to soil. As a consequence the high peaks and ridges are usually composed of bare rock, while their lower slopes are largely made up of talus or "slide rock," as it is locally called. This talus has concealed in a most effectual manner much of the rock once scoured clean by the ice, and has restricted most of the prospecting to the high peaks and ridges where the veins are exposed.

In the vicinity of Red Mountain and Ironton much of the topography has a peculiar hummocky character, which is the result of landslide action. This feature is particularly striking on the northern and western flanks of Red Mountain. Several of the Red Mountain mines, such as the Yankee Girl, Genesee-Vanderbilt, Guston, Silver Bell, Paymaster, and others, are in this landslide area. But the ore bodies of these mines appear to be in place. Apparently the land-slide action is rather superficial, and the siliceous outcrops of the larger ore bodies have remained stationary, while the fractured and mineralized rock around them have slid in considerable masses down the slope. Obviously such sliding renders superficial prospecting very unsatisfactory. It is quite possible that some ore bodies have been covered by landslide material and entirely concealed. For further account and discussion of landslide areas in this and other portions of the quadrangle the reader is referred to the heading " Landslides " (p. 37), in the outline of geology given by Mr. Whitman Cross.

OUTLINE OF GEOLOGY.

By WHITMAN CROSS.

SAN JUAN VOLCANIC AREA.

The complex of volcanic rocks within which the Silverton quad-rangle is situated is one of the most extensive in the Rocky Moun-tains. It consists of a series, several thousand feet in thickness, of tuffs, agglomerates, and lava flows. The more or less distinctly hor-izontal surface volcanics have been penetrated by later stocks of vari-ous rocks, ranging in composition from gabbro almost to granite, and by numerous small dikes of a considerable variety of rocks. The eruptions began with the Tertiary and continued during the greater part of that era.

The area covered at the present time by these volcanic rocks is in general well shown by the Hayden map of Colorada. The main peaks of the San Juan Mountains are situated in the western center of the area. From this point the volcanic series extends northward across the Gunnison River to the West Elk Mountains, eastward to the San Luis Valley, and southeasterly a broad arm passes a considerable distance into New Mexico.

The former extent of the San Juan volcanics was very much greater than at present. There is evidence of enormous erosion, which was greatest along the western and southern sides, and it appears that very possibly the high mountains of the Telluride and Silverton quad-rangles are near what was the center of the volcanic area at its maxi-mum. Certainly the former border of the volcanic pile must have been located many miles farther west and south than at present.

The Hayden map of this region is only of general value. There is great complexity within the area of surface lavas and agglomerates, and the map shows in part quite incorrectly the relations of sedimen-tary and other formations on the borders of the volcanic rocks. The resurvey of this extensive tract has not progressed sufficiently to per-mit the preparation of a satisfactory outline of San Juan geology. In the Telluride folio, however, an attempt has been made to present the results of the more recent work in their general bearing.

From the western front of the San Juan Mountains one looks down upon a great plateau level which reaches for hundreds of miles westward. This level is nearly 2,000 feet below the base of the volcanic series. The isolated mountains which modify this plateau country in the zone bordering the San Juan Mountains are in all cases due to igneous intrusions which have in a measure protected the softer sedimentary beds from erosion.

The removal of the western part of the original volcanic complex has revealed a complicated structure in the underlying older formations, showing that the general center of volcanic activity has also been the site of several great disturbances in earlier geologic times. About a center not yet very well located, because covered by the remaining volcanics, the sedimentary formations of Paleozoic and Mesozoic times have been upturned and eroded at several periods. This structure is well shown in the zone bordering the present western projection of the San Juan Mountains and in the canyons cutting through the volcanics.

The first quadrangle in this area to be resurveyed was the Telluride, lying directly west of the Silverton. The main western front of the San Juan Mountains traverses this quadrangle from north to south in an irregular line, and the fine exposures of the high peaks and deep canyons permit a clear understanding of the principal elements in the mass of the volcanic series, in this portion of the area at least. The sedimentary formations below are also very clearly shown in the Telluride quadrangle. The somewhat detailed text of the Telluride folio [1] must be referred to for much general information which is applicable to certain parts of the Silverton quadrangle.

GEOLOGY OF THE SILVERTON QUADRANGLE.

Location with regard to the volcanic area.—The Silverton quadrangle lies almost wholly within the volcanic area, but two streams, belonging to the drainage of the Colorado River, have cut through the volcanics and disclosed the underlying formations.

On the north the Uncompahgre River with its tributaries, Canyon Creek and Red Creek, reveal the Algonkian quartzites and the Paleozoic and Mesozoic formations underneath the volcanics. The lower members of the section appear in the Silverton quadrangle. From the south the Animas River has penetrated far into the quadrangle, and below Silverton it flows in a canyon excavated below the volcanics. This stream has cut a deep gorge through the Needle Mountains, which rise immediately south of the Silverton quadrangle to elevations of over 14,000 feet. These mountains are composed very largely of Archean and Algonkian rocks. On the west no stream has eaten back into the area of this quadrangle with the exception of the two short tributaries of the San Miguel heading in Ingram and Savage basins. On the eastern border the Lake Fork of the Gunnison River has exposed a mass of granite which is entirely surrounded by volcanics, but as its boundaries are at least in part fault lines, it is possible that dislocations on these planes have produced this outcrop by local upthrust. The structures observable in the formations beneath the volcanics show the Silverton quadrangle to be near the center of an area of repeated orogenic disturbance in Paleozoic and Mesozoic times. Some of the igneous phenomena seem also to center here, but it is as yet premature to conclude that the principal center of the entire volcanic activity was in this vicinity. There may have been several centers of eruption whose products combined to form the great complex known.

All of the principal members of the volcanic series thus far identified are exhibited in the Silverton quadrangle. There are also a number of intrusive rocks, most of them having been previously observed in the Telluride quadrangle. The principal elements of the igneous complex will now be briefly described.

San Juan formation.—The earliest volcanic formation of this series revealed to the present time has been termed the San Juan formation. It is a more or less perfectly stratified series of tuffs and agglomerates of andesitic rocks, no lava flows having been observed within it. The character of the San Juan tuff and agglomerate in its greatest known development is very clearly exhibited on either side of Canyon Creek. It reaches there a thickness of fully 2,500 feet (Pl. IV). In many parts the San Juan seems to be a water-laid deposit, so fine is the stratification. The agglomerates are more or less chaotic, but exhibit distinct stratification when seen in large exposures. These tuffs and agglomerates are also exposed in typical development in the mountain group southwest of Silverton, of which Sultan Mountain and the Grand Turk are the most notable summits. The thickness is here, however, but 1,500 feet. On the east side of the Animas the San Juan tuffs are exposed in very irregular thickness on the slope of gneisses and granites descending from the Needle Mountains. Apparently, erosion was here extremely active in the period succeeding the San Juan fragmental eruptions.

On the eastern side of the Uncompahgre River the floor upon which the San Juan rests and also its upper surface dip rather abruptly southward, bringing the succeeding formation abruptly down to the level of the river, and the San Juan does not reappear in the canyon-like valleys of the streams in the eastern portion of the quadrangle. It is plain that in the entire valley of the Animas from Silverton to Animas Forks the volcanics rest upon a surface much lower than that which constitutes their base in Sultan Mountain. Whether this fact is due to an original depression in this area or to subsequent sinking has not yet been determined.

Silverton series.—The next volcanic series after the San Juan was termed the "Intermediate series" in the Telluride folio. It was very much less developed there than either the San Juan below or the Potosi series above. The relations exhibited by the map (Pl. IV) in Potosi Peak and the mountains south of Canyon Creek are typical of what was observed throughout the Telluride quadrangle. But, as the map clearly shows, this formation increases rapidly in thickness eastward and covers, in fact, much the greater part of the Silverton quadrangle. It will therefore be known in future as the Silverton series, the undesirable term "Intermediate" being discarded. From a thickness of 300 or 400 feet in Potosi Peak the Silverton series increases to 4,000 or 5,000 feet in observed thickness near the center of the quadrangle. This series is a complex of andesitic flows and tuffs alternating with rhyolitic flows, flow-breccia, and tuff. In the Telluride quadrangle, and in most places in the Silverton, its lowest member is a rhyolitic flow or flow-breccia of peculiar character. This is well seen in Potosi Peak, along the crest of the range west of Mineral Creek, and very near the summit of Sultan Mountain.

The relation of the different rocks to one another is very variable in different places. Rhyolite is seldom typical, and its largest masses are reddish flow-breccia—that is to say, lavas of rhyolitic base containing many inclusions of both rhyolite and andesite. Very seldom indeed is the rhyolite free from these inclusions. The andesitic portion of the series consists of augite- and hornblende-andesites, and the relation of massive rock to tuff or agglomerate is extremely irregular. In all the central portion of the quadrangle the Silverton series appears to be principally composed of andesite, either fragmental or in flows. But in the Animas Valley the lowest exposed member of the complex, especially along the eastern side from Cunningham Gulch upward, is a rhyolitic flow-breccia of reddish or almost purplish color, so full of impurities that its character is not easily recognized. It appears either that these rhyolite flows were considerably eroded or that they piled up in quite irregular masses at eruption, for the surface upon which the succeeding andesitic rocks rest is extremely irregular, as will be shown by the final geologic map. The recurrence of rhyolitic material at various horizons within the series may be observed at many points. The Silverton series consists, then, of an irregular alternation of two quite different rocks, erupted doubtless from different centers. They are grouped as one series, because the complex contrasts as a whole so markedly with the products of the San Juan epoch below them, which were purely andesitic and entirely fragmental as far as seen, and with the Potosi series above, which is almost entirely rhyolitic in character.

Potosi series.—The uppermost series recognized in the Telluride quadrangle was termed the Potosi series from its occurrence in maximum development in the peak of that name, situated in the extreme

NAMES OF MINES AND PROSPECTS

1 BI-METALLIST
2 REVENUE TUNNEL
3 WHEEL OF FORTUNE
4 U. S. DEPOSITORY
5 CAMP BIRD, MAIN ADIT
6 CAMP BIRD, UPPER WORKINGS
7 HIDDEN TREASURE
8 HANCOCK
9 JAPAN
10 TOMBOY, MAIN ADIT
11 TOMBOY, UPPER TUNNEL
12 NORTH CHICAGO
13 ABBSTOW
14 SILVER BELL
15 PAYMASTER
16 AMERICAN GIRL
17 WHITE CLOUD
18 GUSTON
19 SCOTCH GIRL
20 ROBINSON
21 YANKEE GIRL
22 GRAND PRIZE
23 MIDNIGHT
24 CARBONATE KING
25 ALEXANDRA
26 NATIONAL BELLE
27 LAKE (RED MOUNTAIN)
28 CHARTER OAK
29 HUDSON
30 WEBSTER
31 GALENA QUEEN
32 MINERAL KING
33 HENRIETTA
34 CARBON LAKE
35 CONGRESS
36 ST. PAUL
37 SILVER LEDGE
38 TOM MOORE
39 TOLTEC
40 SCOTIA AND GOLDEN FLEECE
41 MASTODON
42 SUNNYSIDE EXTENSION
43 SUNNYSIDE
44 GEORGE WASHINGTON
45 LAKE (SUNNYSIDE BASIN)
46 BELLE CREOLE
47 MOUNTAIN QUEEN
48 BELCHER (POUGHKEEPSIE GULCH)
49 BONANZA
50 ROB ROGERS
51 SEVEN THIRTY
52 ALASKA
53 SAXON AND AMAZON
54 TEMPEST
55 POUGHKEEPSIE
56 ALABAMA
57 OLD LOUT, LOWER TUNNEL
58 OLD LOUT
59 FOREST
60 MAID OF THE MIST
61 RED CLOUD
62 EARLY BIRD
63 BUTLER
64 SAN JUAN CHIEF
65 MAMMOTH
66 PRIDE OF SYRACUSE
67 ANNIE WOOD
68 POLAR STAR
69 MONARK
70 FRANK HOUGH
71 SUNSET
72 PALMETTO
73 MICHAEL BREEN
74 SARATOGA
75 BALTIC GROUP
76 A. I. C.
77 GUADALOUPE
78 DOLLY VARDEN
79 JOHN J. CROOKE
80 ANACONDA
81 RED AND BOWIE
82 ADAMS
83 GOLD KING, LOWEST TUNNEL
84 AMERICAN
85 GOLD KING (AND SAMPSON)
86 BEN FRANKLIN
87 QUEEN ANNE
88 WELDRUM TUNNEL
89 HAMMOND TUNNEL
90 GENESEE-VANDERBILT
91 SUMMIT
92 SILVER QUEEN (PLACER GULCH)
93 SOUND DEMOCRAT
94 LITTLE CANDIES
95 COLUMBIA
96 BILL YOUNG

Topography by Frank Tweedy, 1895 and Authur Stiles, 1900
Geology by Whitman Cross assisted by A.C.Spencer and
Ernest Howe. Surveyed in 1899 and 1900.

PRELIM[INARY]
NORTHERN PORTION OF TH[E]

Cont[inued]

LEGEND

SURFICIAL ROCKS

PLEISTOCENE

[Pal]

Alluvium
(cones and valley wash)

[Pls]

Land-slides

SEDIMENTARY FORMATIONS

EOCENE?

[Et]

Telluride
conglomerate

UPPER CARBONIFEROUS

[Ch]

Hermosa

DEVONIAN

[Do]

Ouray limestone

ALGONKIAN

[Asq]

Algonkian schists
and quartzite

IGNEOUS ROCKS

EOCENE AND NEOCENE

[mp]

Monzonite
porphyry
(intrusive masses)

[prh]

Potosi series
(rhyolitic)

[S]

Silverton series
(massive andesite, rhyolitic
flow-breccias, tuff-breccias)

[Sj]

San Juan series
(andesitic breccias)

Lode - fissures

• Mineral deposits other than
  lodes, chiefly stocks

⚒ Mines   No sharp
              distinction
✕ Prospects   between the two

⌐ Strike and dip

⌐ Vertical lodes

Palmetto Gulch
Redcloud Gl.
Rose Cabin
Engineer Mtn
Henson Cr.
Schafer Gulch
Schafer Basin
San Juan Chief
Hurricane Basin
Horseshoe Basin
Poughkeepsie Basin
Focus Mine
Old Lout Mine
Mineral Point
Wood Mt.
Cleveland Gl.
Redcloud Mine
Houghton Mt.
Cinnamon Pass
Uncompahgre River
Tuttle Mt.
Animas Forks
California Gulch
California Mt.
Picayune G.
Handies Pk.
Hurricane
Peak
Mastodon Mill
American Basin
Peak
Burns
Gulch
Sunnyside
Mill
Niagara Mt.
Jones Mt.
Mc Carty
Basin
Sunnyside Mill
Niagara Pk.

OF THE
QUADRANGLE, COLORADO

Geological boundaries are provisional only and
subject to revision upon further field-work
Economic Geology by F. L. Ransome
Surveyed in 1899–1900

Miles
feet.

northwestern corner of the Silverton quadrangle. This series is distinguished as marking the beginning of an epoch of almost exclusively rhyolitic eruptions. For a time, at least, eruptions of more basic rocks were suspended. In Potosi Peak itself, and, indeed, in the Telluride quadrangle, so far as this series was observed, the floor upon which the earliest Potosi lavas were outpoured was approximately level. But either the great development of the Silverton series to the east or other causes have limited the distribution of the Potosi in this first observed simple relation to the other volcanics. No rhyolites which might be correlated with the Potosi series occur to the east in the Silverton quadrangle until the rugged crest of the continental divide is reached. Here several summits, like Sheep Mountain, show light-colored, fluidal lavas at their summits, but they are not sufficiently like the Potosi rhyolites, as known to the westward, to fully justify at this time a certain correlation between them.

The rhyolitic lavas which occur in the higher summits of the northeastern portion of the quadrangle seem more nearly related to the Potosi type, but further examination in the mountains of the Ouray quadrangle is necessary to show the true relations of these lavas to those of the Potosi series.

Stock eruptions.—At a later period than that of the Potosi rhyolite there were in the western San Juan several eruptions of magmas, differing much in composition, and now seen in the form of rather coarsely crystalline rocks penetrating the whole series of surface volcanics to the summits of the highest mountains.

In the Telluride quadrangle a series of these large stocks were observed, in Mount Sneffels, and Stony Mountain, Ophir Needles, Grizzly Peak, and Mount Wilson, all of whose summits are in these stock rocks. There are also several stocks in the Silverton quadrangle, the largest being that of quartz-monzonite, the massive rock of the mountains immediately southwest of Silverton. This stock has a very irregular outline. It extends from Copper Gulch on the west to the base of Kendall Mountain on the east, and the site of Silverton is eroded out of this mass. The northern boundary is very much concealed by talus slopes. The evidence that this stock cuts directly across both sedimentary and surface volcanics is clearly seen in the exposures of the southern contact, on Sultan Mountain and Bear Peak. Smaller stocks of monzonite occur also in the lower part of Cunningham and Maggie gulches. A more basic rock of gabbroitic character appears at the head of Henson Creek, where it is crossed by the wagon road leading to the divide.

Other intrusive rocks.—The Silverton quadrangle is notable for several masses of a peculiar porphyry. This rock occurs in several small stocks; one of them forms a portion of the summit of Red Mountain, another is on Round Hill, while a third is seen at the head of Fullmoon Gulch. A very completely decomposed mass of what is probably the same rock is seen on the north side of Mill Creek, and still another occurs to the west of Mineral Creek, between Mill Creek and the Middle Fork. This rock is characterized by large crystals of orthoclase which give it a very marked porphyritic structure. A very small occurrence of the same rock is found on the wagon road a short distance south of the Yankee Girl mine, and another in the knob north of the National Belle. Intrusions of rhyolitic, and probably of other rocks, are known in the surface volcanics of the northeastern part of the quadrangle, but their character has not yet been definitely ascertained.

A sheet-like mass of porphyry occurs at the eastern base of Sultan Mountain. It has been intruded in quite regular manner in the shales just above the Ouray limestone. Several other masses of porphyry, more or less distinctly in sheet form, with a few dikes, occur at the forks of Mineral Creek west of Silverton. In this vicinity all rocks are so extremely decomposed and are so extensively covered with glacial and other superficial material that the outline of the bodies is very difficult of determination.

A large body of porphyry of rude laccolithic outline occurs on the north side of Canyon Creek on the north border of the quadrangle. The representation of this mass on the map (Pl. IV) is provisional, for the topographic map of this region was revised after the porphyry mass had been examined. This mass is an intrusion in the Triassic beds and extends for some distance into the Ouray quadrangle. The cliffs of this mass are very notable as seen from the wagon road leading up Canyon Creek. The rock is usually very much decomposed.

Formations underlying the volcanics.—As explained above, the volcanic rocks rest upon the greatly eroded surface of many older formations which have been repeatedly elevated and eroded at this general center, producing a complex domal structure. The oldest formation displayed is the gneiss and schist series, in which the Animas Canyon is eroded for several miles, beginning about one mile below Silverton. The rocks of this section have not yet been examined in detail, but will be studied in connection with the Needle Mountains quadrangle.

Associated with these gneisses and schists in the Needle Mountains is a series of quartzites and slates several thousand feet in thickness. These rocks cross the Animas a few miles south of the Silverton quadrangle line. A similar section is displayed in the picturesque canyon of the Uncompahgre from the north base of Abrams Mountain to Ouray. Here, too, there are some thousands of feet exposed, but neither the top nor bottom of the quartzite series is shown. On the south the earliest Paleozoic beds or the volcanic tuffs cover the quartzites and schists unconformably. On the north the Ouray limestone rests upon their upturned edges. This great

NAMES OF
MINES AND PROSPECTS

1 BONNER
2 MAGNET
3 BROBDIGNAG
4 IRENE
5 ZUNI
6 YUKON TUNNEL
7 NORTH STAR
8 BELCHER (SULTAN MT.)
9 MONTEZUMA (FORMERLY BOSTON)
   TUNNEL, HERCULES MINE
10 EMPIRE TUNNEL (HERCULES MINE)
11 KING
12 MOLES
13 MABEL
14 MONTANA
15 LITTLE RAY
16 TITUSVILLE
17 IDAHO
18 LACKAWANNA
19 NEVADA
20 UNITY TUNNEL, (SILVER LAKE MINE)
21 SILVER LAKE MINE (MAIN ADIT)
22 IOWA
23 ROYAL TIGER
24 BUCKEYE
25 NORTH STAR (KING SOLOMON)
26 DIVES
27 HIGHLAND MARY
28 GREEN MOUNTAIN
29 PRIDE OF THE WEST
30 BIG GIANT
31 LITTLE GIANT
32 ASPEN
33 HEIGOLD
34 PHILADELPHIA
35 VETA MADRE
36 ANTIPERIODIC
37 BIG TEN
38 RIDGEWAY
39 HAMLET
40 GOLD NUGGET
41 DEWEY
42 LITTLE MAUD

LEGEND

/ Lode-fissures

● Mineral deposits other than
  lodes, chiefly stocks

☓ Mines
                } No sharp distinction
                  between the two
☓ Prospects

/70° Strike and dip of lodes

/ Vertical lodes

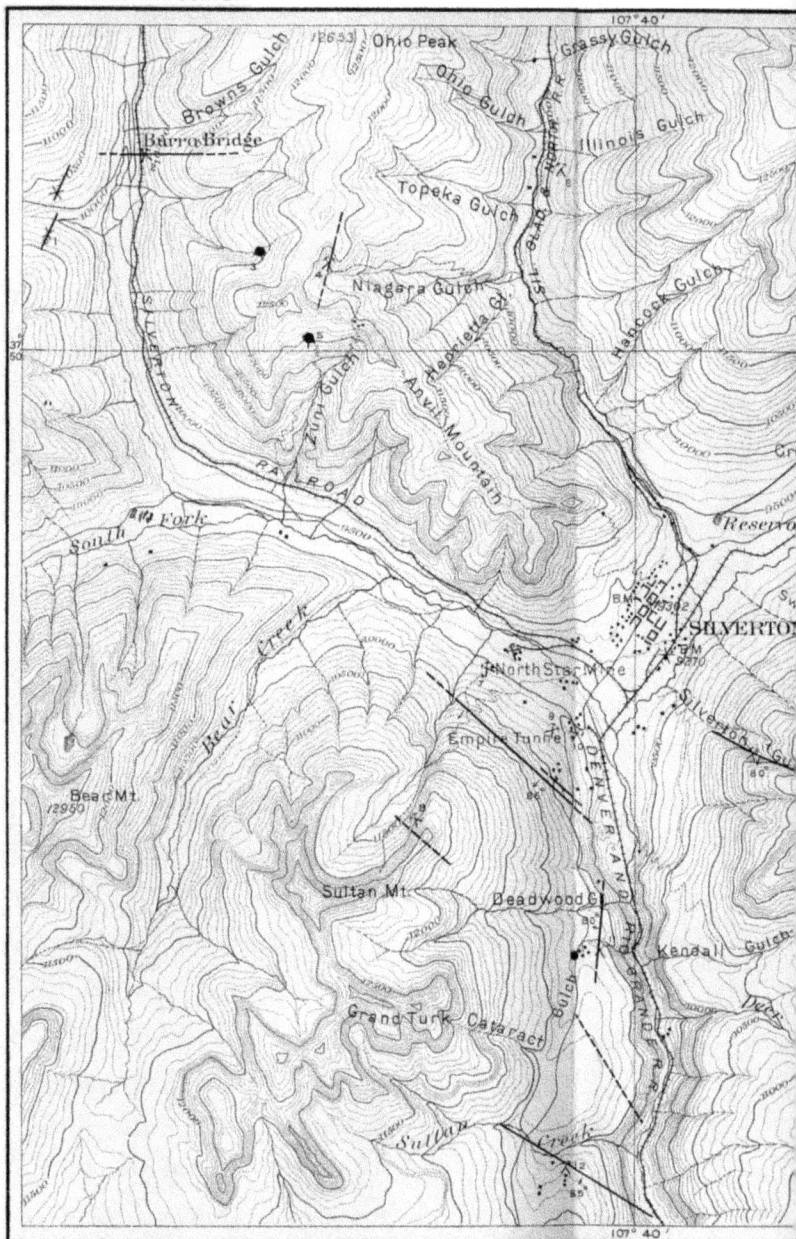

Topography by Frank Tweedy and Arthur Stiles
Surveyed in 1895 and 1900

MAP OF THE SOUTHERN POR

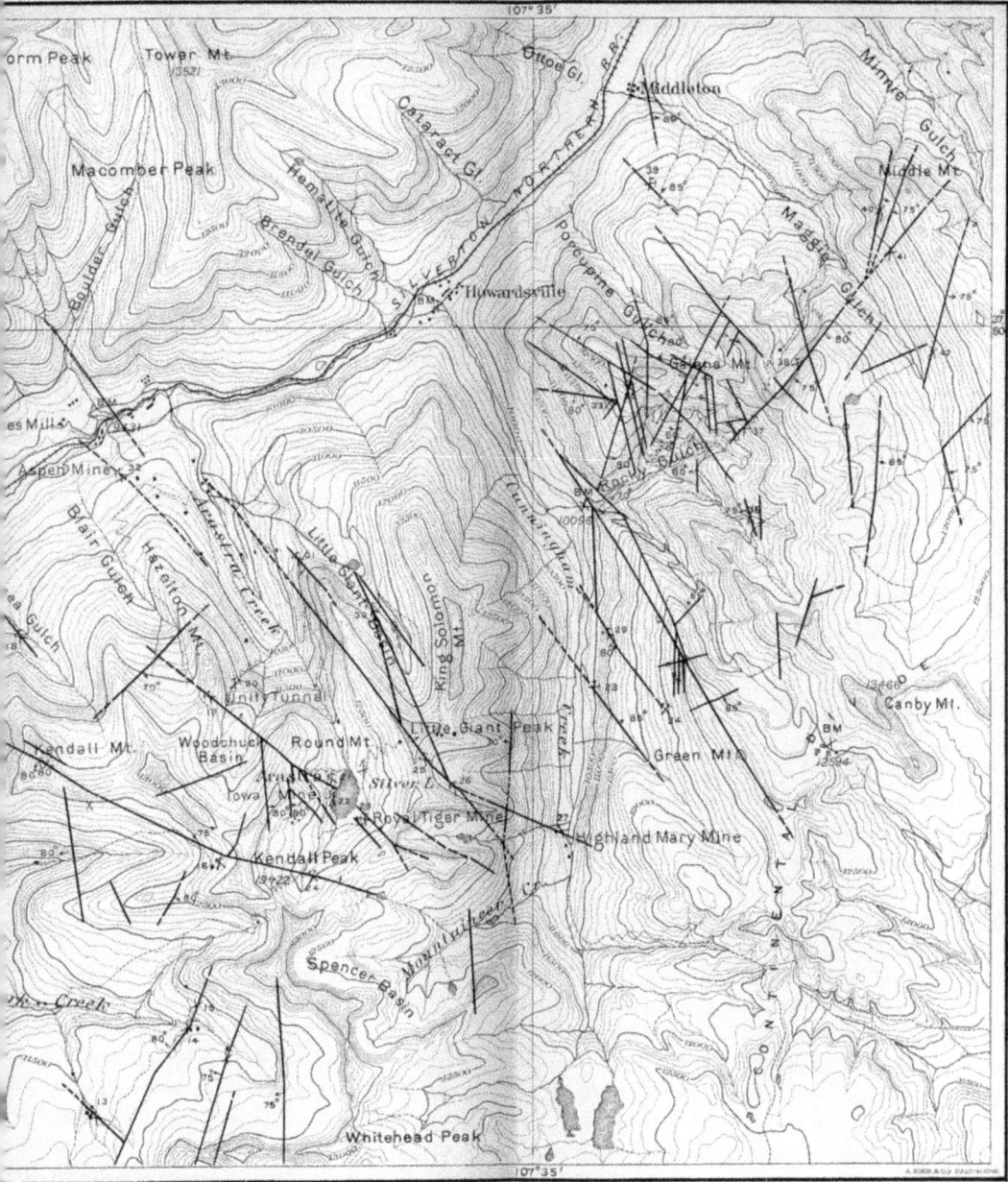

ION OF THE SILVERTON QUADRANGLE, COLORADO

Economic features by F. L. Ransome, 1899-1900

Scale

Contour interval 100 feet.

series of old quartzites and slates has been assigned to the Algonkian period, chiefly from their lack of fossils, and because they may be thus plausibly correlated with the Algonkian rocks of the Grand Canyon of the Colorado.

The earliest Paleozoic formation is a quartzite with some sandy shales 100 to 200 feet thick, which is seen on the west side of the Animas from the monzonite contact to Molas Lake, and imperfectly on the east side of the Animas. This quartzite has been traced down the Animas to a point below Rockwood, and is called the Ignacio quartzite from its characteristic development on the bench where the lake of that name is situated. A southerly dip carries this quartzite onto the southern slope of the Needle Mountains, and there a few indistinct fossils have been found which indicate its Cambrian age.

The rather shaly beds, often calcareous, succeeding the quartzite have as yet yielded no fossils. If there are any Silurian strata in this section they are probably represented by these calcareous shales and sandstones.

The next formation recognized is a heavy white limestone about 200 feet in thickness, which is seen resting directly on the Algonkian quartzites just south of Ouray, and has been called from this locality the Ouray limestone. In the Animas region this limestone rests upon the calcareous shales just mentioned. A few Devonian fossils were found in this limestone by F. M. Endlich during the Hayden survey. At the present time a quite characteristic Devonian fauna has been obtained from this limestone at various places.

Following the Ouray limestone there is a stratigraphic break. Some formation of Lower Carboniferous age should appear at this horizon, and some fossil evidence has been obtained to show that at least remnants of that formation may be found on careful search. But the great series of Upper Carboniferous rocks called the Hermosa formation come nearly or quite in contact with the Ouray limestone in many places. This series of alternating limestones, grits, and sandstones is about 2,000 feet thick and forms the imposing scarp facing the Animas Valley on the west side. The same beds form notable cliffs about the town of Ouray. From the limestone members a very characteristic Upper Carboniferous fauna has been obtained.

At Rico the uppermost division of the Carboniferous was found to be characterized by invertebrate fossils of the Permo-Carboniferous. Three hundred feet of strata have been grouped together as the Rico formation. Evidence is not yet sufficient to demonstrate the presence of the Rico strata in the Silverton quadrangle.

Following the Carboniferous comes the well-known series of reddish grits, sandstones, and conglomerates ordinarily known as the Red Beds. This series has been termed the Dolores formation from its occurrence along the Dolores River, and evidence of its Triassic age has been found in the upper part of the formation. These strata are seen in Canyon Creek and in the southwestern corner of the Silverton quadrangle. All the Paleozoic and Mesozoic formations named exhibit a domal structure by their prevailing dips to the north or to the south, away from the center in the Silverton quadrangle. Beyond the borders of this quadrangle the other formations of the Mesozoic appear.

Prevolcanic surface and the Telluride conglomerate.—The Hayden map represents the volcanic complex of the San Juan as resting on a general plane surface eroded across many different formations. Over considerable stretches the base of the volcanics is represented as approximately level. A detailed examination of the Telluride quadrangle showed that the San Juan tuff was in that region everywhere underlain by a conglomerate free from volcanic material but containing pebbles of very hard sedimentary rocks, Mesozoic and Paleozoic, with schists, granites, and Algonkian quartzites, such as may be seen in the Needle Mountains; that is to say, this conglomerate contained in its pebbles the record of an enormous erosion, such as would be required to produce the surface upon which the volcanic formations in general rest.

This conglomerate below the volcanics was first called the San Miguel conglomerate. It is now necessary to replace that name by another, as it has been found that there is a prior use of that term for a Cretaceous formation in Texas. It is therefore proposed to rename this conglomerate below the San Juan tuff the Telluride conglomerate, or formation, on account of its typical exposures and relations, clearly seen about the town of that name and, indeed, throughout the Telluride quadrangle. The Telluride formation varies greatly in thickness; it is almost 1,000 feet thick in Mount Wilson, on the western border of the Telluride quadrangle. It is 700 feet thick in Sheep Mountain, southwest of Trout Lake, and from that point thins gradually eastward to Sultan Mountain in the Silverton quadrangle, where, on the eastern slope, it has an average thickness of about 30 feet. To the east of the Animas this conglomerate is not continuously developed, but is found occupying hollows in the surface of gneisses and schists. The San Juan tuffs resting upon it bury the small prominences separating these hollows, and the Telluride conglomerate is, as a rule, not conspicuous. It is very distinctly seen, however, beneath the San Juan on both sides of Whitehead Creek. The same thinning of the Telluride conglomerate occurs between Telluride and the exposures in Canyon Creek. The conglomerate is very clearly exposed upon the wagon road leading up the north side of Canyon Creek, and is probably continuous along its southern side to the valley of Red Creek, but beyond that to the east is seen only in disconnected exposures. It is thought probable that the same formation occurs beneath the volcanics in the lower country east of the Uncompahgre River in the Ouray quadrangle.

Structure of the pre-Telluride formations.—The domal structure which has been referred to is clearly seen in the Animas Valley from the Grand Turk Mountain southward. The formations here have a southerly or southwesterly dip, which prevails with some undulations to a point below Durango. Westerly dips prevail in general in the Telluride quadrangle, and northerly dips are seen in the Uncompahgre Valley from Ouray downward. A local structure of interest is seen directly east of Sultan Mountain in the Animas Valley, where the Ignacio quartzites, the Ouray limestones, and the lower Hermosa beds dip rather abruptly northward. The original extent of this fold is not now plain, for these beds are cut off sharply by the great quartz-monzonite stocks, beyond which only volcanic rocks are known for a number of miles. An interesting evidence of the structure of the Paleozoic beds is also seen in Ironton Park. The well-known deposits of the Saratoga mine seem quite clearly to be in the horizon of the Ouray limestone, resting upon Algonkian quartzites, as in the Uncompahgre Valley. They have a westerly dip, which carries them quickly below the level of the park, but along the road on the western side a very few feet of grits may locally be found beneath the San Juan tuffs.

These beds of Ironton Park are supposed to represent the known eastern limit of the sedimentary section which must be continuous beneath the volcanics and which reappears at Telluride, in the drainage of the San Miguel River.

Faulting.—Displacements of the volcanic rocks by faulting are numerous, but it is in most cases difficult to ascertain the amount and character of the displacement. It is supposed that many of the faults seen in the eastern part of the quadrangle, especially in or near the valley of the Lake Fork, have a very considerable throw. In other parts of the area dislocation is, as a rule, comparatively slight. An exception to this statement is afforded by the faults bounding the rhyolite body of Porphyry Gulch, near the head of Mineral Creek. Here there appears to be a sunken block of Potosi rhyolite bordered by faults, some of which are very clearly exposed. Minor faults occur particularly in the southern parts of the quadrangle, on either side of the Animas.

In that region there are faults which seem to be of an earlier date than the volcanic eruptions, but they can be identified as prevolcanic only where they are found to stop at the lower surface of the San Juan. Two such faults are seen in Cunningham Gulch above the Highland Mary mine. Between these faults a block of Ouray limestone has been dropped, and outcrops of that formation occur on each side of the main gulch. These faults do not affect the San Juan tuffs above the Ouray limestone.

Landslides.—The Telluride quadrangle is an area of important landslides, especially on the steep western face of the San Juan, where the Telluride conglomerate rests on soft Cretaceous shales. Two of these slide areas are several square miles in extent and huge blocks have slipped en masse. Minor slides were also noted in several of the gulches of the mountains about Telluride.

The Silverton quadrangle has also been the scene of landslides on an unusually grand scale. The principal area of landslides, shown on Pl. IV, embraces the country on both sides of Red Creek from the lower end of Ironton Park to the head of Mineral Creek, and extends vertically up the adjacent slopes for several hundred feet, or in places 1,000 feet. That landslides have occurred in this region will be evident to all who carefully examine the topographic detail within the areas outlined. This detail is unfortunately too small to admit of expression on the scale of the topographic map. Apparently the movement has been very superficial in comparison with the great slides of the Telluride quadrangle. It has taken place in hundreds of separate small blocks, which are now outlined by trenches or sinks back of the knolls representing the separate masses. When shadows are cast in the right direction the individual knolls of this slide area stand out very clearly. Pl. VI represents a small part of the landslide surface in the vicinity of the Guston mine.

The occurrence of such an area of landslide material is especially interesting in this case because of the fact that the Yankee Girl, Guston, Robinson, Vanderbilt, Paymaster, and other mines are in the heart of the area affected. Since in some cases the ore bodies were traced continuously from the surface downward to depths of over 1,000 feet, it is plain that not all of the surface rock has suffered dislocation, but it is a notable fact that at each one of these mines there is a mound or hill of much-altered rock which stands up as an outcrop in place in the midst of the greatly fractured rock which has slipped downward to an unknown extent through the landslide action. The contrast between a hill of rock in place and the knolls representing landslide blocks surrounding it is very striking in the case of the hill above the Paymaster mine. A critical examination of the mountain side at this point clearly shows that this hill must have been a still more prominent point before the landslide epoch. The great multitude of small slide blocks descending from the higher slope have almost covered Paymaster Hill and have swept down on either side to the level of the main gulch. Whether this landslide material may have entirely concealed mounds of altered rock corresponding to those so prominent at the mines mentioned is an interesting problem to which Mr. Ransome refers.

The evidence of landslide action is chiefly that of the topography, which is entirely irregular and can not have been produced by ordinary erosion. The mounds or knolls can not be regarded as roches moutonnées because of the completely fractured character of the rocks found in them and the fact that many of them are soft tuffs, quite incapable of having resisted moving ice to form a scoriated mound. It is also notably the case that rocks of entirely different character

VIEW OF PRINCIPAL RED MOUNTAIN MINES, FROM THE NORTH, LOOKING UP RED CREEK.

The view shows the character of landslide topography. The Guston and Robinson mines are on the left; the roof and smokestack of the Yankee Girl shaft house appear in the middle distance; while beyond, on the sky line, is the knoll on which the National Belle is located. Photograph by Whitman Cross.

and different states of preservation are found in juxtaposition, breaking the continuity which is normal where the formations were not thus dislocated.

Landside action of similar character has been very prominent about Rico, in the heart of the Rico Mountains.[1] The result is to produce broad topographic forms of rather gentle outline, modified by many small ridges, knolls, and trenches. Such fractured material is now necessarily in process of further disintegration. Indeed, certain parts of the landslide area adjacent to Ironton Park have been so smoothed out by the ordinary creeping down the slopes of the much-fractured rock that the landslide character is rather obscure. Landslide action on so gigantic a scale is at present attributed to violent earthquake shock. Only some such violent local force seems adequate to explain the enormous slides of the Telluride quadrangle, and the explanation is most plausible in view of the volcanic history of the region.

METHODS OF MINING AND TREATMENT OF ORES.

It is not proposed in this section to go into an exhaustive account of mining methods and metallurgical processes. Such a discussion would be outside the scope of the present report and would necessitate the collection of detailed data of a kind not contemplated in planning geological field work. A brief statement of methods and approximate costs may, however, be of use to readers having no practical knowledge of this district or of the modes of procedure followed in it.

Owing to the exceedingly rugged character of the country, the larger mines, with the exception of those in the Red Mountain district, are almost all worked through adit tunnels. Although involving, as a rule, greater initial expenditure, and, in the case of crosscuts, necessitating much "dead work," such adits have for large mines undoubted advantages in this region over shafts. They obviate the need of pumping, and when, as is commonly the case, there is vertical connection with the surface through old workings, they greatly simplify the problem of ventilation. In a region so elevated as the San Juan an adit tunnel possesses great advantage in that it allows the mine entrance and buildings to be placed as low down as possible, where they can be better sheltered from the winter snow, and where water and timber can be more readily utilized. In several of the larger mines, such as the Silver Lake and Tomboy, shafts are sunk from the adit levels, and hoisting and pumping machinery are installed in underground stations.

The levels are usually, although not uniformly, about 100 feet apart, and the ore is worked out by overhand stoping. It is usually allowed to fall on canvas and roughly sorted from the waste. The stopes are sometimes left open, sufficient timbering being carried up

---

[1] Geology of the Rico Mountains, Colorado, by Whitman Cross and Arthur Coe Spencer: Twenty-first Ann. Rept. U. S. Geol. Survey, Pt. II, pp. 129-151.

for working purposes.   But the more approved practice is to fill in the stopes as the work proceeds, blasting out material from the walls if necessary.   In many cases this filling is successfully combined with systematic prospecting of the wall rock on both sides of the vein, often resulting in the discovery of ore which would otherwise have been overlooked.   Underhand stoping is, however, practiced in a few mines, as in the Sunnyside, where the proportion of waste which it is necessary to move is small in comparison with the ore.

Machine drills are not generally employed, save in running long crosscuts.   The progressive Silver Lake mine began, however, to employ several Siemens and Halske electric drills in its stopes in 1899.

In mines of moderate output the ore is taken directly from the ore chutes and run out in single cars pushed by a man or a boy.   In the Silver Lake mine the ore collected from the various drifts and levels is all dumped into ore bins in the main underground station on the adit level, and thence taken out to the mill in a train of cars drawn by a mule.   In the Revenue tunnel, which is over 7,400 feet in length, and from which the daily output is some 200 tons of crude shipping ore and an additional amount of milling ore sufficient to furnish 30 to 40 tons of concentrates, the ore is drawn out in a train of cars operated by an electric locomotive.   The tunnel is furnished with a double track, but practice has shown that even in an adit of this length a single track is amply able to accommodate a much greater output than that now handled.

Many of the mines being situated above timber line in the high basins, frequently from 2,000 to 4,000 feet above the main roads, the proper location of the mill and the mode of transportation of the ore and concentrates to the nearest railway are problems which must be solved for each mine.   Owing to the scanty and intermittent character of the water supply, it is seldom practicable to mill the ores at the altitude of the adit tunnel.   An exception must be made in the case of the Silver Lake mine, where water for the mill is obtained by pumping from the adjacent lake.   But in most cases the mill is built any. where from a few hundred to several thousand feet below the mine. The primitive method of packing the ore down on backs of burros to the shipping point or to the mill is in use by nearly all the small mines and prospects.   But in well-equipped mines, located above timber line, the wire-rope tramway performs an office which makes it indispensable in the mining development of so rugged a country.   These tramways are of various patterns, from the lightly constructed Huson tram, with its small fixed buckets and single rope, to the substantial Bleichert, with large detachable buckets drawn over a fixed rope and supported by well-constructed towers of timber.   The Bleichert tramway of the Camp Bird mine in 1899 was running about 45 buckets, each bucket carrying about 700 pounds of ore.   One man is able to dispatch 350 buckets a day.   This, however, is by no means the full capacity of the tramway, it being restricted in this instance by the limitations of a small Huson tramway which was bringing ore down from upper workings to what is now the main adit. The expense of running these well-constructed gravity tramways is comparatively small, and they are remarkably smooth and efficient in operation.

The ores of the Silverton quadrangle are commercially divisible into shipping ores and milling ores. The former are shipped direct to the smelter as crude ore. They include such high-grade ores as can not be treated economically by simple milling processes and such ores as contain but a small proportion of worthless gangue. The milling ores include such gold ores as can be readily amalgamated or concentrated and those silver-lead ores in which the valuable minerals are associated with considerable quartz or other gangue material. Most large mines produce both grades of ore, while the output of the smaller mines, without mills, is necessarily usually restricted to shipping ore. Exclusive of the ore from the Revenue tunnel, which really comes from the Telluride quadrangle, the bulk of the crude ore at present shipped from the Silverton quadrangle goes to the American Smelting and Refining Company's smelter at Durango. Heavy lead ores preponderate. The treatment charge varies from $2 to $11, the higher rate being for siliceous or "dry" ores. An extra chare of 50 cents per unit (1 per cent of 1 ton) is made for ores running over 10 per cent of zinc.

The treatment of milling ores varies considerably with their character. Formerly many lixiviation plants, usually employing some modification of the original Augustin process, were installed in the region for the treatment of argentiferous ores. This process, involving a chloridizing roasting of the ore and subsequent leaching out of the chloride of silver by the use of strong brine, proved successful in treating the ore of the Polar Star mine, about 95 per cent of the silver being saved. In many cases, however, it was a failure, the ores not being adapted to its employment. At the present time it is no longer used. Mechanical concentration, coupled in some cases with amalgamation, has entirely replaced the various chemical processes, and the concentrates thus obtained are shipped to the smelters.

The silver-lead ores, carrying usually some gold and copper, are crushed by rolls or stamps and, after proper sizing in revolving screens, are concentrated by means of jigs and shaking tables. For this purpose the Wilfley table is most in use, although Woodbury, Bartlett, and Cammett tables and Frue vanners are also employed, and a special form of "end-shake" table is in use in the Silver Lake mill for saving the slimes. Where the ore to be treated is rich in galena the tendency in most of the newer mills is to substitute rolls for stamps. Huntington mills are frequently employed for regrinding tailings.

The richer ores, carrying free gold, require different treatment. In the thoroughly modern mill of the Camp Bird mine the equipment is similar to that which has been found best adapted to the free-milling gold ores of California. The ore passes from the crusher directly to the stamps, weighing 800 pounds and dropping 90 times a minute. The pulp passes through a 40-mesh screen to the amalgamating plates, where from 75 to 80 per cent of the value is caught. It is then passed through classifiers and over Frue vanners, the latter being so arranged that the pulp before being allowed to escape as tailings passes over two vanners. The ore commonly run through this mill averages from $100 to $200 per ton and is frequently much higher. The tailings run from $3 to $5, part of which is recovered in a small auxiliary mill.[1] Considering the peculiar character of the Camp Bird ore, which will be elsewhere described in this report, this showing is rather remarkable. Between mills of this type and the more complicated equipment required for treating lower-grade argentiferous ores there are all gradations, as will appear when each mine is described in detail. At the Tomboy mine, which produces a gold ore, the ore is crushed by rolls and ground in Huntington mills before passing over the plates, no stamps being used.

The various mills will be described in the detailed accounts of the individual mines included in this report.

The conditions under which the several mines operate differ so widely that it is impossible to assign a definite lower limit to the value of ores which can be profitably extracted. Ores ranging in value from $6 to $12 a ton are perhaps as of low a grade as can be now worked, even on a considerable scale and with modern equipment. For many of the smaller mines, however, this limit must be doubled. The Red Mountain mines were expensive to operate on account of the irregularity in the disposition of their ore bodies, their corrosive waters, and the necessity of pumping and hoisting through shafts. In 1888 the New Guston, then beginning operations on an extensive scale, paid dividends on ore mined at a cost of $53.50 per ton. The lowest annual cost per ton was $9.60 in 1894 on ore worth only $12.80 per ton. Location, power, timber, water, kind of country rock, and character and amount of ore form a complex set of factors, which fully accounts for the range in the limiting value between workable and nonworkable ore. Compared with $1.35 per ton, which in 1894 covered the entire cost of mining and treatment of the ore from the Alaska-Treadwell mine,[2] or the $2.50 to $5 per ton which covers the similar expense in many deep California gold mines, the figures given above seem high. But the difference is sufficiently explained by the difficulties inseparably connected with mining in so rugged and elevated a region and by the more involved processes required for the treatment of complex

---

[1] In 1900 a modern cyanide plant was being erected to treat the tailings directly from the mill. It is now in operation.

[2] Reconnaissance of the gold fields of southern Alaska, by G. F. Becker: Eighteenth Ann. Rept. U. S. Geol. Survey, Pt. III, 1898, p. 64.

ores, the concentration of which is often rendered troublesome by the presence of sphalerite and barite, and which almost invariably require smelting.

CLASSIFICATION OF THE ORE DEPOSITS.

The ore deposits of the Silverton quadrangle may be conveniently classified and described under three heads: (1) Lodes, (2) stocks or masses, and (3) metasomatic replacements. To the first class belong by far the greater number of the deposits that are being worked at the present time. To the second class are assigned most of the ore bodies formerly worked in the Red Mountain district, often locally known as "chimneys." In the third class, by far the least important in this quadrangle, are placed a few deposits occurring in limestone or in rhyolite.

In the classification of actual ore deposits it is usually found that a given ore body presents features common to two or more ideally distinct types. Thus a fissure vein may be accompanied by some replacement of the adjacent country rock, or it may be found impossible to draw any definite line between impregnation[1] and total metasomatic replacement. The present field offers no exception to this general difficulty, and the division of the ore deposits here made is not to be regarded as in all cases discriminating between totally different things. The aim is to group the deposits broadly, under most prominent characteristics of occurrence, in order that they may be conveniently and systematically described.

THE LODE FISSURES.

DEFINITIONS.

By fissure, as used in this report, there is meant a somewhat extensive fracture in the rocks. Such a fissure is not necessarily accompanied by recognizable faulting, nor by the production of visible open spaces, nor by mineralization. According to the classical definition of Von Cotta,[2] a vein is the filling of a fissure. As designating a simple type, of frequent occurrence, it seems very desirable that in accurate description the term vein should retain this significance, i. e., a "true vein" or fissure vein. In mining speech, however, the word commonly has a broader meaning, and is made to include not only the filling of a preexisting fissure, but also, in many cases, more or less altered and impregnated country rock alongside the fissure. Frequently the "vein" of the miner is a sheeted zone, embracing several veins in Von Cotta's sense, or it may be a zone of more or less irregular ore-bearing stringers. For this reason, lode will be

---

[1] As used in this report, *impregnation* is merely a structural term indicating the fine dissemination of ore particles through a rock mass. In the majority of cases it involves true metasomatic replacement.

[2] Die Lehre von den Erzlagerstätten, 2d ed., Freiberg, 1859, p. 102.

consistently used throughout this report, as a more general term than vein, to designate either a simple filled fissure or a zone of closely spaced fissuring with possibly more or less impregnated or replaced country rock. Lode thus includes what Von Cotta and Von Groddeck[2] have called complex veins (zusammengesetzte Gänge).[1]

A similar usage has been followed by Emmons and Tower[2] in their description of the ore deposits of Butte, Mont., and to some extent by Purington[3] in describing the ore deposits of the Telluride quadrangle.

DISTRIBUTION OF THE ORE-BEARING FISSURES.

Fractures in the rocks, containing more or less ore, are widely distributed within the quadrangle, but are larger, more abundant, or richer in metalliferous contents in certain limited districts than elsewhere. Prospects occur thickly dotted over the whole area, but the important mines are found in more or less isolated groups. A partial idea of the distribution of the fissures may be gained from the accompanying map (Pl. III), on which but a small proportion of the actual fractures are indicated. Moreover, in plotting the lodes the smaller, poorly exposed, and apparently nonworkable ones must necessarily be often disregarded, and the map can not therefore be taken as showing strictly the true distribution of all the fissures, but only of those which are superficially conspicuous or which carry ore in sufficient quantity to be worked or prospected. As the map suggests, fissures carrying variable amounts of ore are less noticeable features in the southwest and southeast corners of the quadrangle than elsewhere. But they occur in all the rocks which possess any considerable distribution within the area, from the Algonkian (or Archean?) schists, which constitute the basement formation of the region, to the latest monzonitic intrusions that cut the Tertiary volcanic series. By far the greater number of them, however, are found in the volcanic rocks of the Silverton and San Juan series. This is apparently due chiefly to the fact that these rocks occupy the greater part of the quadrangle rather than to any special determining factor in the rocks themselves. It is admittedly difficult, however, to isolate such a possible factor from others more directly associated with the origin and accumulation of these rocks than with their texture and structure. Whether the productive ore bodies that fill some of the fissures are as impartially distributed in the various formations as the fissures themselves is a different question and will be discussed in another place.

The existence of local areas of especially pronounced fissuring and mineralization has already been pointed out. Silver Lake Basin may be cited as a center of one such areal group of fissures. Not only are

---

[1] Die Lehre von den Lagerstätten der Erze, Leipzig, 1879, p. 35.
[2] Geologic Atlas U. S., folio 38, Butte special.
[3] Preliminary report on the mining industries of the Telluride quadrangle, Colorado: Eighteenth Ann. Rept. U. S. Geol. Survey, Pt. III, 1898, p. 772.

the fissures within and around this basin exceedingly numerous, but many of them contain large and productive ore bodies.   Another area is that of Galena Mountain, where the fissures are well exposed and form a remarkable network, but have not hitherto proved to contain ore bodies of great value.   As still other centers of conspicuous fissuring may be cited the heads of Treasure and Poughkeepsie gulches, Ross Basin, and Mineral Point.   These last four local districts of vigorous fracturing are not distinctly separated, and should perhaps be included together as a single area over which fissuring and subsequent veining has taken place on an extensive scale.   Finally, the region embracing Savage, Imogene, and Silver basins, in the northwest corner of the quadrangle, in which are located the Virginius and Smuggler-Union mines, may be regarded as part of a very important district of strong ore-bearing fissures, lying chiefly within the Telluride quadrangle.

COORDINATION OF THE FISSURES.

By the coordination of the fissures is meant their arrangement in natural systems, based upon direction, size, persistency, or other features.   In the present instance the trends of the principal fractures will first be discussed, and afterwards the coordination of minor local groups of fissures occurring in various portions of the quadrangle will be considered.

If fissures were always vertical and were bounded by plane surfaces, it would be possible to represent them on a map, such as Pl. III, by straight lines.   Moreover, a single careful observation along a short exposure of any fissure would afford sufficient data for accurately laying off the direction of the line on the map.   It might then be possible to classify these fissures with mathematical precision and divide them into groups or systems, each system being characterized by a certain common direction.

But actual fissures seldom conform to these conditions.   They stand at various angles to the horizon and are often curved.   The existence of curves of large radii where it is evident that the mapped curvature is not the result of the combination of dip and topography is illustrated by several of the lodes plotted on the accompanying map, Pl. III.   The existence of minor curves may be well seen in almost any plan of the underground workings of a mine exploiting a fissure deposit.   These facts interfere materially with the discussion and classification of the fissures of a given region as a simple problem in plane geometry.   It is obvious that single observations of the course of a fissure in a drift can not be assumed as giving its general strike. Nor is it practicable to tabulate such observations—at least in this region—and thereby to set up various systems of fractures traversing the region, each system narrowly confined and accurately distinguished from other systems by a difference of only a few degrees in trend.

In the Silverton region the important fissures have been mapped in the field with reference to the topography. The lines as shown are approximations to truth, requiring much allowance for necessarily limited observations and for imperfections of the topographic map.[1]

Inspection of the map on which the prominent lodes are thus platted shows that the most conspicuous fissuring has taken place in a general northeast and southwest direction. The dip of these fissures is usually southeast at about 75°. But the angles of dip range from 40° to vertical, and a few of the lodes dip northwest. Perhaps slightly less marked, but thus far more economically important, are numerous strong lodes running in a general northwest and southeast direction. In the southeastern quadrant these usually dip northeast at angles varying from 50° to 90°. Elsewhere they usually, although not invariably, dip southwest at high angles, as observed by Purington in the Telluride area. The existence of these two dominant directions was long ago pointed out by Endlich.[2] Purington,[3] in his description of the fissures of the neighboring Telluride quadrangle, concludes that "there are four general directions of fissuring: (1) East and west, and (2) northeast, best developed in the central and southeastern portions of the quadrangle, and (3) north and south, and (4) northwesterly, best exemplified in the northwestern portion." It will be seen that the two most prominent directions of fissuring in the Silverton area coincide with Purington's second and fourth general directions in the Telluride quadrangle.

There occur in the Silverton quadrangle several approximately north and south, as well as nearly east and west, lodes, and if attention were restricted to certain small areas, such as Galena Mountain, it might be considered advisable to recognize one or both of these directions as characterizing distinct subordinate systems of fractures. Such a grouping, however, while valid for the very limited area named, is lost when the various directions of all the fissures in the quadrangle are taken into account. The possible number of such subordinate systems then becomes so large as to obliterate the divisions between them. For accounts of these minor directions of fracturing the reader is referred to the latter part of this section, where their coordination is discussed, and to the detailed descriptions of the various local groups of lodes in the later pages of this report. It is sufficient in this place to point out that the fissuring of the rocks throughout the quadrangle has been so thorough and has taken place in so many directions that two of these only stand out as dominant. Even these show a certain localization within the limits of the quadrangle, the northeast-southwest fissures notably predominating in the northeast quadrant,

---

[1] The topographic map upon which most of the lodes were originally laid down was partly revised during the year 1900, and these lodes had then to be transferred and adapted to the new map in the office—a procedure which may lead to occasional unavoidable errors of position with reference to minor topographic features.

[2] Hayden Survey, Ann. Rept. 1874, p. 232.   [3] Loc. cit., p. 767.

while the northwest-southeast fissures are the more persistent in the southeast quadrant.

In the accompanying diagram (fig. 1) the lodes represented upon the map have been platted as straight lines passing through a common central point. In the case of moderately curved fissures the line represents the average general direction. A few sharply curved lodes have been treated as if they were formed by two intersecting fissures, and represented in the diagram by two lines. The figure

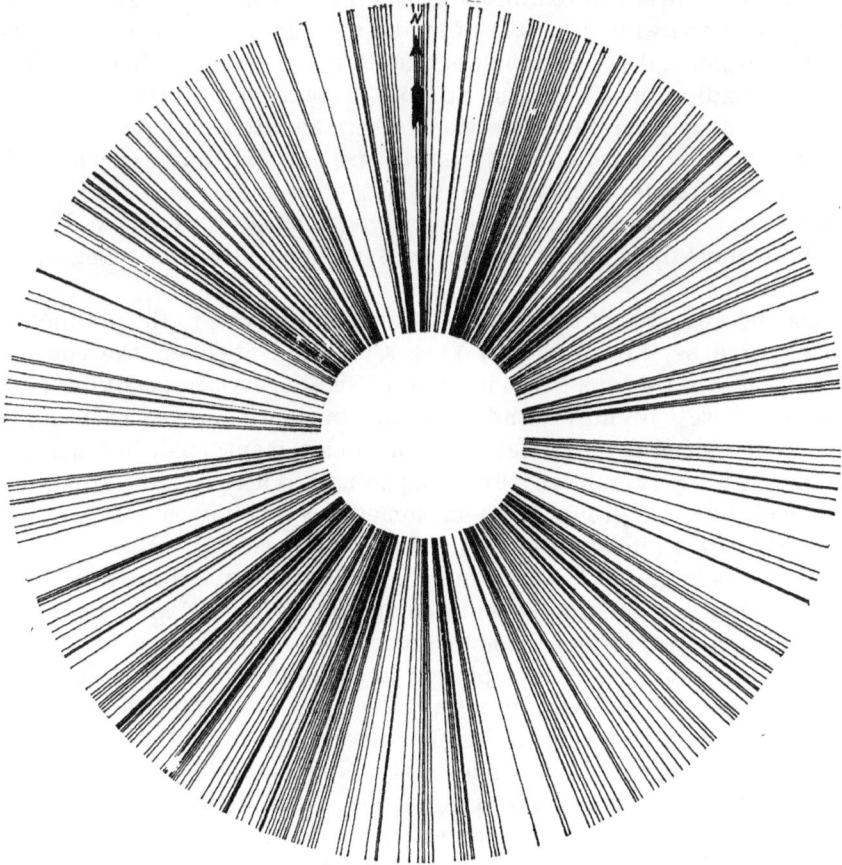

FIG. 1.—Diagram showing courses of some of the lode fissures of the Silverton quadrangle.

brings clearly before the eyes the great variety of direction assumed by the relatively few fissures which have been mapped as lodes. It fails, on the other hand, to express the relative persistency and individual importance of the various fissures represented, and in so far only partially brings out the prominence of the northeast-southwest and northwest-southeast lodes.

In a region traversed by such a great number of fissures, trending in so many different directions, it is not difficult to construct fanciful relationships. Few of the latter are more popular than that which

makes all the important lodes radiate from one or more points—
extinct volcanoes being generally preferred. That there is a certain
radial disposition of the fissures will be presently shown, but this
arrangement is very different from that which has been sometimes
supposed, and which has been described in intendedly serious publications.

Having considered the various directions taken by the fissures in
the region as a whole, it remains to discuss certain mutual relations
in regard to direction exhibited by the fissures of several small areas.
In spite of the wide diversity of strike already dwelt upon, and illustrated in fig. 1, it is nevertheless true that in certain limited portions
of the quadrangular area one or more systems of fissures, characterized by fairly accordant directions of strike, are easily recognizable.
In many cases these parallel fissures are so small and carry so little
ore or vein matter that they have not been indicated on the map.
Moreover, they are at places so closely spaced as to render it impossible to show them on the scale here employed for the topographic
mapping.

On the west side of Silver Lake Basin, in addition to the productive
lodes, such as the Iowa, New York City, and Stelzner, the country
rock is traversed by a great number of smaller fissures, striking from
20° to 40° west of north, and resulting in a pronounced sheeting of
the rock, generally parallel with the lodes mentioned, but without
recognizable regular or rhythmical spacing. These fissures are nearly
vertical, but one group of them sometimes exhibits a steep northeasterly dip, while another group dips southwesterly. Many of them
are ore bearing, and it is not uncommon to see on a clean surface of
rock a vein an inch or so wide composed of quartz and galena in
about equal proportions. Such mineralized stringers may be seen to
branch or to wedge out completely in a distance of 40 or 50 feet and
be succeeded by an overlapping veinlet a few inches to one side. On
a smaller scale they repeat the phenomena observed underground in
the nearly parallel lodes, such as the New York City, belonging to
Group II of the Silver Lake Basin lodes. (See detailed descriptions,
pp. 145–148.) They appear to belong to the same system and to have
been formed at the same time as the latter.

Two prominent sytems of parallel fissures, intersecting nearly at
right angles, are noticeable in the cliffs at the head of Arrastra
Gulch. One system, with a strike of about N. 32° W., and a dip of
from 75° to 80° to the northeast, runs generally parallel with the
trend of the gulch, and divides the compacted breccias of the Silverton series into great plates from a few inches to several feet in thickness. The width of this sheeted zone is from 30 to 40 feet. These
fissures may perhaps be placed in the same system as those just
described west of Silver Lake, and were possibly formed at about the
same time. But they apparently contain neither ore nor quartz, and

may be of later date. The second set of fissures, with a strike of N. 60° E., and a southeasterly dip of about 80°, crosses the gulch near its head, and may be seen also in Blair Gulch, in a considerably broader zone than the first set. Some of these fractures contain quartz, but, so far as observed, are not heavily mineralized and carry only a little pyrite. The relative ages of these two systems of intersecting fissures were not directly determinable in the field. It is not improbable that they were formed simultaneously as conjugate fissures. (See pp. 52–53).

Minor systems of fractures in parallel coordination may be well studied on Galena Mountain, a precipitous peak composed of both massive and fragmental andesite. The mass of the mountain is cut by very numerous fissures, many of them prominently indicated on the surface as lodes or veins. These may be divided into four groups, the members of each group being characterized by approximately the following strikes: (1) N. 45° E., (2) N. and S., (3) N. 25° W., and (4) N. 65° W. In each of these systems there usually occurs one or more fairly strong lodes accompanied by numerous smaller, nearly parallel, fissures, the rock being conspicuously sheeted. These associated parallel fissures are particularly abundant in the N. 25° W. system, as seen on the southern declivity of the mountain, in the vicinity of the Veta Madre mine. They are generally nearly vertical, and rather closely spaced, a foot or less apart. They are usually occupied by small veins which sometimes carry ore minerals, chiefly galena, sphalerite, and chalcopyrite, but which are seldom large enough or rich enough to be worked.

Another area within which the fissures exhibit striking parallel coordination is on Treasure Mountain and at the head of Placer Gulch. The dominant northeast-southwest fissure system of this portion of the quadrangle is exemplified by a series of strong, persistent lodes, represented by the Sunnyside and Scotia, with a general course of about N. 40° E. The dip of these is southwesterly at high angles. A second system of shorter, less conspicuous fractures crosses the first series almost at right angles, with a general course of about N. 55° W. These fissures are nearly vertical, but the majority of them dip southwest at high angles. In their trends these fractures belong with the system of northwest-southeast fissures already recognized as a system characteristic of the quadrangle as a whole. It is interesting to note that here, however, they are not, as in the Silver Lake Basin, the master fissures, but are short, transverse, relatively unimportant fractures, forming a nearly rectangular network with the strong, continuous northeast-southwest lodes. A third local system of transverse veins, represented by the Golden Fleece, has a course of about N. 75° E., with usually steep southerly dip. Still a fourth set of very prominent lodes is exposed at the head of Placer Gulch, with an average strike of about N. 25° E. Their dips are usually steep and

easterly.   There are several other fissures in this vicinity which do not
strictly belong with any of the systems noted.   Several have a nearly
east-and-west strike, and might, perhaps, be grouped together or
possibly included with veins of the Golden Fleece system.

In the canyon of the Uncompahgre, north of Mount Abrams, the
thick masses of San Juan breccia and the underlying Algonkian schists
and quartzites are cut by numerous nearly parallel fissures with a
general trend of about N. 5° W. and a dip of about 80° to 85° to the
east.   These fractures, which do not, as a rule, contain prominent
veins, have divided the San Juan breccia into huge, nearly vertical
slabs (Pl. VII), and have thereby contributed to the preservation of
the lofty cliff faces of this part of the canyon.   A second, less prominent, system of fissures, usually carrying quartz stringers, has a
general strike of about N. 60° W.

At the head of Porphyry Gulch, near the western border of the
quadrangle, the Potosi rhyolite and underlying Silverton series are
traversed by a conspicuous series of nearly vertical fissures, striking
N 8° W., which do not contain workable ore bodies, so far as known.

A similar series of fissures is also very prominently shown in the
steep walls of Canyon Creek northeast of the mouth of Richmond
Basin.   The general strike of these fissures appears to be about N.
35° W., and they usually dip steeply to the northeast.   They belong
genetically with the northwest-southeast system of lodes as developed
in Silver Lake and Richmond basins.   So far as is known they do not
carry any workable ore bodies.

On the eastern side of Ironton Park are several fissures having a
general northeast-southwest course and a steep southeasterly dip.
Their formation was accompanied by considerable normal faulting.

The parallel coordinated fissures or sheeted zones hitherto described
all possess considerable width and are characterized by the division
of the country rock into slabs of relatively great thickness as compared
with the resulting fissures.   But this is not always the case.   The
fissures may sometimes be very small and also very closely spaced, as
in the col separating Lake Como from Cement Creek, where the thorough manner in which the rock, a much-altered andesite or rhyolite,
has been sheeted is beautifully shown.   This minor sheeting is generally parallel in strike with the larger lodes in the vicinity, and at
the particular point where best shown the fracture planes dip westerly at about 60°.   As many as fifty such planes may frequently be
counted within the width of a foot.   There has been some infiltration
of silica along these minute fissures, which causes them to stand out
on weathered surfaces as little parallel ribs.   The local strike of this
sheeting is about N. 35° E., but it is sometimes disturbed by irregular
later cross fractures.   No regularity or rhythm could be detected in
their spacing.

Sometimes several nearly parallel fissures of considerable extent,

SHEETED STRUCTURE IN CLIFFS OF SAN JUAN BRECCIA ABOVE THE SILVER LINK
MINE, CANYON OF THE UNCOMPAHGRE RIVER

The breccia rests upon the eroded surface of Algonkian schists and quartzites.   Photograph by Whitman
Cross.

originally containing open spaces ranging from a few inches to a foot
or more in width, occur closely spaced within a comparatively narrow
zone bounded by less fractured country rock.   Such a sheeted zone
forms a favorable place for the deposition of ore bodies, and, as will
be shown later, the Camp Bird, Tomboy, and other lodes are chiefly
of this type, illustrated in figs. 9 and 14.   Coordinate fractures of this
simple character, in which the individual fissures maintain their individuality and essential parallelism for long distances, may pass without any line of demarcation into groups of fissures of the linked type,
in which the nearly parallel fissures are arranged more or less en
échelon and connected by smaller oblique fractures.   This structure
is exemplified to some extent in the Tomboy mine, but is not a com-

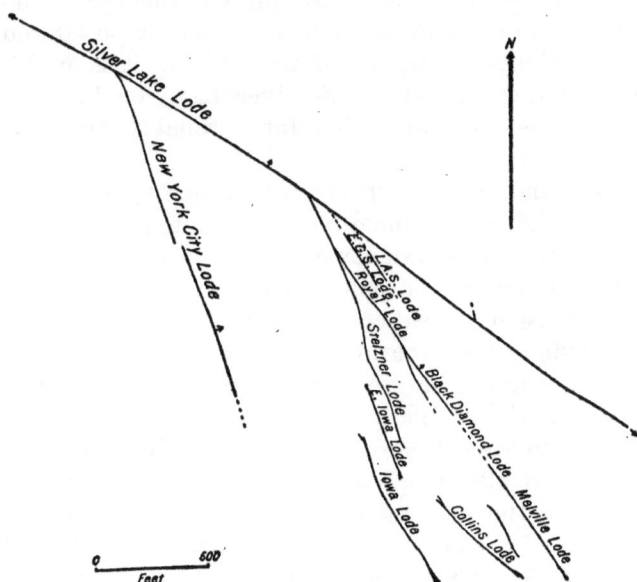

Fig. 2.—Horizontal sketch projection of the productive lodes west of Silver Lake.

mon one in the Silverton quadrangle.   With a still further shortening
of the principal parallel fissures and an increase in the number and
size of the linking fractures the fissure zone becomes a reticulated
aggregation of irregular, curving fractures, which, when filled with
ore, constitute what may be conveniently termed a stringer lode.   Of
this class is the North Star (King Solomon) lode.   Finally, with yet
greater increase in the number and irregularity of the fissuring, this
type of fissure zone passes by insensible gradations into a breccia
zone, such as a portion, at least, of the Polar Star lode.

Thus far the discussion has dealt with the parallel coordination of
fissures into minor systems and with the steps by which such groups
of fissures, when confined to rather narrow zones, may pass into other
less regular structures commonly associated with the deposition of

ore in lodes. It remains to consider another manner in which the fissures of this region are sometimes locally grouped, not in parallel, but in more or less radial, or, perhaps more accurately, branching, arrangement. This feature is well shown by the lodes of Silver Lake Basin. The accompanying diagram (fig. 2) is a horizontal sketch projection of the productive fissures worked in the Silver Lake and Iowa mines, compiled from maps of the underground workings. The relations of the lodes here shown are described on pp. 148–149, and it is only necessary in this place to call attention to the branching arrangement of the fissures and to the fact that all these lodes appear to have been formed at the same time. Another case of conspicuous branching occurs at the head of Placer Gulch, where several lodes having an average strike of about N. 25° E. split off from the main Sunnyside lode. The junctions in this case have not been exposed underground, but from the similar character of their filling it is highly probable that these fissures, like those of Silver Lake Basin, were all filled at the same time, and were therefore probably formed practically simultaneously.

Still a third and interesting case of branching fissures is that at Lake Como, partially and somewhat diagrammatically represented on the map (Pl. III). A view of some of these lodes is shown in Pl. VIII. In this instance, again, the lodes are all of similar character and all appear to have been formed contemporaneously.

Another mode of coordination sometimes observed in groups of fissures is that which Lindgren,[1] following Daubrée, has described under the name of conjugated fissures—i. e., in the particular region described by Lindgren, fissures having generally parallel strikes but dipping symmetrically in opposite directions. In the Silverton quadrangle, however, the fissures usually stand so nearly vertical that such a relation, if it exists, can not be clearly made out. The difference of dip, which would determine to which subdivision of such a conjugated system any given fracture would belong, is so slight, and may be so frequently found in various portions of one and the same fracture, that it does not seem practicable to recognize such systems. The low angle of dip at which most of the Grass Valley and Nevada City lodes lie makes the conjugated character of the fissure systems readily recognizable. It is to be noted, however, that in such conjugated fissures as those above described it is assumed that the rocks, in fracturing under tangential stress, yielded in a vertical direction, and that the prisms consequent upon the rupturing lie horizontal. But it is obvious that if, for some cause, the rocks could yield more easily in a horizontal direction, at right angles to the direction of thrust, then the resulting prisms would stand vertical and the conjugate fissures would, in this case, form two

[1] The gold-quartz veins of Nevada City and Grass Valley, Cal.: Seventeenth Ann. Rept. U. S. Geol. Survey, Pt. II, 1896, p. 164.

VIEW OF LODES EXPOSED ON THE SOUTH SIDE OF LAKE COMO.

intersecting systems of vertical fractures, with their obtuse angles of intersection bisected by the direction of the applied stress. Thus systems of parallel fissures flatly dipping in opposite directions and systems of nearly vertical intersecting fissures may both result from compressive stress, and may both be called conjugated fissures. Further, all gradations are possible between these two extreme cases, dependent upon the relation of the present surface to the original direction of least resistance when the fracturing took place. It is quite possible that some of the steeply dipping fissures of the Silverton quadrangle which intersect nearly at right angles form conjugate systems in the sense of Daubrée. The test of this lies in determining whether the intersecting fissures are of the same or different ages.

RELATION OF FISSURES TO KIND OF COUNTRY ROCK AND TO ROCK STRUCTURE.

It has already been pointed out that fissures carrying ore in greater or smaller amounts occur in practically all the rocks found in the quadrangle. Opportunities are rare, however, for observing directly the character of the fissures as they pass from rocks of one kind to another, and when such observations are possible no marked change is apparent. It has been stated by at least one observer,[1] that the lodes of this region occupy contraction fissures in the volcanic rocks, and do not extend down into the underlying schists or sediments. Nothing could be much farther from the facts than this assertion. Not only does the distribution of the fissures disprove it, but at the Green Mountain and Highland Mary mines, in Cunningham Gulch, the lodes may be seen passing from the Algonkian schists up into the overlying volcanic rocks of the San Juan and Silverton series. In the case of the Green Mountain mine no difference was noted in the fissure itself, either in width, direction, or general character, in the older schists and the younger rhyolite flow-breccia. The exact point at which the fissure passed from one rock into the other was not seen. The Highland Mary lode, where originally worked in Algonkian hornblende-schist, is a simple vein filling a clean-cut fissure 2 or 3 feet in width. Toward the northwest the lode passes into the volcanic rocks of the San Juan and Silverton series, and becomes a rather irregular zone of fissuring about 12 feet wide where observed at a point some 1,500 feet in elevation above the mine. In general it is apparently true that the fissures in the Algonkian rocks tend to be simpler and narrower than those in the volcanic rocks of later age. The fissures as a rule cut the schists cleanly at various angles with the schistosity, which apparently exercised no influence upon the direction of the fracturing.

No very definite differences can be observed in the character of the

---

[1] Philip Argall: Proc. Colo. Sci. Soc., Vol. IV, 1891-1893, p. 353.

fissuring within the various volcanic rocks of the quadrangle. It is very rarely possible to study in any one mine the passage of a given fissure from one rock into rock of a different sort, and thus the kind of observations which could best establish such characteristic differences, if they exist, is wanting. The massive lavas and indurated flow-breccias seems to have a general tendency toward comparatively simple fissuring, such as results in fissure veins of moderate size and regularity or in sheeted zones. In the softer tuffs and volcanic breccias the fissuring tends to be irregular, resulting in stringer or breccia lodes. Irregular fissuring appears to be favored also by alternations of harder and softer members of the nearly horizontal volcanic series, as at the North Star (King Solomon) mine. In this instance the softer rocks were regarded by the miners as the more favorable for ore. According to Purington,[1] in the northeast corner of the Telluride quadrangle the fissures passing from the San Juan formation and massive andesite up into the Potosi rhyolite series contain there less ore, and appear to have originally formed with less open space than in the underlying andesitic rocks. There are no workings in the Silverton quadrangle which afford an opportunity to verify this statement, but the Potosi rhyolite is generally regarded by the miners as unfavorable to ore in the region where it prevails. Near the head of Porphyry Gulch, however, a vein carrying galena and sphalerite ore up to 10 inches in width has been prospected in the Potosi rhyolite. The vein strikes N. 50° E. It can be followed for a short distance to the southwest, over the bare rhyolitic surface, when it either dies out or is cut off by a numerous series of small, nearly east-and-west fractures filled with white quartz and often showing excellent illustrations of linked-vein structure on a small scale.

In the intrusive stocks of monzonite, such as that of Sultan Mountain, the fissures are, as a rule, of simple regular character and moderate width. They are usually occupied by fairly simple fissure veins, subject to local contractions or enlargements, but seldom lose their simple linear character. As examples, may be cited the fissures of the North Star, Hercules, and Little Dora veins in Sultan Mountain, and of the Hamlet vein near Middleton.

The foregoing generalizations are admittedly based upon scanty data, and are to be considered rather as suggestions indicating lines of future inquiry than as authoritative statements of facts.

As a rule, local geological structure has had very little discoverable influence upon the fissuring. The lack of any regular relation between fissures and schistosity has already been pointed out. Contacts between rocks of different kinds and ages also appear to have had but little effect upon the fracturing and subsequent veining. This is partly due to the fact that the contacts are frequently nearly horizontal, while the fissuring took place along nearly vertical planes.

---

[1] Eighteenth Ann. Rept. U. S. Geol. Survey, Pt. III, 1898, p. 774.

Search was made in Cunningham Gulch, along the Uncompahgre, and at other points for any evidence of the diversion of fractures and mineralizing solutions along the contact of the older schistose terrane with the volcanic series, but with negative results.

Lines of fissuring in this region are occasionally determined by igneous dikes. Thus the Magnolia, a superficially prospected north-west-southeast lode just northeast of Silver Lake, follows for some distance an andesitic dike about 6 feet wide which curves across the gulch toward Round Mountain. Both this dike and a larger one which crosses it are irregularly fissured and traversed by poorly mineralized quartz stringers along the greater part of their exposed lengths. A similar occurrence was noted on the south side of Kendall Gulch about a mile a little east of south from Kendall Mountain, where a vein about 6 inches wide lies on the south side of a nearly east-and-west andesitic dike. None of these fissures have yet proved of much economic importance.

DISPLACEMENT OR FAULTING AS AN ACCOMPANIMENT OF THE FISSURING.

In a few cases only, and those of relatively unimportant lodes, has tangential dislocation been detected as a consequence or accompaniment of the original fissuring. The abandoned Moles mine, 3 miles south of Silverton, is on a vein about 2 feet wide which apparently fills a fault fissure of noticeable throw. The head of Deer Park Creek is crossed by a nearly north-and-south fault, with the down-throw on the west side, which is accompanied by some brecciation and veining of the schists along the fracture and with unimportant mineralization.

On the eastern side of Ironton Park, as shown in the Saratoga and Baltic mines, the formation of a parallel series of approximately northeast-southwest fissures has been accompanied by obvious fault-ing. These fissures, as a rule, dip steeply to the southeast. In all cases where it could be made out the faulting is normal, and the maximum throw, as observed on the Mono vein, can hardly be less than 100 feet. This fissure carries a body of low-grade pyritic ore.

In the northeastern portion of the quadrangle the field work of Messrs. Cross and Spencer appears to demand considerable faulting along some of the prominent vein fissures. But as there are no exten-sive underground workings in this portion of the area the evidences of such faulting did not come under my observation.

Notwithstanding the foregoing exceptions, it remains true that such displacement as occurred in connection with the formation of most of the principal productive lodes was slight in amount, although proba-bly not wholly absent. Moreover, the peculiar branching arrange-ment of many of the fissures, described on page 52, and their disposition en échelon, as in the Silver Lake and Iowa mines (p. 157), would seem opposed to the idea that any considerable tangential displacement of the walls had taken place during their opening. In the King lode, south of Silverton, and in other fissures cutting the schists, there is no recognizable displacement or throw of the individual schistose bands by the fissuring. Slickensiding of the fissure walls prior to the deposition of the ore has been nowhere recognized. In the Telluride quadrangle, where the occurrence of workable lodes in nearly horizontal sediments gives better opportunities than are usually available in the Silverton area for detecting and measuring displacement, Purington[1] evidently saw little clear evidence of it. Beyond recording a vertical separation of 1 foot by normal faulting in the San Bernardo mine, and concluding from the shape of the ore bodies of the Virginius and Smuggler-Union mines that the spaces they occupy were probably formed by faulting of the same kind, he makes no further reference to faulting as accompanying the formation of the ore-bearing fissures, although pointing out the desirability of future study in this direction. In nearly horizontal rocks faulting in which the movement is confined chiefly to the horizontal component (offset or heave) is not readily detected when of small amount, and was noted only in the few cases in which later-filled fissures have cut productive lodes, displacing the latter.

The absence of considerable displacement as an accompaniment of the opening of lode fissures is not a special peculiarity of the Silverton quadrangle, but is of very common occurrence. In fact, it seems to be comparatively rare to find more than very moderate faulting in fissures which have formed productive lodes.[2]

INTERSECTIONS AND RELATIVE AGES OF FISSURES.

In a region where the fractures are so numerous and possess such diverse trends intersections must necessarily be of frequent occurrence. But as veins at such points are particularly susceptible to disturbance by later movements of the rocks and to superficial disintegration, the study of these intersections is often difficult. The presence of seams of clay or gouge, due to these later or postmineral movements and the entrance of oxidizing waters, may so obscure the original relationship at the junction of two fissures as to render any conclusive determination of relative age impossible when no conspicuous faulting of one of the fissures has taken place to give the desired information.

In many cases in the Silverton quadrangle intersecting fissures have been formed nearly or quite simultaneously and filled by one process of vein deposition. This is true of the assemblage of branching fissures worked in the Silver Lake and Iowa mines, and probably so of the Placer Gulch and Lake Como fissures. It is also possible that

---

[1] Loc. cit., p. 780.
[2] Phillips and Louis: A Treatise on Ore Deposits. London. 1896.

many of the fissures intersecting nearly at right angles (conjugate fissures) were produced by a single principal stress. In the greater number of instances, however, in which lodes are known to intersect, the exposures are such as leave the actual character of the crossing and the relative ages of the fissures in doubt. They may or may not have been simultaneously formed.

Fairly clear evidence of the intersection of an earlier vein-filled fissure by a later one without recognizable faulting was obtained at several points. The prominent lode outcropping along the crest of Green Mountain crosses and is distinctly younger than a strong vein of barren quartz, about 10 feet wide, which is accompanied by several parallel minor veins and strikes about N. 60° E., dipping 80° to 85° southeasterly. The younger lode has a general strike of about N. 30° W. and a steep southwesterly dip. The intersection is thus approximately rectangular, and is not accompanied by faulting, so far as could be seen. In the complex network of Galena Mountain it was found that at least one of the several north-and-south veins was cut by at least one of the numerous veins striking nearly N. 25° W., and these in turn are apparently intersected in several instances by lodes trending about N. 65° W. One of the latter was also observed to cut a lode striking about N. 30° E. In none of the cases mentioned was the intersection accompanied by visible results of faulting. Other probable examples of northwest lodes cutting northeast lodes occur on Treasure Mountain, but the exposures were not such as to place the question of relative age beyond doubt.

The shaft of the San Juan Chief mine was sunk at the crossing of two lodes, but it is by no means clear that they are of different age. The main lode, striking about N. 40° E., formerly produced some ore from surface workings. The fact that in their underground development the owners of the mine appear to have inadvertently drifted along the barren intersecting lode, striking about N. 60° E., is the only indication that the latter may be younger.

In three places only were lodes found to be faulted by later transverse vein-filled fissures. The King lode, in Cataract Gulch, 2 miles south of Silverton, with a strike of N. 6° W. and a dip of about 80° to the west, is faulted by a small approximately east-and-west quartz vein, dipping south at about 50°. By this faulting the northern portion of the King lode is thrown about 12 feet to the westward. In the Silver Lake mine the important New York City vein, with an average strike of about N. 20° W., and dipping generally northeast at about 80°, is cut by a faulting lode, which, where best exposed, is a breccia zone about 6 feet wide, composed of sharply angular fragments of country rock cemented by quartz carrying a little worthless ore. This fault fissure strikes apparently about N. 73° E. and dips about 75° to the south. The New York City vein is displaced in the same way and to about the same degree as the King lode. In the Ridgway mine the main lode, with a strike of about N. 40° E., is faulted by the Alaska lode, with an average strike of about N. 25° W. The fault here is apparently of the same kind as those just described—an offset of the northern portion of the main lode to the westward, but the developments in the Ridgway mine were not sufficiently extensive in 1899 to clearly show the relations. It is interesting to find that the three observed cases of faulting of one lode by another are of like character. In each case the lode running most nearly north and south has been faulted by barren or low-grade lodes trending more nearly east and west. While the number of observations is at present too small to serve for the formulation of a general rule for the quadrangle, their unanimity is suggestive of the possibility that such a rule may be laid down in the future. Even now, in the event that a productive lode in the southeastern portion of the quadrangle should be found, on drifting northward, to be cut off by a more nearly east-and-west lode with southerly dip, the miner, in the absence of other clues, is not utterly at a loss, but may reasonably hope to regain his ore by crosscutting westward after drifting beyond the interruption.

That, after the original deposition of the important ore bodies, there was at least one period of minor fissuring, followed by fresh deposition of quartz, is abundantly shown in many of the lodes of the region. Thus, in the Royal Tiger, the Dives, the Tom Moore, the Sunnyside, the Red Cloud, the Polar Star, and various other mines and prospects the original ore has been fissured or brecciated, and subsequently healed with generally barren or low-grade quartz. In the Tom Moore the later veinlets are notable in carrying small amounts of native copper in the quartz. This later fracturing is not always conspicuous, but small stringers of quartz, sometimes with rhodonite and carbonates, can usually be detected cutting the ore when the latter is examined in considerable masses. This is by no means a rare phenomenon in lodes in general, which are in most cases complex structures, resulting from successive fissurings and fillings. Such complexity, other conditions being the same, is more likely to be prominent in lodes of which the initial fractures were opened at an early geological period. The lodes of the Silverton quadrangle show much less evidence of repeated movements and filling than do the auriferous lodes of the Sierra Nevada, which were probably initiated at the close of the Jurassic or in early Cretaceous time.

Still younger than any of the fractures hitherto described are numerous fissures, usually noticeable only in mine workings in which neither quartz nor ore has been deposited. These are the "slips," "gouge seams," or "breaks" of the miners. As these names partly indicate, the fissures are generally filled with a wet, plastic, grayish clay, or gouge, which is, as a rule, merely altered and ground-up country rock, resulting from the attrition of the fissure walls. The presence of this unctuous gouge, when sharply limited by harder country rock or vein filling, is taken to indicate relative movement of the walls, not necessarily resulting in any great net displacement. In other words, they are fault planes of greater or less movement, possibly oscillatory. The width of these post-mineral fractures, as they are sometimes conveniently called, may vary from a narrow, scarcely noticeable clay seam, up to zones 12 feet or more in width, in which the gouge is mingled with masses of shattered country rock, and of which the most notable example is found in the zone of so-called "broken ground" in the Silver Lake mine (see p. 154). Occasionally the fine clayey material is lacking and the fracture is the course of a copious stream of water. Such a fissure was encountered in the Silver Lake mine in crosscutting to the New York City vein on level C, and the same or a similar water-bearing fracture was subsequently cut in the Unity tunnel about 400 feet lower. Since these post-mineral fractures can be studied, as a rule, only in the vicinity of the lodes, and as they are frequently directly connected with the latter, they may be conveniently classed as follows:

(1) *Strike fractures generally parallel with the lode.*—These are most commonly seen as gouges next the hanging or foot wall or in the country rock near to and approximately parallel with the lode in strike and dip. Gouge seams of this class, although not generally characteristic of the lodes of the Silverton quadrangle, are frequently well exhibited. They are found on both foot and hanging wall of the Silver Lake lode, particularly the latter, and also in the country rock for a distance of several feet on either side of the main fissure. When such later fractures have followed a previously existing vein or stringer they usually contain crushed vein quartz mingled with the gouge.

(2) *Strike fractures dipping at considerable angles with the lode.*— These are not so common as the foregoing, but sometimes occur. The ore in this case will be found to rest locally on a bench or step of country rock, separated from the latter by a clay seam of varying thickness. An example of this structure was noted in a stope on the Silver Lake lode (see p. 152, fig. 11). Such an occurrence needs to be carefully distinguished from the steps or benches of ore formed by the original opening of a somewhat irregular fissure by thrust faulting. In this case there should normally be no gouge between the ore and the country rock upon which it rests.

(3) *Cross fractures differing from the lode in both strike and dip, and cutting it at an appreciable angle.*—These are observed in nearly all extensive workings, and are sometimes fairly abundant and regular, as in the case of those which cut the Silver Lake lode (p. 151).

Besides the three classes noted, various irregular fractures occur in almost all lodes which can not be definitely grouped.

The post-mineral fractures are simply of local importance to individual deposits, and it has not proved practicable to discuss and classify them with reference to the quadrangle as a whole. Their possible bearing upon secondary enrichment of the primary ore deposits will be discussed in another part of this report.

To briefly recapitulate, while many of the fissures in the Silverton quadrangle, including some which differ widely in direction, were formed at substantially the same time, there have been later periods of fissuring, also followed by vein deposition. The oldest fissures known have a course somewhat east of north, and appear to have been successively cut by later fissures approaching more and more to an east-and-west strike. This generalization, however, should not be taken too rigidly, as it is very probable that whenever a prominent set of nearly parallel fissures were formed, other fissures intersecting the dominant set at various angles were produced at the same time. Lastly, there has been fissuring not followed, as far as known, by any deposition of quartz or ore in the resulting fractures.

The evidence afforded by the intersections as to the possible conjugate character of the dominant northeast-southwest and northwest-southeast fissures is unfortunately inadequate, owing to lack of satisfactory exposures of such junctions to determine the question one way or the other.

GEOLOGICAL AGE OF THE FISSURES.

Beyond the statement that the ore-bearing fissures of the Silverton quadrangle are of Tertiary and probably of late Tertiary age, it is not at present possible to fix the exact geological time at which the first considerable fracturing took place. More precise knowledge waits upon a determination of the exact age of the Telluride conglomerate and the San Juan and later volcanic formations. The Telluride formation, which, when present, underlies the volcanic rocks, has been provisionally referred to the Eocene by Cross.[1] It is difficult to conceive of its being older than this, as it has been found by Cross to rest unconformably on the Colorado Cretaceous shales. The Telluride formation and the younger volcanic series have been cut by several great intrusive masses of monzonite, such as the Sultan Mountain stock, which is thus the most recent rock in the quadrangle and can hardly be older than the Neocene. Since the lodes occur in the monzonite as well as in other rocks of the area, their formation probably does not antedate the latter part of the Tertiary, and may, indeed, have extended into the Pleistocene.

PROBABLE DEPTH AT WHICH THE PORTIONS OF THE FISSURES NOW EXPOSED WERE ORIGINALLY FORMED.

According to Cross,[2] the maximum thickness of the volcanic rocks in the Telluride quadrangle was, in round numbers, 5,000 feet. Owing to the varying thickness of different members of these series

---

[1] Geologic Atlas U. S., folio 57, Telluride, Colorado; also Proc. Colo. Sci. Soc., Vol. V, 1894–1896, pp. 235–241.

[2] Telluride folio.

in the Silverton quadrangle, no very close estimate can be made of the average thickness of the volcanic accumulations over the quadrangle as a whole. It may provisionally be estimated, however, as from 5,000 to 6,000 feet. As no traces of any extensive later deposits have been found in the San Juan, it may be assumed that this thickness represents practically the total deposition from the beginning of Tertiary time. As most of the lodes worked at the present day occur within the volcanic series, and often from 1,000 to 3,000 feet above its base, it is evident that the portions of the fractures now accessible must have been formed at geologically moderate depths—that is, probably under 6,000 feet. When, further, it is remembered that erosion proceeded concurrently with the fissuring and probably made rapid headway during the progress of ore deposition, it appears that many of the ore deposits must have been formed well within depths frequently reached by mining operations. It might be expected that under these circumstances there would be a recognizable, although probably not very intimate, relationship between the ore as originally deposited and the present topographic surface. The full discussion of this subject, however, brings up questions whose solution properly belongs in another place, and which are treated in the section on the origin of the lode and stock ores.

PERSISTENCE OF THE FISSURES HORIZONTALLY AND IN DEPTH.

Fissures are not of indefinite extent either horizontally or vertically. No fixed limit, however, can be assigned to the length which a given fissure may attain. It depends upon the magnitude of the stress that produced the fissuring and upon the relative movement which has taken place between the walls. Great relative movement results in a long fissure, but as profound faulting appears to be usually not favorable to the subsequent formation of a lode, there is a certain variable limit beyond which length is to be regarded as an unfavorable factor in the productiveness of a fissure lode. There are, however, certain notable exceptions to the foregoing general rule that should not be ignored. Thus the Merrifield-Ural lode, near Nevada City, Cal., occupies, according to Lindgren, a thrust-fault fissure of probably over 1,000 feet throw.[1] In the Silverton quadrangle fissures vary greatly in length. Those having a length of 2 or 3 miles are certainly not uncommon, and it is very probable that some of the fractures extend continuously for as much as 6 miles. That great length is not necessary for the formation of a productive deposit is shown by the occurrence of such lodes as the Iowa, Steezner, and East Iowa, which can scarcely be longer than a quarter of a mile, and which die out at their southern ends in small branching fractures. The exact length of a fissure is, of course, rarely determinable, as practical exploitation seldom follows a lode to its total disappearance.

---

[1] Gold-quartz veins of Nevada City and Grass Valley districts; Seventeenth Ann. Rept. U. S. Geol. Survey, Pt. II, 1896, p. 167.

The formation of fractures is limited vertically by that depth below which the rocks are under such pressure that no fissures can form. As the zone of fracture, according to Hoskins and Van Hise,[1] has a depth of about 10,000 meters (33,000 feet), and as fractures may also form in the still deeper zone of combined flowage and fracture, it is plain that this limit will never be reached in mining operations. But many fractures undoubtedly die out long before reaching this ultimate limit. The longer horizontal and the vertical dimensions of the fissures were probably originally nearly the same, and it is not likely that the present depth of any fissure very greatly exceeds its length. The depth will at least be roughly proportional to length. But the depth to which fissures extend is rarely actually determined, as the value of the ore body nearly always falls below the limit of profitable working long before the fissure itself disappears. But that many smaller fissures do die out at moderate depths is a well-attested fact.[2] In the Silverton quadrangle, where mining development is as yet restricted to moderate or slight depths, no well-authenticated case is yet known of the actual dying out in depth of a fissure which carried workable ore at a higher level. The ore may change in character, or may disappear, as in the North Star (King Solomon) mine, but the fissure still continues to an unknown depth.

In such instances as the last many considerations enter into the question as to whether it is advisable to follow the pinched fissure to greater depth in the hope of finding new ore bodies. The first step in such an issue is to determine whether the pinch is merely a constriction in a fissure extending to greater depth or whether it signifies the final diminution of the lode. This can be decided only by a careful consideration of the length and strength of the fissure as exposed above, of the possible faulting which accompanied its formation, and of the behavior of the fissure in those portions already mined. If the length of the croppings be several times the depth attained, if the fissure be usually strong, if it has been opened by faulting, and if it has been found subject to local pinches above, it may safely be concluded that it will persist and open out again with increased depth. But equally important is the question of the character of ore that may be found below, even if the fissure continue, for, as will be fully discussed later on, the ore contents of a given fissure are not constant at all depths, either in kind or in value. Lastly, in connection with these factors must be considered the costs involved in mining the ore from an increased depth.

The foregoing relates to simple fissures. Lodes in general exhibit similar characteristics, but their persistency will in the main be greater than that of a simple fissure. They are subject to the same general laws as the individual fractures of which they are composed.

---

[1] Principles of North American pre-Cambrian geology; Sixteenth Ann. Rept. U. S. Geol. Survey, Pt. I, 1896, p. 593.

[2] Lindgren, loc. cit., p. 162.

ORIGIN OF THE FISSURES.

The attempt to explain the origin of so complex a network of fissures as occurs in the Silverton quadrangle presents many difficulties. No explanation that can be at present proposed is to be regarded as complete or as free from various necessary assumptions. Our knowledge as to the degree of homogeneity of the various rock masses under strain, the depth at which the fissures were originally formed, their exact extent, direction, and distribution, the geological conditions obtaining at the time of fissuring, and numerous other essential data are far too fragmentary to permit of anything like rigid analysis leading to irrefutable results. And yet the problem is not so hopelessly insoluble but that it may be reasonably taken up in the hope of finding at least a working hypothesis to account for major features.

Fissures in which ore deposits occur are very commonly regarded as produced by tangential compressive stress transmitted in nearly horizontal planes by the rocky envelope of the earth, within what Van Hise[1] has termed the zone of fracture, limited by a depth of 10,000 meters. In many districts this is, without much doubt, the immediate cause of the fracturing. Thus, in the Grass Valley and Nevada City region of California the conjugated systems of flatly dipping regular lodes, opened with the accompaniment of more or less thrust faulting, were probably, as Lindgren concludes, "produced by a succession of compressive stresses applied in different directions, chiefly from east to west and from north to south."[2]

In accounting for the fissures of the Telluride quadrangle, Purington[3] arrived at somewhat similar conclusions. He states:

From all observed phenomena it seems probable that the fissuring was made by forces acting at time intervals not far apart and at localities not far removed from one another. Since the fissuring is later than all the rocks of the quadrangle [Telluride], volcanic disturbances whose product is now visible can not be cited to account for it, but it is entirely possible that later disturbances of volcanic nature, which did not result in surface flows of lava, have produced a straining to the point of rupture in the tract under consideration. It is thought, with great reason, that in the area directly east and northeast of the Telluride quadrangle there are centers and necks of volcanic eruption. Such evidence as has been collected in the present investigation points to those quarters as the source from which the pressure came.

Since the fissures of the Telluride and Silverton quadrangles were formed at the same time and undoubtedly had a common origin, Purington's results have an important bearing on the problem in hand, and it becomes necessary to scrutinize them closely. The geological mapping of the Silverton quadrangle and reconnaissances to the north of it, accomplished by Cross and Spencer since Purington's report was written, have failed to identify the actual "centers and necks of

---

[1] Loc. cit., p. 589.

[2] Loc. cit., p. 170.

[3] Eighteenth Ann. Rept. U. S. Geol. Survey, Pt. III, 1898, p. 770.

volcanic eruption " referred to as the possible sources of the compressive
stresses supposedly effective in fissuring the rocks of the Telluride
quadrangle.   There is, however, ground for believing that there may
have been local manifestations of volcanism in the Red Mountain
region.   It is not impossible that the plug-like masses of porphyry
mapped by Mr. Cross (Pl. IV) may occupy conduits through which
volcanic materials reached a former surface some thousands of feet
above the present one.   The character of the ore deposits, the intensity
of the associated metamorphism, and the geological distribution of
the Tertiary volcanic rocks all strongly suggest that one or more centers of former volcanic activity underlie the Red Mountain region.
This, of course, does not mean that any trace of volcanic form remains
in the present topography.   Cones or craters, if they existed, have
been obliterated in the evolution of the erosional forms of to-day.

In the Silverton, however, no less than in the Telluride area, the
greater part of the fissuring is not only post-volcanic, but is later
than the monzonitic intrusions which cut the volcanic series.   Thus,
although the vast volcanic accumulations of the San Juan must have
issued from local vents, the reference of the forces which produced the
fissures to one or more centers of eruption immediately east or northeast of the Telluride quadrangle as their source is as yet scarcely
warranted by the facts.   In this connection it may be noted that the
intrusion of the great stocks of mozonite into the rocks of both quad-
rangles brings up the rather perplexing problem of the mechanics of
such intrusions and the disposition of the rocks which formerly occupied the space now filled by the intrusive mass.   He would be rash
who would assert that the intrusion of these stocks was not accompanied by strains and fissures in the surrounding rocks, and yet the
fact is clear that the principal ore-bearing fissures of the region were
formed after the monzonite had solidified, since they cut the latter as
well as the adjacent formations.

But leaving aside for the present the question of their source, it is
necessary to investigate more fully the extent to which tangential
stresses of the kind indicated may be regarded as an effective cause
of the fissuring of the Silverton quadrangle.   Normally, in homogeneous rocks, such stresses might be expected to produce systems of
conjugated fissures of parallel strike, with dips not far from 45°, and
accompanied by some, although not necessarily great, thrust faulting;
i. e., reversed faulting.   The actual fissures do not conform even
approximately to these conditions.   Assuming that the direction of
least resistance was vertical, it is inconceivable that fissures so nearly
vertical and showing so little evidence of thrust could have been
formed by tangential stress.   Moreover, such evidence of faulting as
Purington was able to obtain in the Telluride quadrangle indicated
normal faulting and not thrust faulting.

If, on the other hand, the direction of least resistance was not vertical, but horizontal, then tangential stress might produce systems of nearly vertical fissures intersecting nearly at right angles conjugate fissures. As shown by Daubrée's [1] classic experiments, such fractures would be undulating, not strictly parallel, and might branch as do the fissures of the Silver Lake Basin. Moreover, as shown by Daubrée's illustrations, the walls of such fissures will not necessarily be always slickensided, nor need the faulting be in all cases perceptible. Such faulting as took place would result mainly in horizontal displacement, or offset, and, not being necessarily great in amount, might be easily overlooked. It is possible in this way to account for much of the fissuring of the Silverton quadrangle, and perhaps this view best explains the two dominant directions of fissuring intersecting nearly at right angles.

This hypothesis that the dominant northeast-southwest and northwest-southeast fissures were formed by compressive stress, acting nearly horizontally, although the direction of least resistance which it presupposes is not that which would appear most probable, seems to offer fewer serious objections than any other that has been devised. Fissures of contraction they certainly can not be, for they cut indifferently rocks of most diverse age, character, and origin. Fissures originating from tensional strains would probably be more irregular and would not be associated with close parallel sheeting of the rock. Tension would find relief in a single fissure rather than in a series of closely spaced parallel fractures. Shearing strains due to differential elevation or subsidence might culminate in fissures, but these could not be expected to show the general regularity and the persistency of the principal fissures of the Silverton quadrangle, and would probably reveal evident faulting. Faults probably due to such forces occur within and about the quadrangle, but they rarely contain ore deposits. Torsional stress, which by Daubrée [2] is regarded as a system of pressures, and by Becker [3] as a system of tensions, is capable of producing systems of parallel fissures intersecting at nearly 90° as well as radial fissures. It may at times be impossible to distinguish such fissures from those formed by simple pressure. But the fissures produced by torsion are in general decidedly curved, with a more strongly marked radial disposition. Their characteristics are given in detail by Becker in the paper cited. In a homogeneous block subjected to pressure, there are formed, in addition to the main conjugate fractures, various more or less irregular fractures, making angles of less than 45° with the direction of pressure, and frequently branching from the main fissures (see Daubrée, loc. cit., Pl. II). Such fractures are comparable with the branching fissures of Silver Lake Basin and Placer Gulch. For this reason, and on

[1] Études Synthétiques de Géologie Expérimentale, Paris, 1879, Pl. II.
[2] Loc. cit., p. 321.
[3] Torsional theory of joints: Trans. Am. Inst. Min. Eng., Vol. XXIV, 1894, p. 137.

account of the preponderance of nearly north-and-south fractures over nearly east-and-west fractures, it is thought that the effective stress producing the principal fissuring of the quadrangle may have acted in a north-and-south direction. Further, since there is no known source of such stress exterior to the quadrangle and immediately adjacent region, it is thought probable that the stresses were generated chiefly within the quadrangular area, by slight vertical movements following the enormous transfer, in Tertiary time, of volcanic material from an intratelluric to a superficial position, and that the surrounding country merely acted as a relatively passive buttress against which the thrust was directed. In other words, it is believed that the stresses were due principally to local gravitative readjustment. Some genetic connection between the volcanism and the subsequent fissuring, mineralization, and veining can scarcely be doubted, as these phenomena rapidly diminish in intensity away from the volcanic district of the San Juan Mountains. It does not seem possible at present to do much more than suggest that hypothetical relation which seems to adhere most closely to the known facts.

In addition to the principal stress outlined, there were doubtless many minor directions of stress set up at various times, some contemporaneous with the principal north-and-south stress and others earlier or later, frequently leading to the production of minor local fissure systems. These also were probably in part compressive tangential thrusts, and it is the existence of these widely differing directions of effective stress that furnishes one of the strongest arguments against seeking sources for the latter outside of the region itself. It is exceedingly improbable that there should have been set up at various widely separated points in the relatively undisturbed regions inclosing the volcanic area, stresses which were to find their most energetic expression in the fissuring of the area itself. It seems likely that torsional strains also had their place in the readjustments following volcanism. Such a complex of fissures as that around Lake Como seems hardly to admit of any other satisfactory explanation. The probability that torsion has been a factor in the minor fissuring of this region has been suggested by S. F. Emmons.[1] It seems not unlikely that earthquake shocks, to which there is good reason to suppose the region was subjected during the faulting and principal fissuring accompanying postvolcanic readjustments, may have, in some instances, transformed torsional strain into actual fissuring, as suggested by Crosby.[2]

In conclusion, it may be well to recall the complex forces and results that enter into the fissuring under simple pressure of a block of what is commonly called homogeneous material. The geological conditions under which rocks are fractured are vastly more complex, and, as Dr. Becker[3] has remarked, "it is entirely safe to presume that every pos-

[1] Structural relations of ore deposits: Trans. Am. Inst. Min. Eng., Vol. XVI, 1888, p. 804.
[2] The origin of parallel and intersecting joints: Am. Geologist, Vol. XII, 1893, p. 368.
[3] The torsional theory of joints: Trans. Am. Inst. Min. Eng., Vol. XXIV, 1894, p. 130.

sible mode of deformation and rupture is exemplified," and, it may be added, frequently in a single limited field of fracture. The best that can be hoped for in any case is to ascertain the *dominant* mode in which fracturing has taken place.

STRUCTURES OF THE LODES.

The larger structural features of the various lodes depend mainly upon the character of the fissures in which they were deposited. Where the original fracture was a simple, clean dislocation, the resulting lode is a fissure vein. Most of the lodes of the Silverton quadrangle are of this character—nearly vertical plates of gangue and ore confined between definite walls. They sometimes show local irregularities and may divide into numerous branching stringers (stringer lodes) at their edges, but in the essential character of their workable portions they are veins, in the original sense of Von Cotta. Such are the veins of the Empire group on Sultan Mountain, the New York City (fig. 3), Stelzner, Royal, and Iowa veins of Silver Lake Basin, the Green Mountain vein, most of the veins of Galena Mountain, the Hamlet vein, and many of the lodes in the northeast portion of the quadrangle. The width of the workable veins usually varies from a few inches up to 10 or 12 feet. Lodes attaining greater width than this are rarely simple veins, although some of those near Sunnyside Basin, with widths of from 30 to 50 feet, appear to have filled simple open fissures. A width of 2 or 3 feet

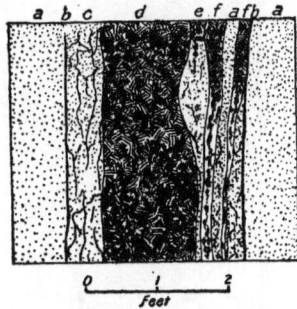

FIG. 3.—Cross section of the New York City lode, Silver Lake mine. *a*, country rock; *b*, miners' wall; *c*, broken country rock; *d*, ore, chiefly galena, chalcopyrite, and sphalerite; *e*, ore and country rock; *f*, ore and quartz.

is perhaps a rough characteristic average of the productive veins of the Silverton region. The vein filling usually fits snugly to the fissure walls and is frequently adherent to them—"frozen," as the miners say. Quite commonly, however, there has been sufficient movement along the fissure to cause the ore to come away readily from one or both walls, and sometimes there is a gouge or selvage present. This is rarely thick or extensive. Fissures sometimes contract, or pinch, and the vein then becomes much reduced in width and may be entirely absent. Such pinches, where the fairly solid walls are separated by a mere crack, were encountered on the New York City, Stelzner, and other veins, and on the Camp Bird lode. It is often difficult, when such a pinch is passed through by a crosscut, to believe that it really represents a lode which is elsewhere wide and productive. Many crosscut tunnels have on this account overshot the vein sought for. Careful systematic surveying and mapping is the safeguard against such mishaps and is the prerequisite of all intelligent and extensive prospecting.

As a rule the country rock adjoining the veins is not strikingly altered and retains practically the original form of the fractured surfaces. In some veins in rhyolite, however, there has been some replacement of the rock by ore, as may be seen in the Tom Moore mine.

Veins of the simple type described are connected by many transitional forms with lodes occupying closely spaced sheeted zones and consisting really of several parallel veins. Such are the small lodes of the Micky Breen and the important lodes of the Camp Bird and Tomboy mines. In both the Camp Bird and Tomboy, however, the parallel veining resulting from sheeting of the country rock is asso-

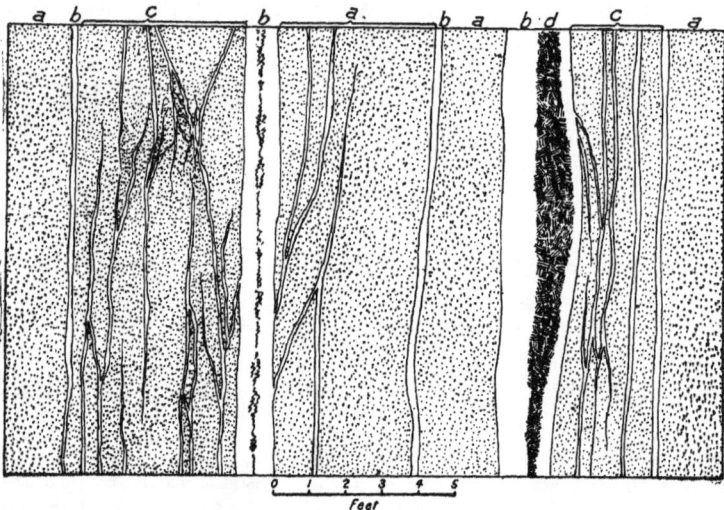

FIG. 4.—Sketch section of the Silver Crown lode, showing stringer-lode structure. a, andesite; b, quartz; c, andesite and quartz stringers; d, ore.

ciated with the less regular linked-vein structure, in which the lode is made up of nearly parallel or slightly diverging veins connected by linking stringers (Trümmer) and with the yet more irregular stringer-lode structure, in which the lode consists of a mass of stringers without noticeable parallelism among themselves. In these mines the more regular structure is usually found with the gold ore separated from the hanging wall by a little gouge, while the more irregular structures characterize the foot-wall portion of the lode, carrying a low-grade galena ore and sending off irregular stringers into the country rock, and being thus without regular foot wall.

The stringer-lode structure is perhaps best exemplified in the North Star (Solomon) lode, in portions of the Royal Tiger and of the Pride of the West lodes, in the upper part of the Forest lode, and in the Alabama lode. But it may frequently be well studied in surface out-

crops, notably those which stand out on the steep slopes of Dome
Peak, north of Howardsville, and in particular near the head of Mill
Creek, west of Chattanooga, where the Silver Crown lode is beautifully exposed in the bed of the stream, and the details of its structure
may be seen through the clear water. Fig. 4 is a sketch of the lode
as seen at the bottom of a clear pool in the creek. In minor development it is to be seen in nearly every fissure deposit in the quadrangle.
The most regular of the veins will sometimes locally split up into a
stringer lode, particularly where the fissure is seen about to die out
in the country rock.

Breccia lodes—that is, lodes in which the ore and gangue originally
filled the spaces in a zone of brecciated country rock, do not appear
to be common. The only deposit
seen which would appear to be
characterized chiefly by this structure is that of the Silver Queen
mine on Bear Creek, near the northern edge of the quadrangle. Here
the pay streak, from 3 to 6 feet
wide, lies in a brecciated zone about
12 feet wide in the San Juan formation. Judging from the materials
seen on their dumps, the Polar Star
and Red Cloud lodes also were in
part deposited in breccia zones.

To a certain extent the lodes as
originally formed have had their
structure modified by later movements, resulting in fractures in the
ore already deposited, followed by
a secondary veining, usually of
quartz. It could not be proved
that this process has exerted any

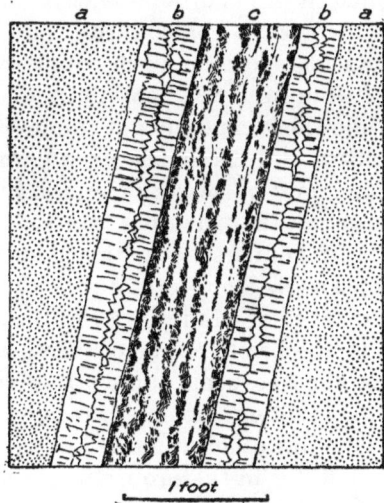

FIG. 5.—Cross section of the Japan lode,
showing structure produced by successive
openings of the original fissure. *a*, country
rock; *b*, quartz; *c*, ore.

important influence upon the ore. As a rule the secondary stringers
are small and relatively barren. In the Tom Moore lode the later
stringers carry a little free copper, but the amount is insignificant
from a commercial point of view. In the Polar Star and Red Cloud
mines an earlier deposition of quartz carrying pyrite was certainly
brecciated and followed by renewed veining. But there was no means
of determining in 1899 to which period of deposition the ore of these
mines belonged. The Japan lode, a section of which is shown in fig.
5, appears to have been formed by successive openings of the original
fissure. The main ore streak of banded galena and sphalerite ore was
first deposited. Later, nearly barren white quartz, with well-marked
comb structure, was deposited in fresh openings on each side of the
original vein.

The internal structure of the fissure fillings, such as banding and comb structure, and the size and shape of pay shoots, while, strictly speaking, embraced under the head of lode structure, will be more conveniently described in the following sections on the ores of the lodes.

THE ORES OF THE LODES.

MINERALOGY OF THE ORES.

GANGUE MINERALS.

Under this head are included those mineral constituents of the lodes which commonly make up the matrix of or are intimately associated with the metallic ore minerals. The term is a somewhat relative one. Thus pyrite is often a matrix for free gold, and has sometimes been treated, with chalcopyrite, etc., as a gangue mineral.[1] But in the present report the term gangue will be restricted to the so-called non-metallic minerals,[2] chiefly oxides, carbonates, sulphates, and silicates, while the compounds of the heavy metals, largely sulphides, sulphar-senites, sulpharsenates, sulphantimonites, or sulphobismuthites, all commonly possessing metallic luster, will be treated as ore minerals, although they may not in all cases be valuable. This is but following the convenient distinction laid down by such classical writers on ore deposits as Von Cotta and Von Groddeck, and is in accordance with the usage of Lindgren,[3] Phillips,[4] and others.

In the brief résumé of physical properties given with each mineral, the aim has been simply to give those features of color, form, etc., which will enable one to recognize the minerals as they occur in this particular quadrangle. The same minerals occurring in other regions may differ in the properties named.

Quartz.—SiO_2. Rhombohedral. Massive, or in hexagonal prisms terminated by rhombohedrons. Usually white or colorless. Hardness, 7. Specific gravity, 2.6.

This, as a rule, exhibits the usual character of vein quartz common in most mining districts. It varies from semiopaque, milk-white varieties to those which are vitreous and transparent. The latter occur in the Camp Bird and Tomboy lodes carrying free gold, usually in minute particles. In the Tomboy the occurrence of the gold in quartz of that particular vitreous appearance which is elsewhere commonly regarded as a sign of worthless vein matter has been commented upon by Purington.[5] The quartz of the veins is generally massive, and in thin section under the microscope is seen to be com-

---

[1] Mining industries of the Telluride quadrangle, Colorado, by C. W. Purington: Eighteenth Ann. Rept. U. S. Geol. Survey, Pt. III, 1898, p. 781.

[2] Von Groddeck, Lagerstätten der Erze, p. 58, Leipzig, 1879.

[3] Gold-quartz veins of Nevada City and Grass Valley districts: Seventeenth Ann. Rept. U. S. Geol. Survey, Pt. II, 1896, pp. 114–119.

[4] Treatise on Ore Deposits, 2d ed., pp. 1–2 and 85, London, 1896.

[5] Loc. cit., p. 840.

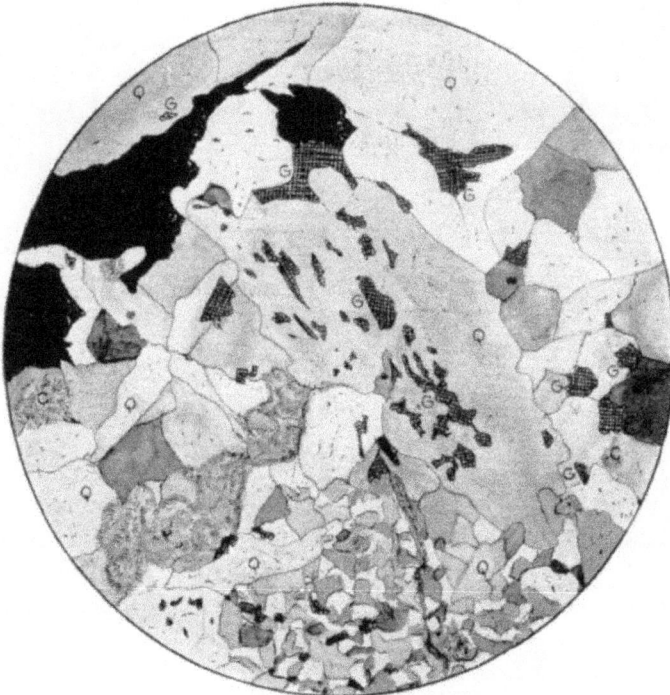

A. Stelzner lode.

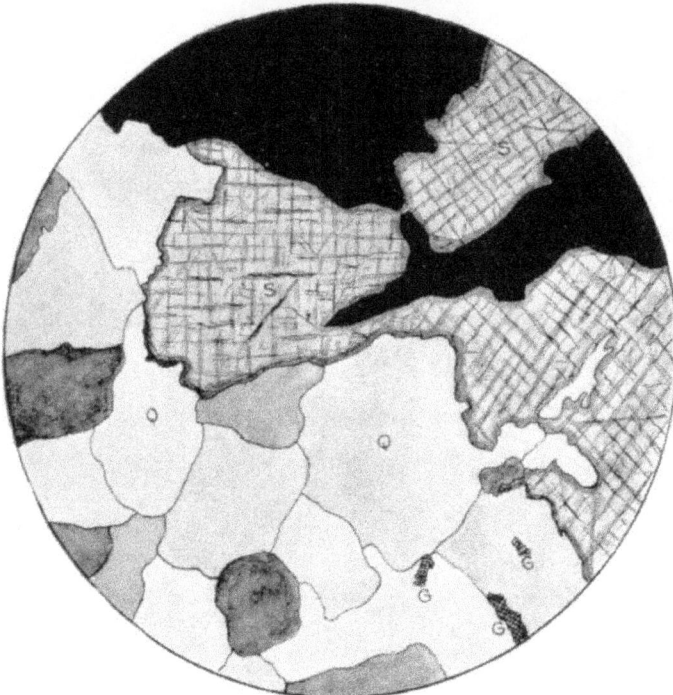

B. Sunnyside lode.

NORMAL VEIN-QUARTZ STRUCTURE.

Magnified 17 diameters; nicols crossed; black areas, chalcopyrite; Q, quartz; C, chlorite; G, galena;
S, sphalerite.

posed of interlocking grains with incomplete crystal boundaries (see Pl. IX, *A* and *B*). Distinct crystals are found only in small vugs and are never of large size. A radial arrangement of the imperfectly formed crystals was noted in the Dives, Magnet, Camp Bird, and Tomboy mines. The quartz of the Silver Lake lodes frequently incloses chlorite, which gives it a green color, usually regarded as an indication of good ore. In some of the lodes the presence of minute included crystals and grains of galena, sphalerite, tetrahedrite, argentite, various sulphobismuthites, and other ore minerals results in a dark-clouded quartz in which the ore minerals are not recognizable with the naked eye. Such is the richest ore of the Ridgway mine (Pl. X, *A*). Microscopic fluid inclusions are common, usually with gaseous bubbles. They are frequently arranged along irregularly disposed curved surfaces in the quartz.

In addition to quartz of the foregoing character, which has crystallized in open spaces, there occurs within and adjacent to many of the lodes, and especially in connection with the stocks of the Red Mountain district, a much more finely crystalline quartz, sometimes resembling a fine-grained quartzite, which has resulted from a partial or complete replacement of country rock by silica. As seen in thin section under the microscope, this quartz is associated with more or less sericite or kaolin, and sometimes alunite, in a cryptocrystalline mosaic. Such a mosaic often reveals the outlines of the former crystals of feldspar, which have been metasomatically replaced, as more fully described on pages 114–131.

As vein filling, with other gangue minerals and ore, quartz occurs in nearly all the productive lodes of the quadrangle, but the relative amount of quartz and ore minerals varies widely between the highly siliceous gold ore of the Tomboy, showing to the eye insignificant mineralization, to the coarsely crystalline, heavy lead ore of the Royal Tiger and Iowa mines. In a few of the lodes occurring in rhyolite, vein quartz may be practically absent. It did not occur to any considerable extent with the ore stocks of the Red Mountain district, where the quartz is chiefly the result of metasomatic replacement of the country rock.

Barite.—$BaSO_4$. Orthorhombic. Massive, or in groups of diverging tabular crystals. Cleaves perfectly in three directions. White. Transparent to opaque. Hardness, 3. Specific gravity, 4.5; whence common name "heavy spar."

This is not nearly so important a gangue mineral in the Silverton quadrangle as quartz, but it occurs massive with the latter mineral in the veins in the Sultan Mountain monzonite mass, in the Royal Tiger, Melville, and probably other lodes in Silver Lake Basin, as a heavy vein in the Dives and Potomac claims, in the veins of Galena Mountain, where it has sometimes been replaced by pseudomorphous quartz, and in the Bonanza, Alaska, Tempest, Alabama, Old Lout, and other lodes in Poughkeepsie Gulch.   In practically all the stocks of the Red Mountain district barite was found in close association with the argentiferous copper ores.   Thus in the Guston and Yankee Girl the bornite or chalcocite often incloses numerous isolated crystals of translucent barite, and it is to be found on nearly every mine dump in this vicinity.   It also occurs in the Zuñi mine, on Anvil Mountain, embedded in kaolinite, and with zunyite in guitermanite. Where occurring in lodes, the barite is very commonly associated with gray copper (tetrahedrite).   In the Mastodon claim, on the Sunnyside lode, barite forms with quartz a finely crystalline aggregate which constitutes the gangue of a highly argentiferous lead sulphobismuthite.

Calcite.—$CaCO_3$.   Rhombohedral.   Commonly massive or in rhombohedrons, scalenohedrons, or prisms.   Cleaves perfectly in three directions, affording rhombohedrons.   White.   Transparent to opaque. Hardness, 3.   Specific gravity, 2.71.   Effervesces freely with cold dilute acid.

Calcite is less abundant than barite in the productive lodes, but is nearly always present in small amount, particularly in small vugs and in minute veinlets cutting the ore.   As the principal filling of mineralized veins it was noted at the Oneida, a prospect at the head of American Basin, where it carries sphalerite with a little pyrite, chalcopyrite, and galena, and at the Yellow Jacket claim on Bear Creek, on the northern edge of the quadrangle, where also it contains abundant sphalerite.   In both cases the ore is too poor to work.   As an important gangue constituent it was noted in some of the copperbearing lodes north of Hanson Creek.   At the Osceola claim in Cunningham Gulch it occurs pseudomorphous after tremolite.   In the Camp Bird mine it is fairly abundant, filling vugs and spaces in the quartz left by comb structure.   As a microscopic constituent calcite is found almost invariably in the wall rocks near the lodes, except where the rock has been silicified, in small amount with rhodonite and rhodochrosite in the so-called "pink spar" (chiefly rhodonite) of the Sunnyside and other mines, and with fluorite and quartz in a palegreen cryptocrystalline aggregate accompanying and forming part of the rich streak in the Camp Bird lode.

As limestone, impure massive calcite occurs at several points in the quadrangle, and on the east side of Sultan Mountain and at the Saratoga mine in Ironton Park is directly associated with deposits of ore.

Dolomite.—$CaMg(CO_3)_2$.   Rhombohedral.   Usually massive or in curved rhombohedra.   Cleavage like calcite.   White to brownish. Milky to opaque.   Hardness, 3.5.   Specific gravity, 2.88.   Effervesces with cold dilute acid only when finely powdered.

This mineral is not abundant or important in connection with the ore deposits, but occurs as a microscopic constituent with rhodonite and rhodochrosite.

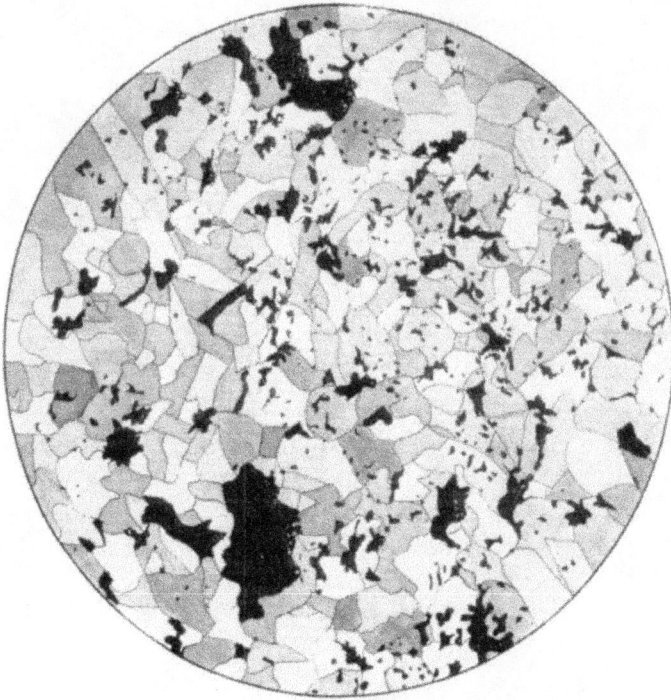

A. Finely crystalline ore from the Ridgway mine. Magnified 17 diameters; nicols crossed; large black areas, pyrite; small black areas, argentites, with a little sphalerite and galena; shaded areas, quartz.

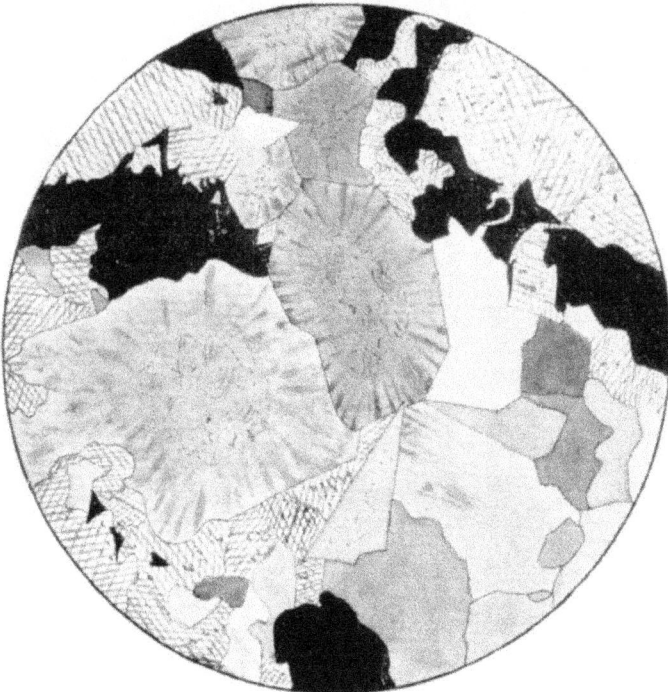

B. Gangue of rich ore, Camp Bird mine. Magnified 17 diameters; nicols crossed; black areas, fluorite; shaded areas, quartz; cross-hatched areas, calcite.

VEIN-QUARTZ STRUCTURE.

Rhodochrosite.—$MnCO_3$. Rhombohedral. Massive or in small rhombohedra in vugs. Cleavage like calcite. Usually some shade of pink. Hardness, 3.5–4.5. Specific gravity, 3.5. Effervesces freely in powdered form with dilute acid.

Occurs abundantly in massive form as gangue in the Titusville lode, and in small amounts in the veins of the Empire group on Sultan Mountain, and in most of the lodes in the northeast quarter of the quadrangle, where it may usually be observed as minute rhombohedral crystals lining small vugs. In small quantity it always occurs in the veins containing rhodonite. In the Golden Fleece vein it forms, with quartz, the gangue of the rich free-gold ore, and also occurs in small amount in the Camp Bird lode. It is found in beautifully colored rhombohedral crystals in the Grizzly Bear mine on Bear Creek, just within the northern boundary of the quadrangle.

Kaolinite.—$K_4Al_2Si_2O_9$. Monoclinic. Massive, or in a loose powder consisting of microscopic crystalline scales. White. Hardness, 2–2.5. Specific gravity, 2.6. Can be scratched with the finger nail. Smooth to the touch.

This mineral occurs in very pure form in the National Belle mine as a snowy white powder, made up of minute crystalline scales. As seen in the upper workings in 1899 it occurs filling fissures in the country rock or as a filling between the fragments of brecciated zones near the ore bodies. It was apparently in these cases deposited later than the ores, partly taking the place in this mine of the clay gouge commonly associated with post-mineral movement in other localities.

As an original constituent accompanying the ores, kaolin occurs abundantly in the stock deposits of the Red Mountain district, as shown by the materials now visible on their dumps. This is usually a firm, compact variety, intimately associated with pyrite. A characteristic form is that in which the kaolin is traversed by numerous anastomosing veinlets of pyrite, giving it a spotted and sometimes a schistose appearance (fig. 6). It was not possible to investigate the occurrence of kaolin in any of the deeper workings of the Red Mountain mines, but from what could be seen it appears to have accompanied the ores to the greatest depths there attained—about 1,300 feet. It was evidently derived from the country rock adjacent to the ore bodies as a product of its alteration by the thermal waters.

In the Zuñi mine, on Anvil Mountain, compact kaolinite is abundant and is directly associated with pyrite, barite, and small nests of enargite to form the lower exposed portion of the ore body. The pyrite is often beautifully crystallized in very slightly modified octahedra, which are thickly embedded in the white kaolinite.

As a soft white powder kaolinite occurs with the gold quartz of the Tomboy and Camp Bird mines. In the Tomboy the kaolin is sometimes mixed with sericite.[1]

---

[1] Purington, loc. cit., p. 840.

As an alteration product of the wall rock, more particularly of the feldspathic constituents, kaolinite is extremely common; but it is often impossible, without a chemical analysis, to distinguish it from sericite. Both minerals occur in similar fine aggregates, and they sometimes occur together. In such cases the microscope will not always serve to distinguish them.

Fluorite.—$CaFl_2$. Isometric. Massive, or in variously modified cubes and octahedra. Cleaves in four directions, forming octahedra. Colorless, or pale green or lilac. Hardness, 4. Specific gravity, 3.1.

Pale-green fluorspar is abundant on some of the dumps of the Aspen

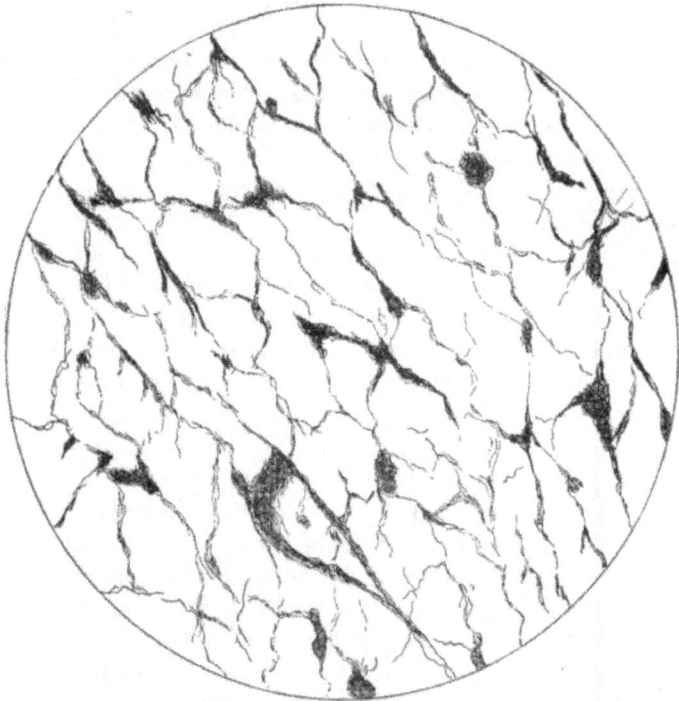

FIG. 6.—Diagram illustrating common occurrence of kaolin in the Red Mountain region. The kaolin is traversed by a network of pyrite. From Zuñi mine. Natural scale.

mine, but was not seen in place. In a massive condition it forms a vein which has been superficially prospected on the north side of Picayune Gulch, near its mouth, and was seen in a prospect near the Mountain Queen mine, in California Gulch, and in another prospect just east of Lake Como. It was also noted in the dump of the upper tunnel of the Micky Breen mine. Well crystallized with quartz, fluorite occurs in a prospect just north of the Old Lout mine, and in the Indiana tunnel in Grey Copper Gulch. In the Sunnyside lode, fluorspar, usually of a lilac tint, accompanies the best ore and is used by the miners as an indicator for free gold. Pale-green fluorite occurs with calcite in the quartz of the Tomboy lode and as a microscopic constituent in some of the rich Camp Bird ore. It probably occurs also in visible masses in portions of the latter lode, although none was seen at the time of visiting the mine. It is abundant in the Morning vein of the Japan mine, and in the Empire-Victoria vein with hübnerite. This latter association, however, is much more strikingly shown on the Adams claim, near Gladstone, as described on page 256. On the whole, however, it is not a common gangue mineral in this quadrangle, and when not too abundant is frequently associated with free gold.

Rhodonite.—$MnSiO_3$. Triclinic. Occurs only in cryptocrystalline massive form in Silverton quadrangle. Color, rose pink, fading and then turning black on exposure to the weather. Hardness, 5.5–6.5. Specific gravity, 3.5.

The silicate of manganese, comprising much of the so-called "pink spar" of the miners, occurs in many of the larger lodes of the northeast quarter of the quadrangle as a fine-grained pink material, very hard and tough, forming partitions between the ore-bearing portions of the lodes, as elsewhere described. It is a conspicuous and abundant constituent in the Sunnyside lode, in the neighboring lodes in Placer Gulch, and on Treasure Mountain, particularly in the dominant northeast fissures, and occurs much less abundantly in other lodes of the northeast quadrant. It is also found in the Saratoga mine, near Ironton, where it has partly replaced limestone. It is often difficult to distinguish this mineral by inspection and ordinary physical tests from the pink carbonate of manganese, rhodochrosite, when both are massive. This is due to the fact that the rhodonite is seldom pure, but is associated with quartz, calcite, and rhodochrosite in grains or crystals of microscopic dimensions. The massive rhodonite, when powdered, often effervesces slightly with acids, due to this slight admixture of carbonates. The greater part of the pink powder remains insoluble in dilute, boiling hydrochloric acid, and this fact and an examination of thin sections under the microscope show that the so-called "pink spar" of the Sunnyside and neighboring mines is the manganese silicate, rhodonite.

Zunyite.—A very basic orthosilicate of aluminum. $(Al(OH,F,Cl)_2)_6Al_2Si_3O_{12}$. Isometric, tetrahedral. Colorless. Hardness, 7. Specific gravity, 2.8.

This mineral, which was first described and named by Hillebrand,[1] from the Zuñi mine, is limited in its occurrence, so far as known, to the Zuñi and to one or two adjacent prospects on Anvil Mountain. It occurs as small tetrahedral crystals up to 5 millimeters (three-sixteenths of an inch) in diameter, embedded in guitermanite or its oxidation product, lead sulphate, and associated with pyrite, enargite, bournonite, kaolin, and barite. The chemical composition of

---

[1] On zunyite and guitermanite, two new minerals from Colorado: Proc. Colo. Sci. Soc., Vol. I, 1884, pp. 124–129. Also Bull. U. S. Geol. Survey No. 20, 1885.

zunyite given by Hillebrand as the mean of eleven partial analyses
is as follows:

Analysis of zunyite.

Constituent.	Per cent.	Constituent.	Per cent.
SiO_2	24.33	P_2O_5	.60
Fe_2O_3	.20	F	5.61
Al_2O_3	57.88	Cl	2.91
K_2O	.10		102.76
Na_2O	.24	Less O	3.02
Li_2O	Trace.		
H_2O	10.89		99.74

Although of much scientific interest, the mineral at present has no
economic importance.

Other minerals.—Sericite, epidote, chlorite, and zircon occur in
some of the ores, often merely as microscopic constituents. In the
mines of Silver Lake Basin, chlorite, with the usual green color and
minute radially foliated structure, occurs as nests in the vein quartz.
The spots of chlorite are commonly regarded as an indication of good
ore. It is abundant also in ore from the Silver King tunnel, on Mill
Creek, and in vein quartz carrying pyrite in the Barstow mine. Sericite is often associated with kaolinite in ores formed by partial replacement of rhyolite and related potash-bearing rocks. A little chalcedony
was noted in the Camp Bird ore. Alunite, although occurring in
the National Belle mine and as an alteration product of some of the
volcanic rocks of the Red Mountain range, has not been noted as a
true gangue mineral. Gypsum occurs occasionally as small crystals
in vugs of the Red Mountain ores.

In a fissure in the Silver Ledge mine there is a considerable quantity of a soft white material which, when moist, has a beautiful
faint-green tint. When dry the substance is a snow-white powder,
apparently of great purity. A partial analysis of this substance by
Dr. Hillebrand gives the following approximate chemical composition:

Constituent.	Per cent.	Constituent.	Per cent.
SiO_2	14.83	Alkalies (largely K_2O)	.20
Al_2O_3 (with trace of Fe_2O_3)	39.42	CaO	0.68
H_2O	40.25	CO_2 (calculated for $CaCO_3$)	.53
SO_3	4.26		100.17

This may be calculated as a mixture of allophane, aluminite, gibbsite, alunite, and calcite, in the following proportions, by weight:

Per cent.
Allophane (hydrous silicate of alumina) 58.0
Aluminite (hydrous sulphate of alumina) 15.5
Gibbsite (hydrate of alumina) 16.2
Alunite (hydrous sulphate of alumina and potash) 2.0
Calcite 1.2

92.9

There remains over 5 per cent of superfluous water, which is not included in the foregoing calculation. The chemical analysis, however, does not pretend to be more than approximate, and the mineralogical composition deduced from it is of similar character.

ORE MINERALS.

Under this head are included minerals generally mined as ores, together with some compounds of the heavy metals not of commercial value in this region. Unless otherwise stated, these minerals are characterized by a metallic luster.

Pyrite.—Iron disulphide (FeS_2). Isometric, pyritohedral. Massive, granular, or in pentagonal dodecahedrons (pyritohedrons), octahedrons, or cubes, or in combinations of these forms. Pale-brass yellow. Hardness, 6. Specific gravity, 5.

The isometric sulphide of iron is common in all the ores of the district, and impregnates to a varying extent all of the country rock in the vicinity of the ore bodies. In some cases this impregnation has involved huge masses of rock, as in the Red Mountain range, which owes its color to the oxidation of the pyrite that is scattered in minute crystals throughout the altered andesites and other rocks which compose the range. Large bodies of pyrite, formed in part by replacement of country rock, are known to exist in the lower workings of many of the abandoned mines of the Red Mountain district. Other large masses of crumbling granular texture occur near Ironton, in the Saratoga and Baltic mines, in great part replacing limestone. Pyrite, in beautiful octahedra embedded in kaolin, is abundant in the Zuñi mine, on Anvil Mountain. As a rule the pyrite, when occurring in large bodies with little or no quartz, is not of sufficient value to pay for working. When, however, as in the Henrietta mine, it is associated with a considerable amount of chalcopyrite, it can sometimes be mined at a profit. Associated with true vein quartz, especially when the latter carries some free gold, the pyrite itself is usually sufficiently auriferous to repay treatment. In the Tomboy and Camp Bird mines the gold occurring in the pyrite is less in amount than that occurring free in the quartz. In the Gold King, however, the reverse is true. A large part of the gold in the Silver Lake mine is known to occur in pyrite.

Pyrite in radially fibrous spherules, and at first supposed to be the orthorhombic sulphide of iron, marcasite, was noted in the dumps of the Red Cloud and Old Lout mines, and also occurred as small stalactites in cavities in the Genesee-Vanderbilt ore body. It was noted in radial spherules embedded in calcite at the Yellow Jacket claim on Bear Creek, near the northern edge of the quadrangle. Dr. H. N. Stokes, in the course of a chemical investigation on the iron sulphides, has shown that the foregoing radial forms are not marcasite but common pyrite. Dr. Stokes has kindly furnished the following note in regard to his method of determination:

Supposed marcasite from the Red Cloud mine, Colorado.—Carefully selected portions of the mineral, as free as possible from foreign substances, gave the "oxidation number" 65, that of pyrite being 60, and of marcasite 17. There can be no question, therefore, that the substance is pyrite, the excess of 5 over the normal figure being accounted for by small quantities of As, Pb, and Cu, which were found on analysis, the effect of these being to raise the "oxidation number."

The "oxidation number" may be thus defined: When a sulphide is boiled with a solution containing 1 gram ferric iron per liter, Fe''' is reduced to Fe'', and iron, if present in the mineral, goes into solution. At the same time more or less S is oxidized to SO_3. In the case of FeS_2 this is expressed by

$$\frac{0.333\, b}{a} - 25 = \text{per cent of sulphur oxidized} = \begin{cases} 60 \text{ for pyrite} \\ 17 \text{ for marcasite} \end{cases}$$

where a=amount of iron dissolved,
b=amount of Fe'' produced by reduction.

When other sulphides are present the effect is to contribute to b but not to a, thus increasing the apparent percentage of sulphur oxidized, as in the present case.

Tetrahedrite (gray copper).—Sulphantimonite and sulpharsenite of copper. The composition of tetrahedrite proper is $4Cu_2SSb_2S_3$, corresponding to 52.1 per cent of copper. For tennantite, the arsenical variety, it is $4Cu_2SAs_2S_3$, corresponding to 57.5 per cent of copper. These two varieties are connected by various intermediate compounds. Isometric, tetrahedral; commonly massive. Color, gray. Streak, brownish or reddish. Hardness, 3–4.5. Specific gravity, 4.4–5.1.

This mineral rivals galena in this region in importance and abundance as an ore constituent. Its high percentage of copper, with the fact that it often carries a large amount of silver replacing part of the copper, gives it its value. Most of the varieties seen contain both arsenic and antimony, but the latter predominates, especially in the highly argentiferous varieties sometimes known as freibergite (the typical freibergite from Freiberg contains over 30 per cent of silver), which is usually of a lighter gray than ordinary tetrahedrite. Moderately argentiferous tetrahedrite occurs in the North Star, Belcher, Empire, Little Dora, and other lodes on Sultan Mountain, in the Aspen mine on Hazelton Mountain, and in the Royal Tiger and other mines in Silver Lake Basin, associated with galena, chalcopyrite, sphalerite, pyrite, quartz, and barite. An antimonial variety rich in silver, probably to be referred to freibergite, constituted the principal ore of the North Star (King Solomon) mine. Tetrahedrite has been found in the Pride of the West, Philadelphia, and Highland Mary mines, in Cunningham Gulch, and usually constituted the richest ore. It occurs in most of the lodes of Picayune and California gulches, in the Tom Moore lode, in the Mickey Breen, Poughkeepsie, Bonanza, and other mines in Poughkeepsie Gulch, and in many other mines and prospects in various parts of the quadrangle. In some, such as the North Star (King Solomon), Tom Moore, and Mickey Breen, it is the principal constituent of ore bodies of some size and continuity, but in others, especially in those lodes carrying much galena, the tetrahedrite often occurs as bunches in the other ore. A little lead may sometimes replace part of the copper, as was found to be the case with a tetrahedrite carrying both arsenic and antimony, from the Black Diamond claim, on California Mountain. Tetrahedrite was present also in the ores of the Yankee Girl, Silver Bell, and other Red Mountain mines.

Enargite.—Sulpharsenate of copper ($3Cu_2S.As_2S_5$). Contains 48.3 per cent of copper. Orthorhombic. One perfect cleavage. Grayish black to iron black. Hardness, 3. Brittle. Specific gravity, 4.4.

Enargite is of somewhat frequent occurrence in the ores of the Red Mountain range. It was found abundantly in the Zuñi and Congress mines, and formed the principal ore of the National Belle. Handsome clusters of prismatic crystals incrusted with malachite and quartz were seen from the now inaccessible workings of this mine. According to Mr. Emmons, enargite occurred in the·Yankee Girl.

The enargite of the Zuñi mine is said to have carried over 200 ounces of silver per ton. In the National Belle it was probably of much lower grade.

Chalcocite.—Cuprous sulphide (Cu_2S), corresponding to 79.8 per cent of copper. Orthorhombic, commonly massive. Blackish lead-gray. Hardness, 2.5–3. Specific gravity, 5.5–5.8.

Stromeyerite.—Sulphide of silver and copper (($AgCu_2$)S). Form and physical properties like chalcocite, except slightly higher specific gravity—6.1–6.3.

These two closely related species do not always admit of sharp distinction. Stromeyerite may be regarded as chalcocite in which about half of the copper is replaced by silver. Chalcocite is itself a valuable ore mineral on account of its high percentage of copper, but stromeyerite, which may contain over 50 per cent of silver, is particularly valuable. The two minerals are probably connected by intermediate varieties.

Chalcocite, more or less argentiferous, was noted on the dump of the Frank Hough mine, and bornite in the Silver Link and John J. Crooke. Stromeyerite is reported to have been formerly an abundant ore in the Yankee Girl and Guston mines, where it occurred above a depth of 600 feet, between galena above and bornite below. One lot of 6 tons of this rich ore from the Yankee Girl contained over 5,300 ounces of silver per ton,[1] while one small lot from the sixth level of the Guston is stated to have contained 15,000 ounces of silver per ton, which corresponds roughly to the percentage of silver in pure stromeyerite. None of this ore could be seen in 1899 or 1900.

Some of the richest ore in the New York City lode, in the Silver Lake mine, contains small quantities of a black amorphous substance,

---

[1] Schwarz: Trans. Am. Inst. Min. Eng., Vol. XVIII, 1889-90, p. 145.

associated with chalcopyrite and pyrite. It was at first supposed to
be black oxide of copper, but upon closer investigation proves to be
a copper sulphide, probably chalcocite.

Bornite.—Sulphide of copper and iron ($3Cu_2S.Fe_2S_3$). Contains
55.5 per cent of copper. Isometric. Commonly massive. Copper
red or pinchbeck brown on fresh fracture, but soon becomes irides-
cent from tarnish, whence the name "peacock ore." Hardness, 3.
Brittle. Specific gravity, 4.9–5.4.

This mineral, often highly argentiferous by the replacement of a
portion of the copper by silver, was a very important constituent of
the large ore bodies formerly worked in the Red Mountain district.
According to T. E. Schwarz, it was the "principal ore for large
masses"[1] in the Yankee Girl. It formed large solid masses in the
Guston mine, associated with barite, the crystals of the latter being
often embedded in the bornite. It also occurred in the Genesee-
Vanderbilt mine. It is usually intimately associated with chalcocite
and chalcopyrite. With quartz it occurs in bunches in the Silver Link
mine and also in the John J. Crooke, a prospect north of Hensen
Creek, associated with chalcocite.

Chalcopyrite (yellow copper).—Sulphide of copper and iron ($Cu_2S.
Fe_2S_3$), corresponding to 34.5 per cent of copper. Tetragonal, sphe-
noidal. Commonly massive. Brass yellow. Hardness, 3.5–4. Brittle.
Specific gravity, 4.1–4.3.

This is a common ore mineral throughout the quadrangle, and is
sometimes auriferous, as in the Sound Democrat, or argentiferous
(carrying also a little gold), as in the Yankee Girl, Guston, National
Belle, and other Red Mountain mines, and in the Guadaloupe mine
on Abrams Mountain. It is very abundant in the Titusville mine,
in the New York City lode of the Silver Lake mine, and in the Hamlet mine. Associated with pyrite, it forms the ore of the Henrietta
mine, and with tetrahedrite much of the ore of the Tom Moore
lode. It is always present in ores carrying galena, sphalerite, and
pyrite, although the amount varies widely in different occurrences.
In the Saratoga mine it occurs with pyrite and galena, replacing
limestone.

Galena.—Lead sulphide (PbS). Isometric. Rarely showing external crystal form, but characterized by its perfect cubic cleavage.
Lead gray. Hardness, 2.5. Specific gravity, 7.5.

This very important ore mineral, which when pure contains 86.6
per cent of lead, occurs in nearly every ore deposit in the quadrangle, although in extremely varying amounts. In the lodes of
the Silver Lake Basin and vicinity it occurs in coarsely crystalline
masses, associated with quartz, sphalerite, chalcopyrite, pyrite, and
sometimes barite and tetrahedrite. Such coarsely crystalline varieties carry in this region relatively low silver values. In other lodes,

---
[1] Personal letter.

in Maggie, Picayune, and Placer gulches, and around Mineral Point, the galena sometimes occurs so minutely crystallized as to give merely a dark color to the quartz in which it is inclosed. This finely crystalline galena is often, but not invariably, highly argentiferous, as in the Sound Democrat mine in Placer Gulch. In the Sunnyside mine the foreman, J. James, informed me that he had seen free gold embedded in galena.

Sphalerite (*zinc blende*).—Zinc sulphide (ZnS). Isometric, massive, and tetrahedral. In variously modified tetrahedral forms. Perfect dodecahedral cleavage. Luster resinous. Color, various shades of yellow or orange to dark brown or black. Hardness, 3.3 to 4. Specific gravity, 4.0.

This is a very common mineral in this region and always accompanies galena. It is not, however, worked as an ore of zinc, and its presence in the ore often involves additional cost in concentrating and smelting. The light-yellow varieties, commonly termed "rosin zinc" by the miners, are often associated with the occurrence of gold. An orange or red sphalerite is common in the ores of the Red Mountain range. In the Yellow Jacket, a prospect on Bear Creek near the northern edge of the quadrangle, sphalerite occurs abundantly in a peculiar form, viz, as cylindrical aggregates of radial structure, thickly embedded in a gangue of calcite. These aggregates, which occur in all sizes up to an inch or so in diameter, frequently contain small specks of galena and chalcopyrite. Spherules of marcasite also occur occasionally embedded, like the sphalerite, in the calcite. When the calcite is dissolved in acid the sphalerite remains as a loose mass of cylindrical rods lying in all directions. In nearly all cases these rods are broken sections, as if they originally formed a loose mass of small shattered stalactites which were subsequently cemented by calcite.

Bournonite.—Sulphantimonite of lead and copper (3(PbCu$_2$)S.Sb$_2$S$_3$). Orthorhombic. Steel gray, lead gray, or black. Hardness, 2.5 to 3. Rather brittle. Specific gravity, 5.8.

According to Mr. S. F. Emmons's unpublished notes, this mineral is reported as having been found in the Yankee Girl mine. A lead-copper sulphantimonarsenite, probably bournonite with some arsenic replacing the antimony, occurs at the Zuñi mine, in small vertically striated prisms with pyrite and zunyite.

Zinkenite.—Sulphantimonite of lead (PbS.Sb$_2$S$_3$). Orthorhombic, but occurs chiefly massive. Steel gray. Hardness, 3 to 3.5. Specific gravity, 5.3.

Described and analyzed by Hillebrand [1] from the Brobdignag claim, where it is said to occur sparingly with barite. None could be seen in 1899 or 1900.

---

[1] Mineralogical notes: Proc. Colo. Sci. Soc., Vol. I, 1884, pp. 121-123.

Guitermanite.—Sulpharsenite of lead ($3PbS.As_2S_3$). Massive, compact. Bluish gray. Hardness, 3. Specific gravity, 5.9.

This mineral, with zunyite, was first described and named by Dr. Hillebrand[1], and occurs, as far as known, only in the Zuñi and adjacent claims. It always incloses the minute, sparkling tetrahedrons of zunyite, and is intimately associated with pyrite, enargite, kaolin, and barite. It alters superficially to anglesite. As it contains about 66 per cent of lead, and probably some silver, it is a valuable ore.

Stibnite.—Antimony sulphide (Sb_2S_3). Orthorhombic. Usually in clusters of radiating prisms. Perfect cleavage in one direction. Lead gray. Hardness, 2. Specific gravity, 4.5.

A single specimen of this mineral was seen, which was stated on reliable authority to have come from the North Star mine on Sultan Mountain. It is readily distinguished by its lustrous cleavage surfaces and softness, being easily scratched with the finger nail.

Polybasite.—Sulphantimonite of silver ($9AgS.Sb_2S_3$), with part of silver replaced by copper and part of antimony by arsenic. Orthorhombic. In characteristic, short, six-sided tabular prisms, with triangular striations and beveled edges. Iron-black. Hardness, 2 to 3. Specific gravity, 6.1.

This rich silver ore is known to have occurred in the upper workings of the Yankee Girl, and probably in other mines of the Red Mountain district. Well-crystallized specimens from this region are preserved in the museum of the Bureau of Mines, Denver, and in various private cabinets.

Proustite (*ruby silver*).—Sulpharsenite of silver ($3Ag_2S.As_2S$), corresponding to 65.4 per cent of silver. Rhombohedral, hemimorphic. Luster, adamantine. Transparent to translucent. Color, scarlet-vermilion, but somewhat masked by brilliant luster. Hardness, 2.25. Brittle. Specific gravity, 5.5 to 5.6.

Few of the ores now mined show this mineral, although it was seen in the Ben Butler mine, associated with galena and perhaps argentite. It is also known in the Ridgeway mine and has been reported on good authority from the Red Cloud, Polar Star, Mammoth, Annie Wood, Palmetto, and Wheel of Fortune mines. It is known to have occurred in the Yankee Girl, and a specimen was seen which was said to have come from the Genesee-Vanderbilt mine. It characterizes the upper portions of argentiferous ore bodies and is not known at depths over a few hundred feet.

Bismuthinite.—Sulphide of bismuth (Bi_2S_3). Orthorhombic. One perfect cleavage. Lead-gray. Hardness, 2. Somewhat sectile. Specific gravity, 6.5.

This sulphide of bismuth occurs in slender prismatic crystals, with specularite in quartz at the Neigold claim on the south slope of Galena Mountain.

---

[1] On zunyite and guitermanite, two new minerals from Colorado: Proc. Colo. Sci. Soc., Vol. I, 1884, pp. 129-131. Also Bull. U. S. Geol. Survey No. 20, 1885.

Argentite.—Silver sulphide (Ag_2S), containing 87.1 per cent of silver. Isometric. Blackish lead-gray. Hardness, 2 to 2.5. Sectile. Specific gravity, 7.3.

This valuable ore of silver, which is readily distinguished by the ease with which it can be cut with a knife without breaking or splintering, is not abundant in the Silverton quadrangle. It was distinctly recognized only at the Ridgway mine, where, curiously enough, it is called "brittle silver." Here it constitutes the richest ore, and is evidently one of the most recently formed of all the ore minerals, as it occurs characteristically in vugs, incrusting or filling the interstices between the quartz crystals. It probably occurs also in the Gold Nugget and other prospects in Maggie Gulch, and is said to have constituted a large part of the rich ore formerly extracted from the Polar Star mine, on Engineer Mountain, and from the Palmetto mine, in American Flats.

Molybdenite.—Molybdenum disulphide (MoS_2). Probably hexagonal. Occurs in scales, with eminent basal cleavage, or compact. Laminæ flexible, but not elastic. Lead-gray. Leaves a bluish-gray mark on paper. Feel, greasy. Hardness, 1 to 1.5.

This mineral occurs in the Sunnyside Extension mine, where it has been mistaken for graphite. Here it frequently contains free gold. It has not been recognized elsewhere within the quadrangle.

Hematite (specularite).—Ferric oxide (Fe_2O_3). Rhombohedral. Lamellar or in thin scales. Color, steel-gray or iron-black. Streak, cherry-red or reddish-brown. Hardness, 5.5–6.5, but often apparently much softer on account of its scaly structure. Specific gravity, 5.2.

The variety of hematite known as specular iron is frequently met with in small quantities in the lodes of the Silverton quadrangle. It occurs, for example, in the Crown Point lode on Silverton Mountain, in the Little Giant lode (on the dump), in the Neigold claims on Galena Mountain, and in the Daniel Webster prospect in Maggie Gulch. In all cases observed it is a vein mineral inclosed in quartz. It is of no economic importance, but is frequently mistaken for more valuable minerals—in one case for "brittle silver." It is easily recognized by its scaly structure, brittleness, and red color when crushed.

Sulphobismuthites.—These include cosalite ($2PbS.Bi_2S_3$), galenobismuthite ($PbS.Bi_2S_3$), alaskaite (($PbAg_2)S.Bi_2S_3$), beegerite ($6Pb-S.Bi_2S_3$), kobellite ($2PbS.(Bi,Sb)_2S_3$), and possibly other species.

Sulphobismuthites of lead, carrying usually considerable silver and sometimes a little copper, are of frequent occurrence, particularly in the northern half of the quadrangle. As most of the above-named minerals resemble one another very closely in physical properties and as a rule occur massive, their specific identification is impossible without a quantitative chemical analysis in each case. Nor is this method often available, for the association of these sulphobismuthites with other ore minerals is usually such that pure material suitable for chemical analysis can not be obtained.   They are nearly all lead-gray in color, with hardness varying from 2.5 to 4.

The argentiferous galenobismuthite named by Koenig[1] *alaskaite* occurs abundantly in the Alaska mine and in the Acapulco claim adjoining it on the east.   Its occurrence and accompanying minerals are described on page 195.

Cosalite, from the Yankee Girl mine, with more or less of the lead replaced by silver, has been analyzed by Low,[2] and probably occurs in other mines of the vicinity.   It has been reported also from the Alaska mine.   Kobellite has been analyzed and described by Keller[3] from the Silver Bell mine, and beegerite, from Poughkeepsie Gulch, by Koenig.[4]   The naming and identification of these species from the Silverton quadrangle rests upon chemical analysis of massive, more or less impure material, and these results have not yet been confirmed by any investigation of crystal form.

In the course of the present investigation sulphobismuthites have been met with in many of the ores, but never with external crystal form nor in such amount or purity as would warrant complete chemical analysis, whereby alone they can be specifically determined.

At the Barstow mine the richest ore is a bright lead-gray mineral, of hardness 3 to 3.5 and specific gravity 7.04 at 30° C. (Hillebrand). It shows indistinct crystal forms, suggestive of isometric symmetry, and is very intimately intergrown with pyrite.

The purest material that could be picked out proved to be a mixture of a lead sulphobismuthite with pyrite and a telluride.   According to Dr. Hillebrand the following may be taken as a rough approximation to the chemical composition of this mixture:

Approximate chemical composition of material from Barstow mine.

Constituent.	Per cent.	Constituent.	Per cent.
Pb	45.5	Fe	2.8
Bi	22.5	Te	3.5
Ag	6.0	S	14.1
Au	.06	Insol	1.5–2.0
Cu	.3		

Had pyrite been the only mineral present with the sulphobismuthite it might have been easily deducted from the completed analysis. But the presence of tellurium shows that there is present also some telluride of unknown composition which can not be thus allowed for. The idea of a quantitative analysis was therefore abandoned.   The above rough approximation indicates the probable presence of a mineral of the composition of beegerite, mixed with pyrite and perhaps an auriferous hessite (telluride of silver).

---

[1] Proc. Am. Philos. Soc., Vol. XIX, 1881, pp. 472–477.

[2] Proc. Colo. Sci. Soc., Vol. I, 1883–84, p. 111.

[3] Zeitschr. für Kryst., Vol. XVII, 1890, pp. 67–72.

[4] Loc. cit.

In the Sunnyside Extension mine a sulphobismuthite of lead and silver, showing no distinct crystal form, occurs intimately associated with molybdenite and free gold in an ore containing quartz, barite, sphalerite, pyrite, and galena. In the Mastodon, which adjoins the Sunnyside Extension on the north, an argentiferous sulphobismuthite of lead showing indistinct prismatic crystallization occurs so finely disseminated in quartz as to give the latter a dark clouded appearance.

In the Custer claim on California Mountain an argentiferous lead, sulphobismuthite, forms the richest ore, and similar bismuthiferous compounds occur in the ores of the Sound Democrat and Silver Queen mines in Placer Gulch.

A lead-copper sulphobismuthite, probably argentiferous, occurs in small, ragged prisms, frequently forming star-like radial clusters, in quartz with pyrite at the Uncompahgre Chief, near Mineral Point.

At the Silver Bell mine, between Ironton and Red Mountain, the richest ore was a massive argentiferous lead sulphobismuthite carrying up to 1,000 ounces of silver per ton. Massive argentiferous lead sulphobismuthites, of probably more than one species (one of them cupriferous), occurred in the Genesee-Vanderbilt ores, intimately associated with barite and pyrite. A massive lead sulphobismuthite (with, perhaps, a little copper) has been found in the Silver Queen mine on Bear Creek. In the Neigold group of claims on Galena Mountain a lead sulphobismuthite, carrying a little silver and perhaps some copper, occurs as small indistinct prisms and specks in quartz with pyrite, specularite, and bismuthite. Lastly, it is known that rich bismuthiferous ores, probably argentiferous lead sulphobismuthites, were formerly mined in the Old Lout and Poughkeepsie mines, and are probably present in varying amounts in most of the lodes of Poughkeepsie Gulch.

Tellurides.—The occurrence of tellurides of gold or silver has been noted at only four localities within the quadrangle, and at each of these in small amounts only. In some of the rich ore of the Camp Bird mine a little tellurium, probably occurring as a telluride, has been detected chemically by Dr. Hillebrand, although no tellurium mineral has been recognized. In the Barstow mine small amounts of an unknown telluride are intimately associated with an argentiferous lead sulphobismuthite and pyrite. Its presence can be detected only by chemical means. At the Silver Ledge mine a small amount of a telluride resembling calaverite was found, with free gold in a single small pocket. At the Magnet mine telluride of silver, probably hessite, occurs, intimately associated with argentiferous galena and a little free gold in a quartz gangue.

Gold.—Isometric, but rarely showing crystal form. Usually in irregular hackly particles. Gold yellow. Hardness, 2.5–3. Very malleable and ductile. Specific gravity, 15.6–19.3.

Free or native gold occurs in arborescent sheets in quartz and rhodonite in the Golden Fleece vein (see p. 183). In the Sunnyside

Extension mine it has been found, scattered through masses of spongy quartz, as implanted crystals on the faces of quartz crystals in vugs, and embedded in yellow sphalerite and molybdenite. It is also sometimes present in the ores of the Sound Democrat and Silver Queen mines of Placer Gulch. In the Sunnyside mine it occurs intimately associated with quartz, rhodonite, fluorite, yellow sphalerite, and galena, and has been found embedded in the latter. In the Camp Bird it is inclosed as small particles with galena, sphalerite, pyrite, chalcopyrite, and traces of some telluride, in quartz and fluorite (see p. 90). In the Tomboy it is found with pyrite in quartz, but only rarely in visible particles. Visible particles are also met with in bunches in the quartz in the Gold King mine. In the mines of the Silver Lake Basin free gold is very rarely seen, and only one specimen has been noted in the Royal Tiger.

One small bunch of free gold, associated with a telluride (probably calaverite), has been found in the Silver Ledge mine. Rich pockets of free gold are reported to have been mined in early days in the Whale and Argentina lodes in Savage Basin, and it was the free gold of the Little Giant in Arrastra Gulch which in the early seventies first called attention to the San Juan region.

Silver.—Isometric. Commonly filiform (wire silver). Silver white, but sometimes black through tarnish. Hardness, 2.5–3. Ductile and malleable. Specific gravity, 10.1–11.1.

Native silver is very rarely seen in the ores worked at the present day. But in the form of wire silver it was formerly found in the Pride of the West, Aspen, Ben Franklin, and Sunnyside Extension mines. It occurs occasionally in the Aspen and in the Antiperiodic mines, and, as small hackly particles and plates solidly embedded in limestone (probably Devonian), on the Fairview claim on Sultan Mountain. It is probably in all cases of secondary origin, resulting from the oxidation of other silver minerals.

Copper.—Isometric. Occurs in irregular plates and branching forms. Copper red. Ductile and malleable. Hardness, 2.5–3. Specific gravity, 8.8–8.9.

Occurs in small amounts in the Royal Tiger, Tom Moore, and Sunnyside Extension mines. It is confined, as far as known, to the superficial portions of the lodes and is of later formation than the bulk of the ore. In the Tom Moore it occurs in a stringer of quartz which contains small crystals of hübnerite and cuts the ore.

Hübnerite.—Tungstate of manganese ($MnWO_4$), usually with some iron. Monoclinic. One perfect cleavage. Usually in bladed prisms; often radial. Brownish red to black. Luster, submetallic or metallic-adamantine. Hardness, 5–5.5.

This mineral occurs rather widely distributed over the Silverton quadrangle as a vein mineral associated with quartz and fluorite. It is most abundant at the Adams lode on Bonita Mountain, where it forms striking radial clusters of brownish-red crystals embedded in

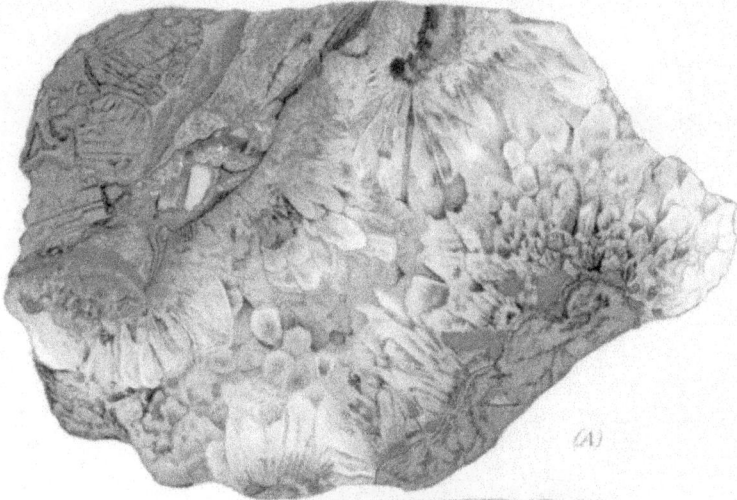

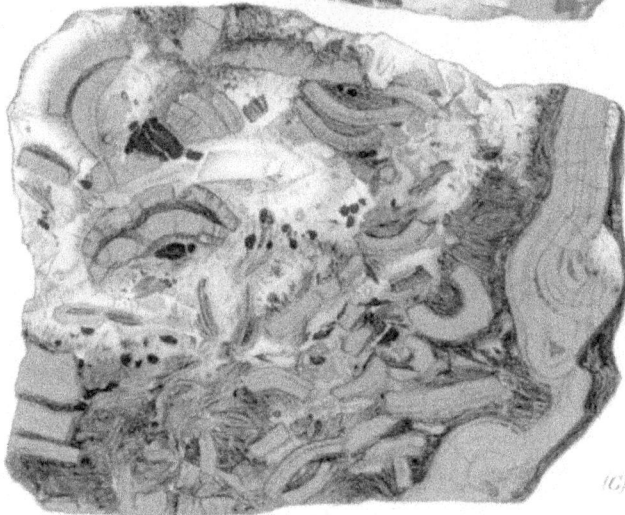

A. HOEN & CO. LITHOCAUSTIC, BALTIMORE.

ORE - STRUCTURES

quartz and pale-green fluorite.   It is also found in a quartz lode in Dry Gulch, but in smaller crystals.   In the Tom Moore lode it was noted as minute brown prisms in quartz.   On Sultan Mountain it has been obtained from the North Star mine, and occurs in moderate abundance in the Empire-Victoria lode.   The mineral from the last-named locality is nearly black and probably contains considerable iron, thus approaching wolframite in composition.   It is embedded in quartz and pale-green fluorite.

This mineral, if found in sufficient quantites, would be valuable as an ore of tungsten.   It is doubtful, however, whether any of the above-mentioned deposits contain it in sufficient abundance to render possible its commercial extraction.

PRODUCTS OF SUPERFICIAL DECOMPOSITION.

Owing to the vigorous erosion to which the Silverton region has been subjected, oxidized ores have seldom had opportunity to accumulate to any great depth, and play a small part in mining operations.   Small amounts of the carbonates of copper, malachite, and azurite, and the carbonate of lead, cerussite, can be found in the croppings of most of the lodes, and cerussite was mined to some extent from the upper levels of the Silver Lake mine.   Anglesite, or sulphate of lead, also occurred near the surface in the Silver Lake and Whale lodes and formed an important part of the ore body of the Zuñi mine, where it resulted from the oxidation of guitermanite.   Anglesite, resulting from the oxidation of galena and often containing kernels of the sulphide, occurs in the Anaconda mine about three-fourths of a mile south of Cinnamon Pass, whence four or five carloads, carrying 55 per cent of lead and 14 to 18 ounces of silver per ton, have been shipped. In the Saratoga mine the greater part of the ore has been mined comparatively near the surface, and consisted of a soft mass of ferruginous clay carrying carbonate of lead and silver, probably native or as chloride.

Under some circumstances partial oxidation has extended to depths of several hundred feet.   In the Silver Lake mine more or less oxidation has taken place where the lodes of Group II join the Silver Lake lode. Subsequent movements have allowed surface waters to descend at the intersections of the planes of weakness determined by the lodes.   The partial oxidation of the Tomboy and Camp Bird lodes at considerable depth, and the deposition within them of black oxide of manganese, are apparently due to fracturing of the original lode filling, thus allowing the downward seepage of oxidizing waters.

STRUCTURE OF THE LODE ORES.

MEGASCOPICAL STRUCTURES.

Six kinds of ore structure have been recognized in the lode ores of the Silverton quadrangle:

1. *Massive structure.*—The quartz, galena, sphalerite, chalcopyrite,

pyrite, and other minerals are all crystallized irregularly in the fissures without external crystal form and without definite arrangement. As a rule the constituent minerals appear to have crystallized practically simultaneously.

This structure is exceedingly common and is the characteristic one of the quadrangle. It is typically exhibited by the ores of the Silver Lake, Royal, Stelzner, New York City, Iowa, and East Iowa lodes of the Silver Lake Basin (Pl. XI, *B*). The quartz sometimes incloses empty spaces, either as minute interstices between interlocking quartz prisms or as small vugs lined with quartz crystals. Occasionally, instead of being perfectly massive, the ore exhibits indistinct traces of the structure next to be described. The two are connected by intermediate forms.

2. Banded structure by deposition.—The ore and gangue minerals have been deposited in more or less parallel sheets, distinguishable

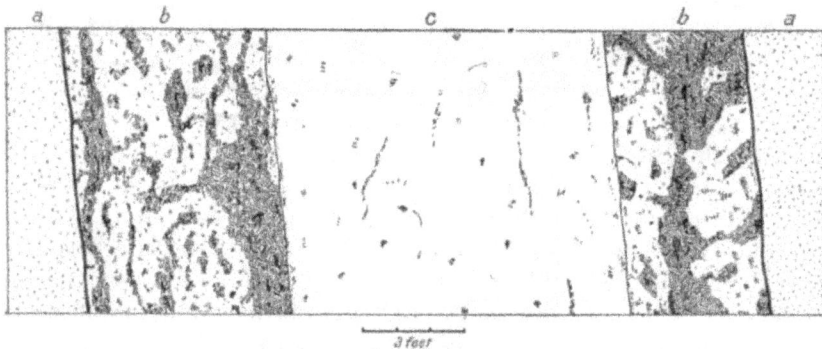

FIG. 7.—Cross section of No-Name lode, Sunnyside mine. *a*, country rock; *b*, ore, chiefly galena with quartz gangue; *c*, rhodonite.

from each other by the fact that they contain the constituent vein-forming minerals in different proportions.

This is far less common than the preceding, but it is found in many small and unproductive veins and in some of the more important lodes, such as the Gold King, where the pyrite has been deposited in bands alternating with white quartz. On the large scale it is perhaps exemplified in the Sunnyside lode, where the ore streaks, themselves of massive structure, consisting of galena, sphalerite, chalcopyrite, pyrite, tetrahedrite, and free gold in a gangue of quartz, rhodonite, and a little fluorite, are separated by plates or lenses of relatively barren rhodonite. This structure is illustrated by fig. 7, which is a section of the No-Name vein in the Sunnyside ground.

In the smaller veins and stringers carrying galena and quartz the galena frequently fills the medial suture formed by the opposing pyramidal ends of the quartz crystals which have grown out from the walls of the fissure. This is well seen in some of the veinlets of the Silver Lake Basin. Even the larger lodes, with prevailingly massive

(A)

(B)

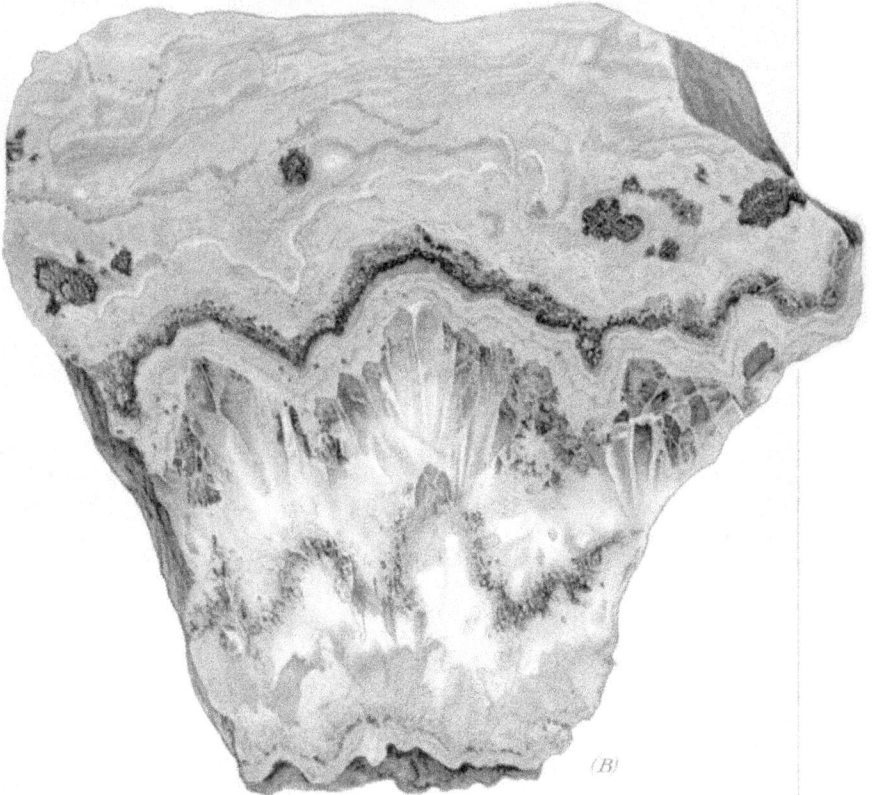

ORE - STRUCTURES

structure, not infrequently reveal a tendency toward a final crystallization of the galena in the middle portion of the vein, as may be seen in parts of the Iowa lode and in the Idaho claim, on the Titusville lode. This same tendency is also illustrated in fig. 8, which is a sketch of a small regularly banded vein prospected by the roadside half a mile north of Houghton Mountain and close to the London shaft.

Remarkably fine and regular banding was observed in the ore thrown out on the dump of a small deserted tunnel on what is probably the Osceola claim, in Cunningham Gulch, about half a mile above Stony Gulch. In its most perfect form this banding consists of dark sheets of finely crystalline sphalerite and galena about one-half millimeter in thickness, separated by plates of vitreous quartz about 2 millimeters in thickness. The result is a remarkably regular and striking fine banding (Pl. XII, A). The little sheets of quartz frequently show comb structure and have apparently crystallized in

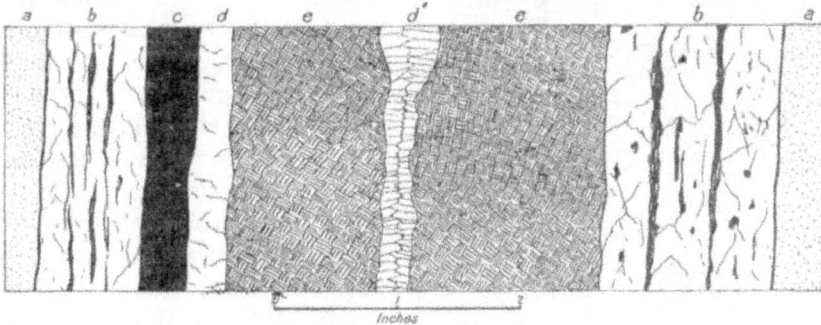

FIG. 8.—Cross section of banded vein near the London shaft, Mineral Point. a, country rock; b, quartz and chalcopyrite; c, tetrahedrite; d, d', quartz; e, galena.

open spaces. In some facies with rather wider banding chalcopyrite occurs, chiefly in the quartzose bands. The cause of such fine and regular banding is not known, but is probably connected with metasomatic replacement. The ore is apparently of too low grade for profitable working and was not seen in place.

Another form of original banding is that exhibited by the rich gold ore of the Camp Bird mine (Pl. XII, B). As elsewhere described, the rich portion of this lode is a sheeted zone in San Juan andesitic breccia and lies next the main hanging wall of the lode. Within the individual fissures of this zone the quartz has crystallized freely in open spaces. Thus, in the specimen illustrated, which represents a characteristic cross section of one of the richest ore streaks, the quartz crystals are implanted at their bases on a thin seam of chalcedony and carbonates carrying a little finely divided galena and sphalerite and associated with pulverulent white kaolin. The thickness of this seam of chalcedony and kaolin varies from 1 to 2 centimeters. From this as a foundation the quartz crystals grew out into the fissure. But after they had attained a length of 2 or 3 centimeters there was an enrichment of the depositing solutions, as is shown by an undulating dark band traversing all the crystals in a general direction at right angles to their direction of growth and parallel to the walls of the fissure. This dark band, which is usually about a quarter of a centimeter in thickness, is made up of free gold, galena, sphalerite, and a minute amount of some unknown telluride, all in very small particles embedded in the quartz. After this the quartz crystals continued to grow outward for another 2 or 3 centimeters, containing in this portion only scattered specks and bunches of ore minerals. The direct outward growth of the quartz crystals now stopped and there was deposited upon the wavy surface defined by their pyramidal terminations a succession of at least 6 distinct coatings, differing in color, composed of cryptocrystalline aggregates of quartz, calcite, fluorite, sericite, and a little chlorite, aggregating in thickness about one-fourth centimeter. Upon these coatings there was next deposited a layer, about one-half centimeter thick, consisting chiefly of finely divided particles of sphalerite, galena, and possibly other ore minerals. This is the most conspicuous dark band seen in the illustration. This band is succeeded by a delicately mottled pale-green aggregate of quartz, the grains of which sometimes contain free gold, calcite, sericite, and fluorite, with dark patches of ore. This aggregate does not exhibit the characteristics of material that has crystallized freely in open spaces and is probably, in part at least, altered and replaced country rock. Variations from the foregoing structure are met with in various parts of the Camp Bird lode, but the specimen described is thoroughly characteristic of the richest ore and illustrates well its curious and beautiful banding (fig. 9). The conspicuous dark band, locally designated "the worm," is the miners' sign for good ore in this mine.

In the Tom Moore mine the pay streak frequently consists of chal-

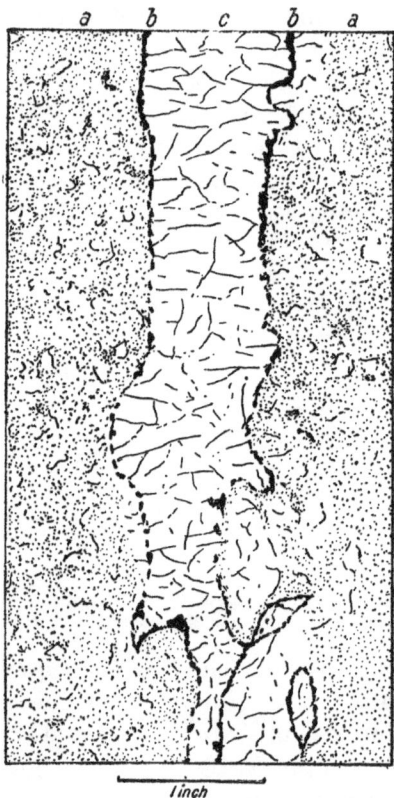

FIG. 9.—Section through a rich ore band in the Camp Bird lode. *a*, quartz, fluorite, calcite, etc.; *b*, ore particles including free gold; *c*, quartz.

copyrite in the middle with tetrahedrite on each side, the latter fading off into the altered country rock.

3. Brecciated structure.—Ore formed at an earlier period of deposition has been broken into angular fragments by renewed movement along the fissure, and recemented by fresh deposition of the ore or gangue.

Ore so shattered as to come under this head has not been noted as a conspicuous feature of any important deposit, although some brecciation can often be detected. The structure, however, is best exemplified by material from the dump of the Red Cloud mine, in which an original deposit of radially fibrous pyrite has been broken up and held together by a younger accumulation of vein quartz carrying a little sphalerite and galena (Pl. XI, *C*), and also by vein filling from the Polar Star and Palmetto mines. In the Red Cloud occurrence not only have the concentric shells of pyrite been broken up, but in some cases exceedingly thin coatings or sheets of the sulphide have been floated or moved out into the quartz, and in the polished specimen from which Pl. XI, *C* was made these delicate curved films embedded in the quartz resemble the finest cloisonné work of the Japanese.

4. Cellular structure.—The quartz or other vein material has crystallized as a mass of irregularly intersecting septa, leaving numerous drusy cavities separated by relatively thin walls.

This is not a structure characterizing any important ore deposit, and was not seen in place. But it is of interest in view of the possible light that it may throw on the processes by which the vein-forming materials are deposited. The structure is best exemplified by fragments on the dump of the Pride of Syracuse mine and on an upper dump of the Empire· group, Sultan Mountain. In the latter case the septa are composed chiefly of sphalerite and galena, and the cells are lined with drusy quartz and in some cases filled with rhodochrosite. Beautiful little crystals of amber-colored sphalerite and of tetrahedrite were observed in some of the cells, implanted upon quartz. It is possible that this cellular structure may have resulted from some kind of pseudomorphous replacement coupled with the solution and removal of some unknown mineral.

5. Spherulitic structure.—The quartz of a vein has crystallized in prisms, often imperfectly formed, which show a marked tendency to group themselves in radial fashion about local centers of crystallization. It is not intended to treat specially under this head such minerals as tremolite, which exhibits characteristically a radial structure.

A rough spherulitic crystallization, previously described by Purington,[1] was noted in the gold quartz of the Tomboy mine and, in connection with banding, in the Camp Bird lode. It is also a notable feature of the North Star lode, where worked in the Dives mine (see

---

[1] Preliminary report on the mining industries of the Telluride quadrangle, Colorado: Eighteenth Ann. Rept. U. S. Geol. Survey, Pt. III, 1898, p. 840.

Pl. XI, *A*), and of the Magnet lode.   The quartz in this case has often grown radially outward from kernels of tetrahedrite.   Less conspicuous radial structure was noted in the Silver Queen, at the head of Placer Gulch.

MICROSCOPICAL FEATURES.

In the larger lodes, with massive structure, particularly those carrying much galena, the crystallization is, as a rule, so coarse that microscopical study is of use only in furnishing details of mineralogical structure.   The quartz of such lodes shows the usual character of vein quartz, and has crystallized in large, irregular grains, interlocking with one another and with the principal ore constituents.   The structure, when sufficiently fine to be embraced in a thin section, is typically hypidiomorphic granular (Pl. IX).   Microscopic fluid inclusions with gaseous bubbles, often arranged along curved surfaces within the quartz, are always present.   No special investigation of the hermetically inclosed liquids of these minute cavities has been made for this region, but they are probably solutions of calcium and alkaline sulphates, with some chlorides, and very likely carbonic acid.[1] Interstitial or miarolitic cavities into which the quartz crystals sometimes project are common, and are often filled with calcite, or, in the Silver Lake Basin, with chlorite (Pl. IX, *A*).

Vein quartz varies widely in the size of the constituent crystal grains.   In the Sunnyside and neighboring lodes the galena, sphalerite, and chalcopyrite are distributed through the gangue in smaller particles, and the quartz grains themselves are correspondingly smaller than those in the Silver Lake veins.   Pl. IX, *B*, shows a typical thin section of Sunnyside ore, in which chalcopyrite and sphalerite happen to be the dominant ore minerals.

Still other ores, of which that of the Ridgway mine may be taken as a type, exhibit throughout much finer crystallization.   The quartz occurs in small interlocking grains, and the ore particles, which in hand specimens are often so small as to render the quartz cloudy or black, without being themselves individually noticeable, are seen under the microscope to be thickly sprinkled through the gangue and to be very irregular in shape.   They occur both in the sutures between the quartz grains and within the grains themselves (Pl. X, *A*).   Such ore is almost invariably said by the miners to contain brittle silver.   But neither chemical tests nor the microscope reveal stephanite.   In the Ridgway ore the particles are chiefly argentite with some pyrite, a little sphalerite, and sometimes a little galena.   These finely crystalline ores are worked for silver and gold, as the galena is rarely abundant enough to pay for extraction.   But they are by no means equally rich.   In some cases the finely disseminated ore parti-

[1] Gold-quartz veins of Nevada City and Grass Valley districts, by Waldemar Lindgren: Seventeenth Ann. Rept. U. S. Geol. Survey, 1896, Pt. II, p. 131.

cles are chiefly argentiferous galena, sometimes a lead sulphobismuthite, and occasionally tetrahedrite. Proustite (ruby silver) was
noted in the Ben Butler mine, with finely crystalline galena, which
is erroneously called brittle silver by the miners.

The gangue of the rich ore of the Camp Bird mine shows several
interesting variations in its microstructure, and differs from that of
any other lode known in the quadrangle. A characteristic structure
and mineralogical composition is that shown in Pl. X, *B*, where quartz,
calcite, and fluorite are crystallized together. This is a common
appearance of the pale-green, finely crystalline aggregate which has
crystallized concentrically outside of the dark, wavy ore band shown
in Pl. XII, *B*. The quartz here shows a distinct tendency toward idio-
morphic form, and in sections transverse to the prism usually reveals,
in polarized light, a zonal structure the outer shells of which are
composed of shadowy sectors differing in their interference tints (Pl.
X, *B*). These quartz grains contain inclusions of chlorite and a mineral in acicular crystals too small for identification. The inclusions
are most abundant near the center of the crystal. The fluorite is not
always present, and the gangue is frequently much more finely crystalline and often shows nests of sericite. This portion of the gangue
has evidently crystallized in part in open spaces, but it has also
undoubtedly in part resulted from a metasomatic replacement of the
San Juan breccia adjacent to or included within the main fissure
zone. The microscope shows the free gold to occur in nests of minute
hackly particles solidly embedded in quartz. It is apparently most
abundant in the coarsely crystalline, vitreous, radial quartz (see Pl.
XII, *B*), but also occurs in the quartz just outside of the dark ore band
already referred to. As seen under the microscope this ore band consists of particles of galena, light-yellow sphalerite, and chalcopyrite
in quartz. The gold was not seen in the dark band itself, but is
nearly always within a few centimeters of it, usually associated with
a little galena and sphalerite.

Although as a rule the ores occupy filled fissures, yet in a few cases
the gangue in which the ore occurs is composed wholly or in part of
country rock which has been altered to an aggregate of quartz, and
usually one or more of the following minerals: Calcite, sericite, kaolinite, chalcedony, fluorite. Such aggregates generally differ from true
vein filling in showing under the microscope a much finer crystallization of their constituents into a mosaic which is more or less cloudy
when seen between crossed nicols. The grains of quartz seldom
exhibit the clear, definite boundaries seen in vein filling, such as is
shown in Pl. IX. Moreover, the sericite and calcite are often
grouped in areas whose form is that of former feldspathic phenocrysts. This structure, however, will be more fully described in the
section on the metamorphism of the country rock.

AREAL DISTRIBUTION OF THE LODE ORES.

The various mineralogical and structural types of ore are not distributed at random over the quadrangle, but a given portion of the area is often characterized by ore of a certain kind, differing from those which preponderate in other portions. These areal subdivisions are not sharply bounded, and it often happens that in a district in which most of the lodes contain a certain type of ore there will occur other ores which are more characteristic of some other district in the quadrangle. But, notwithstanding these transitions and exceptions, the localization of ore types is a striking feature of the region.

The lodes traversing the mass of Sultan Mountain are quite commonly filled with rather coarsely crystalline vein filling, carrying galena, tetrahedrite, and chalcopyrite, with some sphalerite and pyrite, in a gangue of quartz and barite. Their chief value is in silver. The most productive of these lodes has been the North Star, which produced an ore running about 40 per cent lead and 70 ounces of silver. This lode is in monzonite, but mineralogically similar ore is found in the King lode in Algonkian schists, said to carry about 0.1 ounce of gold, 55 ounces of silver, and 15 per cent of lead. The ore of the Empire group of veins is generally similar in character.

Southeast of Silverton, in the vicinity of Deer Park, are numerous mineralized fissures cutting the schists and granites, and in some cases the overlying San Juan breccia. These are occupied, as a rule, by quartz veins carrying pyrite, with a little galena, sphalerite, and chalcopyrite. They are prospected for gold, which occurs free in small pockets. Compared with most of the lodes in the quadrangle these veins are not heavily mineralized, and are not mined on an important scale.

The lodes of Silver Lake Basin are characterized by coarse, massive structure and are heavily mineralized. Galena often exceeds all other constituents of the vein filling in amount, and lead is an important factor in the output. Sphalerite and chalcopyrite accompany the coarsely crystalline galena, and there is usually some pyrite present. The gangue is quartz, often containing chlorite. Tetrahedrite occurs sporadically in many of these lodes, and barite was noted in the Royal Tiger workings. These are generally low-grade concentrating ores, averaging from $8 to $15 per ton. Free gold is rarely seen, but the gold tenor in some of the lodes of Group II (see p. 147) may rise to 2 or 3 ounces per ton. In the Silver Lake mine about half the value of the output is from the gold; the remainder being in silver, lead, and copper. The other mines contain less gold, and the Royal Tiger mine depends chiefly upon its lead. Most of the ore of Hazelton Mountain appears to have been of the general type just outlined. A thousand tons from the Aspen mine is said to have produced 110 ounces of silver per ton and 60 per cent lead, although the average tenor was somewhat less than this. An exception of the sort referred to at the beginning of this section is afforded by the Little Giant lode, on the north side of Arrastra Gulch, which produced a free-gold ore.

In Maggie Gulch the greater number of the lodes contain siliceous ores, of which that of the Ridgway mine may be taken as the general type (Pl. X, *A*). In these ores the quartz is finely crystalline and the ore minerals occur as minute dark specks, which, when abundant, give the gangue a dark, clouded appearance. In the Ridgway, the only mine working these ores on a commercial scale, the ore particles are chiefly argentite. But in other cases they are galena, and apparently not always very argentiferous. The erroneous belief of the prospectors that these minute ore particles are nearly always "brittle silver" has in some cases resulted in much fruitless labor. In one instance a prospecting tunnel of considerable length was being run on expectations excited by a few minute bunches of "brittle silver," which were in reality worthless specularite. Ores generally similar to those of Maggie Gulch occur at various points in the northeastern part of the quadrangle, particularly in the vicinity of Mineral Point.

In Sunnyside Basin and at the head of Placer Gulch the ores are commonly of moderately coarse crystallization and carry abundant galena, sphalerite, and chalcopyrite. They are almost invariably associated with abundant rhodonite. Of low grade as a whole, they sometimes contain bunches of rich ore, which, in the Sunnyside Extension mine, are known to have carried, in carload lots, as much as 74 ounces of gold per ton, mostly free. The ore of the Sunnyside mine may be considered as typical of this district.

In Poughkeepsie Gulch the ores partake somewhat of the characteristics of those of the Sunnyside Basin and of Mineral Point. But their most characteristic feature appears to have been in the occurrence of the lead and silver in some combination with bismuth, usually as an argentiferous sulphobismuthite of lead. The ore of the Alaska mine is a well-known example and has been described on page 195.

In the extreme northeast corner of the quadrangle many of the prospects and some of the mines formerly worked are characterized by argentiferous copper ores, usually chalcocite or bornite, with some galena and chalcopyrite.

What may conveniently be termed the Red Mountain Range, embracing the region bounded by Cement, Grey Copper, and Mineral creeks, is characterized by the predominance of the stock deposits and their peculiar ores, which will be described in subsequent pages.

Lastly, in the northwest corner of the quadrangle, ores carrying much galena, and recalling in some of their features those of Silver Lake Basin, are associated with gold ores such as those of the Camp Bird and Tomboy mines.

As far as could be determined, the distribution of ore types as briefly sketched in the foregoing paragraphs is purely areal and is not dependent upon kind of country rock or direction of fissuring. Finally, emphasis should again be laid upon the fact that there is much over- lapping and intermingling of types, and that ores of many kinds may occur in any given limited district. Free-gold ores, in particular, may be expected to occur in any part of the quadrangle.

DISTRIBUTION OF THE ORES WITHIN THE LODES.

Lodes are seldom equally well mineralized in all portions. Those parts of a lode or vein which are sufficiently large and rich to be profitably worked are commonly known as pay shoots or ore shoots, sometimes spelled ore chutes. The first term is regarded as prefer- able, inasmuch as it expresses concisely the thing meant, and can not be confounded with the ore chutes or mill holes of timber through which ore is drawn from the stopes into the mine levels. A pay shoot is obviously not a thing capable of exact delimitation. Its size and shape depend to some extent on fluctuations in the metal market (unless purely a gold ore) and on variations in the operating expenses of the mine. But the concentration of ore in certain portions of a lode separated by unprofitable lode matter is a very important phe- nomenon, and one that can, as a rule, be studied and discussed only as it is revealed in the exploitation of pay shoots by actual stoping.

A complete study of pay shoots requires good stope maps, carefully revised to date, and an advanced stage in the development of the lodes. Except for a few localities, these favorable conditions are not found in the Silverton quadrangle. The advantage in all future development of having a record of past work is not always fully realized, and the miner is too often content simply to follow his ore where he finds it, adapting his work merely to the needs of to-day. In simple lodes with continuous pay shoots this rough-and-ready mining may succeed well enough. But in complex lodes, or in groups of lodes, such as those of the Silver Lake mine, systematic mapping and the adaptation of the development to future probabilities may mean success where the more ordinary method would result in failure. On account of the general absence of accurate stope maps and the small depth of most of the mines operating in the quadrangle the following observations on the pay shoots are fragmentary and often unsatisfactory.

In most of the workable veins and simple lodes, such as those of Sultan Mountain and Silver Lake Basin, pay ore is usually, although not invariably, found wherever the fissure is wide enough to hold an ore body. This appears to be particularly true of the New York City, Stelzner, Royal, and Iowa veins. Thus, in these veins, distinct pay shoots, separated by stretches of barren lode, do not properly occur, or have not been demonstrated by mining operations. The pay shoots take their shape, as a whole, from the form of the fissure. A partial exception to this is found, however, in the impoverishment of the lodes of Group II as they approach the Silver Lake lode, which is itself of lower grade, and in occasional spots where the lodes retain their width but are too low grade to work. The pay shoots of these lodes also grow longer with depth, which is apparently due to the increasing length of the fissures in connection with the dip of the Silver Lake lode. In the Silver Lake lode the pay ore is found in bunches of varying size; but as these bunches usually occur wherever the filling of the fissure is wide and firm they probably do not constitute a marked exception to the foregoing general rule. In the Royal Tiger mine not enough work has been done to reveal the real shape and extent of the ore body. In the North Star mine, on Little Giant Mountain, the ore occurred irregularly in a stringer lead, but the old stope maps show the pay shoot as a whole to have been about 700 feet in length on the levels and to have pitched to the southeast. Other small ore bodies were found farther in the mountain, to the northwest. In the Sunnyside and Sunnyside Extension mines the pay shoots are mainly lenticular bodies 30 or 40 feet in length, lying on the hanging or foot wall side of the lode, or between plates of rhodonite. Owing to the lack of maps, the exact sizes and shapes of these ore bodies are not known.

In the Silver Queen mine, on Bear Creek, the ore occurs in three separate pay shoots, pitching south and separated by from 12 to 15 feet of lean quartz. The northern pay shoot consists of low-grade chalcopyrite, with subordinate galena and tetrahedrite. The middle ore body is rich in an argentiferous sulphobismuthite of lead associated with galena, but changed to chalcopyrite in stoping upward from the tunnel level. The south pay shoot is almost wholly galena. The development of this mine is very superficial and the three pay shoots may, with more extensive workings, be found to be simply portions of one ore body.

In the Camp Bird mine the pay shoots, so far as known, appear to be coextensive with the lode, disappearing only where the fissures locally contract to a mere crack. What is known as the west pay shoot is about 1,200 feet in length, and others occur along the lode, separated by pinches. The pay shoots are stated to be increasing in length with depth. As previously pointed out (p. 89), the auriferous portion of the lode lies next the hanging wall, with a silver-lead streak near the foot wall.

In the Tomboy, also, the pay shoots, as indicated by the stopes, are long, and are practically identical with the lode itself, disappearing only where the latter is pinched. According to Mr. John Herron, superintendent, the pay shoots lie flat in the lode, one above another, and extend diagonally across from wall to wall. They are said, moreover, to be often connected with certain post-mineral fractures which cross the vein diagonally in the same direction (see p. 209). There was no means of verifying these statements, and such minor pay shoots, with the intervening lower-grade quartz, are all stoped out in working and would not appear in an ordinary stope map. It is beyond question, however, that in this mine the best ore occurs in quartz which has been crushed by movements subsequent to its original deposition. The solid, unbroken, unstained quartz is invariably of somewhat lower grade. As in the Camp Bird, the auriferous ore is next the hanging wall, a silver-lead streak lying near the foot wall.

From large, regular pay shoots, such as are found in the New York City, Camp Bird, and Tomboy lodes, all gradations may be found through smaller, more irregular ore bodies down to the occasional small pockets of rich ore encountered in some of the lodes of Deer Park and Maggie Gulch.

The changes which take place in pay shoots with depth are exceedingly important in mining operations. Unlike the auriferous lodes of California, which rarely show any progressive or regular change in the character of their ores even to depths of 2,500 feet, the pay shoots of the Silverton quadrangle are less constant. It is a common belief among mining men of this region that lead ores do not go down to great, or even to moderate, depths. In part this opinion is plainly traceable to the experience gained in exploiting the stocks of the Red Mountain district, which will be described in subsequent pages; but it is pertinent here to see how far it is justified by the known facts in regard to the lodes.

In the Aspen mine there are said to have been several veins near the surface which came together at moderate depth. Below a depth of 800 feet the galena, which was very abundant in the upper levels, is said by Mr. W. H. Thomas to have so diminished in amount as to make further work unprofitable. According to the same informant the ore of the Belcher mine, on Sultan Mountain, carried abundant galena of good grade in the upper workings, but this decreased in amount and value in the lower levels, while sphalerite grew more abundant. Actual verification of these statements by observation is not now practicable, but Mr. Thomas, who had been connected with both mines, was undoubtedly himself convinced of the truth of his statements. In the Royal, Stelzner, New York City, Iowa, and East Iowa lodes there is no present indication of any diminution of galena or of any other regular change in the ore with increasing depth. The ore of the New York City lode where cut by the Unity tunnel, about 1,200 feet below the croppings, shows abundant galena, and is similar to the ore in the same vein on levels C and I, 400 and 700 feet, respectively, above the Unity level. If anything, it is of higher grade. The Buckeye lode (part of the Titusville lode) has not been exploited to great depth, but tetrahedrite was undoubtedly more abundant in the upper than in the lower tunnel. In the North Star mine, on Little Giant Peak, a change in the character of the ore is well authenticated by Mr. W. Crooke, the manager. For 200 feet below the croppings the ore was largely sulphate of lead (anglesite) with some galena, and did not carry much silver. Below that level it changed to unoxidized galena. With increasing depth the place of the galena was largely taken by argentiferous tetrahedrite, and in the lowest workings, about 600 feet below the cropping, there was very little galena found. The bottom of the pay shoot is said to have been reached in the lowest workings, although the fissure zone continues to an unknown depth and probably contains other pay shoots, which, however, may not be large or rich.

In the Polar Star mine, on Engineer Mountain, the ore, consisting of argentite and proustite, is said by Mr. W. Crooke to have changed abruptly to iron pyrite, carrying about 12 ounces of silver per ton and no gold. According to Mr. Crooke, this change was generally encountered at about the same level in all of the mines on this portion of the mountain. The exact depth at which the change took place is not known, but it must have been moderate, probably not more than 500 feet. In the Palmetto mine, on the northeastern extension of the Polar Star lode, the ore near the surface ran as high as $500 per ton, and contained ruby and so-called "brittle" silver (probably proustite and argentite), resembling that of the Polar Star mine. The lode struck in the shaft at a depth of 400 or 500 feet was of low grade, and, as the dump shows, contained much pyrite. In this, as in the case of the Old Lout and Pride of Syracuse, it is sometimes asserted that the lode found in the deeper workings was not the same as that which had been productive above. While a mistake of this kind is not impossible in individual cases, the threefold repetition of such a mischance in the Silverton region may be set down as highly improbable. The histories of the Pride of Syracuse mine on Engineer Mountain and of the San Juan Chief mine are almost repetitions of that of the Palmetto. The Old Lout lode, which produced rich bismuthiferous ore to the extent of several hundred thousand dollars from a shaft sunk on the lode, proved too poor to work when crosscut by a long tunnel from the bottom of Poughkeepsie Gulch, about a thousand feet below the croppings. The ore from this lowest level shows galena, chalcopyrite, sphalerite, and pyrite, about equally distributed in a gangue of quartz with a little barite. The U. S. Depository lode, in Richmond Basin, produced some galena ore in the upper workings, but where cut by a tunnel, at a depth of something over 500 feet, the lode proved worthless. In this case the lode is well defined, but shows scarcely any mineralization in the lower workings. In the Uncompahgre Chief, a prospect near Mineral Point, a shaft 50 feet in depth has passed through ore containing an argentiferous sulphobismuthite of copper and lead, into an ore of which pyrite is the chief mineral constituent. In the Dolly Varden mine, north of Rose's Cabin, according to Mr. James Abbott, good ore containing tetrahedrite and "brittle silver" (?) was found in the shaft, but no pay ore was encountered in a lower crosscut tunnel.

From the foregoing it appears that in this region pay shoots containing chiefly ores of lead, silver, and copper may change in mineralogical character and in value within very moderate depths. But while galena in one case has been succeeded by tetrahedrite and in another case partly by sphalerite, in other lodes containing silver-lead ore tetrahedrite (Royal Tiger and Buckeye mines) and sphalerite (Iowa lode) have been more abundant in the upper than in the lower workings. Thus the facts, as at present known, do not justify the general statement that galena passes into gray copper with depth, or vice versa. The change from galena to sphalerite with increasing depth is a well-known phenomenon in the lead and zinc deposits of southwestern Wisconsin, and is explained by Van Hise [1] as a result (1) of a first concentration by ascending waters and (2) of a second concentration by descending waters. It is possible that the change recorded in the Belcher is of this character. But the occurrence of abundant argentiferous galena and tetrahedrite in the Virginius lode at the Revenue Tunnel level, over 2,000 feet below the croppings, and of galena in the deepest workings on the New York City lode, indicate that both tetrahedrite and galena may occur abundantly and retain their values at far greater depths than were reached in the Aspen, North Star, and other mines of the Silverton quadrangle, in which the pay shoots are reported to have diminished in value or to have disappeared entirely. As applied to the lodes, therefore, the often-heard arbitrary dictum that galena and tetrahedrite "do not go down," in so far as it relates to the depths ordinarily met with in mining, is a generalization based on insufficient premises, although, as will appear in the discussion of the origin of the ores, there are theoretical reasons for supposing these minerals less likely than certain other sulphides to be found in abundance at great depths. With regard to the rich silver ores argentite and proustite, the known facts, incomplete as they are, seem to warrant the conclusion that these minerals occur in the upper portions of the lodes only, and do not extend to great depths, probably rarely over 500 feet in this region. This is in general accord with observations in other districts where these and other rich silver minerals occur, and, as will be later shown, this fact has an important bearing on the question of the genesis of the ores.

The vertical range of the argentiferous sulphobismuthites of lead occurring in the Silverton region is not known, but there are some indications that they do not extend to great depths.

The free gold ores have not yet been worked to a sufficient extent in the Silverton quadrangle to furnish any very reliable data as to

---

[1] Some principles controlling the deposition of ores: Trans. Am. Inst. Min. Eng., Vol. XXX, 1900, pp. 104-109.

their possible changes in value with depth.   The indications, however, are in favor of a varying, but on the whole more nearly constant, tenor as they are mined to greater depths, than in the case of silver-load ores.   It is fair to assume that the considerable depth to which the lodes of California have been mined without revealing any regular increase or decrease in the tenor of the pay shoots below the zone of oxidation is something more than a mere local phenomenon, and holds for gold-quartz lodes somewhat generally, provided that the fissures are continuous and the country rock is uniform.

RELATION OF ORES TO KIND OF COUNTRY ROCK.

As far as known, the influence of the country rock upon the deposition of the lode ores in the Silverton quadrangle is small.   Extensive productive ore bodies occur in monzonite and in the volcanic rocks of both the Silverton and San Juan series; and, although none are worked in the Algonkian schists, there is no indication that, given a suitable fissure, ores would not be deposited as readily in these as in other rocks.   The influence of the wall rock upon ore deposition appears to be limited to a tendency toward impregnation and replacement in the deposition of ores in some of the rhyolitic and andesitic flow breccias and tuffs.   This, however, is a phenomenon relating to the form of the deposit, and to the subject of the alteration of the wall rocks rather than to the character of the ores themselves.

SECONDARY ALTERATION OF THE LODE ORES.

Owing to the fact that erosion in this high region is effected by mechanical disintegration rather than by secular decay, the oxidation of the ores is not, as a rule, very extensive, and is sometimes negligible.   In the North Star mine, on Little Giant Peak, oxidized ore, consisting chiefly of anglesite, is reported to have extended to a depth of 200 feet.   Ordinarily, however, complete oxidation is very superficial, although traces may be found at much greater depths than 200 feet. Carbonate and sulphate of lead and carbonates of copper, often associated with the unaltered sulphides, occur in the croppings of most of the lodes carrying galena and tetrahedrite or chalcopyrite.   In the Silver Lake mine sulphate and carbonate of lead occurred abundantly in the highest level (No. 4) with unoxidized ore.   At the level below, however, the secondary ores were unimportant.   As is commonly the case in this region, there is no sharp line separating an upper zone of oxidation from the sulphide ores.   The decomposition of the original ore minerals has penetrated irregularly downward wherever oxidizing waters found opportunity to descend, and is seldom limited by a distinct water level.   Very commonly the level of permanent water lies far below the oxidized portion of the lode.   In the Tomboy and Camp Bird mines black oxide of manganese occurs in the deepest workings and usually indicates good ore.   In these cases the oxide appears to be associated with post-mineral fracturing and crushing and to have been deposited later than the bulk of the ore. In the New York City lode of the Silver Lake mine a sooty, amorphous sulphide of copper occurs with pyrite, chalcopyrite, galena, and sphalerite. This is probably of secondary origin, and although not a product of ordinary oxidation is yet considered as due to descending waters percolating downward from the zone of weathering above.

ORIGIN OF THE LODE ORES.

Although the ores of the lode present many features not found in the stock deposits of the Red Mountain district, yet in the essential points of their origin they possess so much in common that it would be impossible to treat the two forms of deposit separately without much overlapping. Accordingly the question of the origin of the lode ores will be left in abeyance until the stocks have been described.

VALUE OF THE LODE ORES.

Generally speaking, the lode ores of the Silverton quadrangle are of low grade and require careful mining and milling to yield profitable returns. The ores which are or have been worked vary in value from a probable minimum of about $6 to several thousand dollars per ton. The extremely high values, however, are for ores carrying free gold occurring only in small amounts in pockets in otherwise low-grade lodes, or in very small veins, such as the Golden Fleece. From the Sunnyside Extension mine, however, ore was obtained in carload lots carrying $1,500 per ton in gold alone, as shown by the smelter returns. In the Silver Lake Basin the ores worked probably vary from $8 or $10 to $60 or $70 per ton, according to the proportion of lead, silver, or gold which they contain. The average of the ore from the Iowa mine is not far from $12 per ton, although in places it may contain from 2 to 3 ounces of gold, thus affording much higher returns. The general average of the Silver Lake mine is probably higher, as half the value of its output is in gold. The ore of the Royal Tiger is, on the other hand, somewhat lower, the product being chiefly in lead. The North Star mine, on Sultan Mountain, was producing quantities of ore in 1882 carrying, according to the Mint reports, 40 per cent of lead and 70 per cent of silver. At the commercial values for that year this would mean an ore of about $120 per ton, gross value, or, at present prices, about $72 per ton. The Aspen mine is said to have produced one lot of 1,000 tons of ore containing 110 ounces of silver and 60 per cent of lead, which at present prices would correspond to about $110 per ton. The average value of the ore from this mine was, however, probably less than half this. The shipping ore from the Ridgway mine, containing finely disseminated argentite, is reported to average about $110 per ton, of which about 3 ounces is in gold and the rest in silver. The free-gold ore of the Tomboy mine averages about $20 per ton, of which $19 is in gold. The most uniformly rich ore now worked in quantity in the quadrangle is undoubtedly the gold ore of the Camp Bird. This is all milled, and averages from $40 to $200 per ton as delivered at the stamps. The gold ore of the Gold King mine is generally of low grade and can be worked to about as low a limit as any in the quadrangle.

THE STOCKS OR MASSES.

DEFINITION AND GENERAL DESCRIPTION.

As defined by Von Cotta,[1] ore stocks are irregular bodies of ore possessing distinct boundaries. They differ from veins or lodes in the absence of a characteristic tabular form, and from impregnations in being fairly solid masses of ore with definite limits, or, in the case of stockworks (Stockwerke), of a great number of small veins or ore bodies constituting a mass which is worked as a whole. Phillips[2] has translated Von Cotta's *Stöcke* as "masses," but this word is so general in its scope and has been used in so many classifications with varying significance that it is here discarded as a definite descriptive term. The nearly vertical ore bodies of the Red Mountain district, sometimes locally known as "chimneys," are substantially what Von Cotta has described and defined as standing (or upright) stocks.

The stocks of the Silverton quadrangle are nearly vertical, irregularly lenticular or spindle-shaped bodies of almost solid ore, surrounded by an envelope of much altered, partly silicified country rock, which is usually impregnated with pyrite. Neighboring ore bodies are commonly connected by fissures, usually filled with wet clay gouge or kaolin. The sizes of the individual ore bodies vary greatly. In plan, lengths of 40 or 50 feet and widths of 10 or 15 feet appear to have been not uncommon. Accurate sections of actual stocks are not now available, but individual ore bodies appear to have been followed in many cases through several levels, and therefore to have had a nearly vertical maximum dimension of several hundred feet. Although sometimes nearly circular, the ore bodies generally showed a more or less elongated or elliptical plan, the longer axis of which usually lay more nearly north and south than east and west. The direction of the longer axis was not, however, always constant for all depths of a given stock.

The ore bodies, particularly in the deeper workings, were usually found within nearly vertical zones of fractured, altered country rock, locally known as "ore breaks." Mr. T. E. Schwarz, who appears to have originated this term, says:

The ore bodies of the Red Mountain district occur as chimneys or chutes, having great persistence in depth and varying greatly in dip and cross section, but ramifying toward the surface. The outcrops of these ore chutes occur along lines of alteration of the andesite or inclosing rock, and such lines were undoubtedly fracture planes of greater or less extent. These fracture planes or belts of metamorphosed andesite were in the early days of the district termed by me "·ore

---

[1] Erzlagerstätten, p. 191, Freiberg, 1859.   [2] Op. cit., p. 4.

breaks," to signify the character of the rock along the line of which the ore chimneys occurred or might be expected to occur. They were not fissure veins, in that the ore bodies did not confine themselves to any given plane, and in following them no definite course either laterally or vertically could be counted upon ahead of exploration. The course of an ore break I have frequently found to change 90˚ in a depth of several hundred feet.[1]

DISTRIBUTION.

The stocks do not occur in all portions of the quadrangle, but are limited to a small, well-defined area, which corresponds mainly to the crest and westerly slope of what may be termed for convenience the Red Mountain Range. This district is bounded on the northeast by Grey Copper Creek, on the northwest by Red Creek, on the west and south by Mineral Creek, and on the east, with one or two exceptions near Red Mountain, by the ridge crest extending from Anvil Mountain northward to Red Mountain and thence northeast to the saddle at the head of Grey Copper Gulch. The area thus outlined has certain characteristic features, apparent to the most unobservant eye. There is scarcely an exposure of rock within it which has not been bleached, silicified, or otherwise altered, and the whole has been more or less thickly impregnated with fine crystals of pyrite. The iron originally contained in this pyrite, through weathering and oxidation, has given to the Red Mountain Range the beautiful coloring for which the region is noted, blending from the deepest red, through vermilion and orange tints, to the most delicate yellows and grays.

The greater part of the deposits, and all of the more important ones, are closely grouped in the northwestern part of this area, in a belt less than a mile wide and about 4 miles long, extending from Ironton to a point about a mile south of Red Mountain village.

Although no continuous fissures or veins can be detected on the surface, it is noteworthy that the St. Paul, Congress, Senate, Hudson, Enterprise, Charter Oak, Genesee-Vanderbilt, Yankee Girl, Robinson, Guston, White Cloud, and Silver Bell mines all lie close to a straight line bearing about N. 21° E. The National Belle, Paymaster, and Grand Prize mines, and a host of less-noted claims, are apparently irregularly disposed on either side of this line. (Fig. 10.)

South of the vicinity of Red Mountain the stock deposits are scattered and have not proved of much importance. The only one which has produced ore in important amounts is that of the Zuñi mine, on Anvil Mountain, which, however, is interesting chiefly as the source of the minerals guitermanite and zunyite.

ORIGIN OF THE SPACES NOW FILLED WITH ORE.

In seeking the origin of the ore spaces, there are, in spite of the present obstacles in the way of complete investigation, certain facts

---

[1] Trans. Am. Inst. Min. Eng., Vol. XXVI, 1896, p. 1057.

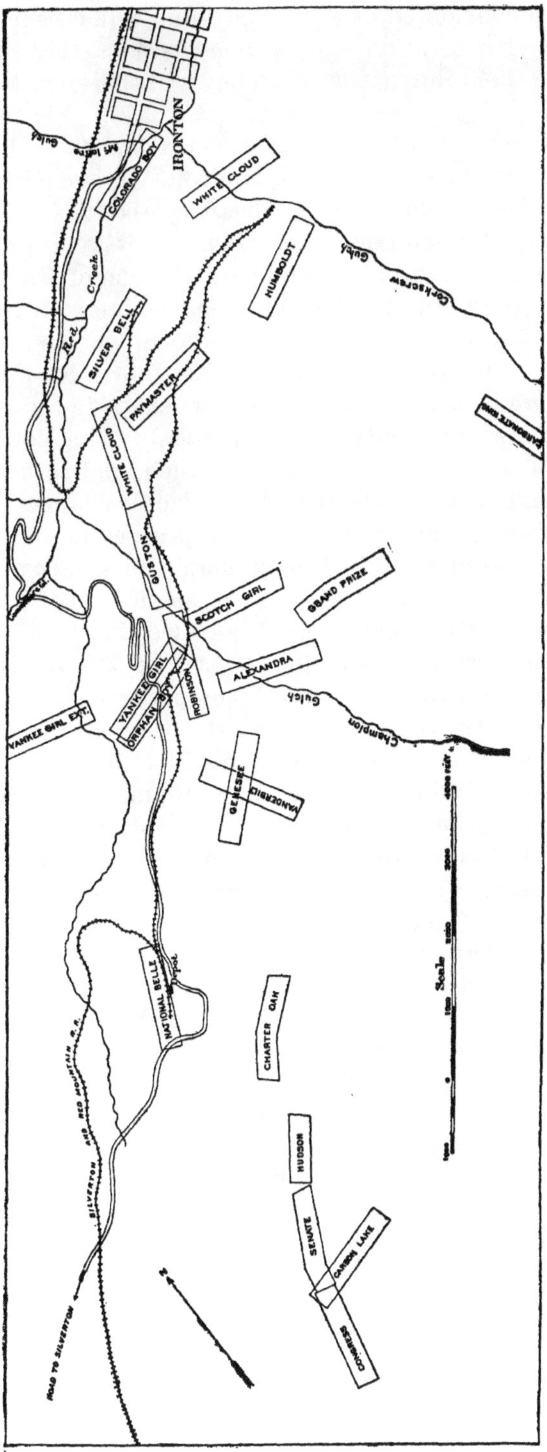

FIG. 10. Map of the principal mining claims in the Red Mountain region.

which may be considered as established and which bear directly upon the problem. No solution can be considered as satisfactory which does not take them into account. They will therefore be briefly summarized.

The greater number of the productive ore bodies in the Red Mountain district, such as the Congress, Hudson, Yankee Girl, Guston, and Silver Bell, lie on or close to a line bearing about N. 21° E., which is probably a line of dominant fracturing. It is very probable, if not demonstrable, that a second nearly parallel zone of fractures extends through the ground of the National Belle mine southwesterly, past Summit and down Mineral Creek, to the Silver Ledge mine, near Chattanooga. The general course of this line is N. 30° E. The ore spaces are more or less irregularly spindle-shaped or ellipsoidal, with the longest axis commonly nearly vertical. The country rock in the vicinity of the ore spaces is usually much, and rather irregularly, fissured. Neighboring ore bodies are commonly connected by fissures, and this fact was taken advantage of in prospecting. When the ore spaces are elongated or elliptical in plan, the longer axis generally, although not always, lies a little east of north. In the Guston claim the ore spaces usually occur in what is called the "ore break," a zone of fissured country rock striking N. 20° or 30° E. Similar "ore breaks" were recognized in the Genesee-Vanderbilt and in other mines. The country rock adjacent to the ore spaces is much altered, as described under metamorphism, pages 124–131, and is often thickly impregnated with fine pyrite. Mr. S. F. Emmons has recorded what seemed to him to be undoubted replacement of country rock, especially of the Silverton breccia, by ore, and even the limited observations now possible show that some replacement undoubtedly took place in both the Yankee Girl and Guston mines.

It is believed that the explanation most consistent with the facts above stated and with others presented in the detailed descriptions of the mines is that the ore spaces were formed primarily by complex, intersecting fissuring, and were enlarged both by solution and by metasomatic replacement. The dominant fissuring seems to have been in a direction between 20° and 30° E. of N., and is well shown along two principal zones. But there was also much minor fracturing in various other directions. At the intersections of two or more of these fissures the country rock was usually brecciated and furnished channels for the ascending currents of warm or hot mineral-bearing waters which effected the pronounced hydrothermal metamorphism so conspicuous in the rocks of this particular district. That such waters were capable of considerably enlarging the irregular conduits through which they circulated, either prior to or during the deposition of the ore, seems likely when the shape of the ore spaces and the extent to which the rocks in their vicinity have been altered are taken into account. It must be admitted, however, that in the case of the small Zuñi ore body no such solvent action was detected. The foregoing explanation is essentially that proposed by Emmons[1] in 1888, when he pointed out the similarity existing between the Yankee Girl mine and the Bull Domingo and Bassick mines in the Silver Cliff district, Colorado. As described by L. R. Grabill,[2] the Bassick ore body fills an irregular opening nearly elliptical in plan and varying from 20 to 30 feet on the shorter diameter to nearly 100 feet on the longer. It extends vertically downward for nearly 800 feet. No walls have been found, and there is no distinct boundary between ore and country rock. Emmons[3] concluded that the ore body was due to fumarolic action in the throat of a volcano after all explosive action had ceased. He remarks, however:

It would appear * * * that it was the intersection of certain fracture planes that determined the course of the ore-bearing channel, and that the ore body is not necessarily the center of the volcanic vent, but that, as a second ore chimney has already been discovered on one side of the first, it is by no means impossible that other chimneys or ore shoots may exist in the body of the agglomerate, which might be discovered by judicious and systematic exploration in the direction of the principal fracture planes.

In the Bassick mine the ore was deposited in the agglomerate, replacing the fine interstitial tuff and partly replacing the larger volcanic fragments.[4] It will be recalled that a somewhat similar replacement of the San Juan breccia by ore was noted by Mr. Emmons in the Yankee Girl mine (p. 216). In the Bull Domingo mine "the ore occupies a nearly vertical chimney-like channel in a conglomerate or breccia mass, and the minerals are deposited in concentric scales around the bowlders or rock fragments."[5] This ore channel "was primarily formed by a complicated intersection of a number of fracture planes which produced a zone of broken country rock, in which the included fragments may have been somewhat rounded by attrition, but whose final rounding was more likely completed by the solvent action of percolating solutions."[6]

It is evident from the descriptions given of the Red Mountain stocks that the term "chimney" applied to them is somewhat misleading. They are not simple vertical pipes of ore extending from the surface to indefinite depths, but are separate bodies of irregular, lenticular, or spindle-shaped form, often completely inclosed by country rock, but linked with neighboring stocks by fissures which are often small and which carry little or no ore. There is no evidence whatever to indicate that they occupy the necks of former volcanoes. Although they occur in volcanic breccia, it is the

[1] Structural relations of ore deposits: Trans. Am. Inst. Min. Eng., Vol. XVI, 1888, pp. 833-834.
[2] On the peculiar features of the Bassick mine: Trans. Am. Inst. Min. Eng., Vol. XI, 1882, p. 110.
[3] The mines of Custer County, Colo.: Seventeenth Ann. Rept. U. S. Geol. Survey, Pt. II, 1896, p. 436.
[4] S. F. Emmons, loc. cit., p. 438.
[5] Emmons, loc. cit., p. 445.
[6] Emmons, loc. cit., p. 446

normal andesitic breccia of the Silverton formation, which covers hundreds of square miles of the San Juan region, and is in no sense an agglomerate filling a volcanic neck.

Another common explanation is that the Red Mountain stocks occupy the throats of extinct geysers and hot springs. The supposition that the ascending heated waters which were primarily instrumental in forming ore bodies may have issued as hot springs, or even geysers, from a former surface, is of course perfectly tenable and difficult to disprove. But to argue that the knolls of silicified and impregnated andesitic breccia conspicuous in the topography of to-day are really mounds of siliceous sinter is to be carried away by an analogy of the most unessential and superficial kind. The knolls exist merely because the bleached and altered rock of which they are composed resists erosion better than the crumbling, more easily weathered breccia, which has been less affected by the mineralizing solutions and is more readily disintegrated and carried away by rains and streams.

It seems most reasonable to regard the ore spaces of the Red Mountain district as a local modification of the general fissuring of the region. It is possible, however, that much of the minor, very irregular fissuring which is characteristic of this region may be due to contraction within the rock mass consequent upon the prevalent alteration of the volcanic rocks to aggregates of quartz, kaolin, and pyrite, with the removal of certain constituents, as more fully discussed on pages 114–131. Not only was the fissuring locally complex, but the ascending thermal waters had more chemical and probably more physical activity than elsewhere within the quadrangle, whereby solution played an important part in enlarging zones or aggregations of fractures into ore spaces and in metamorphosing the country rock to an extent not elsewhere observed in this region. The cause of this local intensity of chemical and physical activity is not known, although it is probable that the circulating intratelluric waters were here hotter and more copious than elsewhere. Some indication of the latter is afforded by the abundant strongly mineralized springs which issue from the surface in this district at the present day. There is, moreover, some geological grounds for believing that the Red Mountain region may have been formerly a local center of volcanic activity.

There has probably been fracturing of the rocks, or at least movement along preexisting fissures, since the ore bodies were formed, but there are not at present facilities for studying the relation of the older to the later fissures. It may even be considered doubtful whether the so-called fault found below the seventh level in the Guston and the sixth level in the Yankee Girl is entirely a postmineral dislocation, or whether it is in part a fracture antedating the deposition of the ore.

CHARACTER AND STRUCTURE OF THE ORE.

It is very difficult, on account of the present condition of the mines, to give a satisfactory account of the mineralogy and structure of the ores occurring in the stocks. The chief difference between these ores and those occurring in the lodes seems to have been the fact that the former occurred in nearly solid masses, with very little gangue. The occurrence of enargite, moreover, is apparently peculiar to the stock deposits in this region, the mineral not being known in the lodes. It does not, however, occur in all of the stocks. The more common ore minerals of the latter are galena, sphalerite, tetrahedrite, enargite, chalcocite (stromeyerite), bornite, chalcopyrite, and pyrite. While the galena ore of the upper levels was argentiferous, the highest silver values occurred in connection with the copper-bearing minerals, especially the chalcocite, bornite, and chalcopyrite. These three ore minerals were almost invariably accompanied by pyrite. In the richer portions of the ore bodies the pyrite was subordinate, and appears to have been found chiefly in the peripheral portions of the stocks. But in the poorer ores the argentiferous copper minerals occurred as nodular bunches in masses of low-grade iron pyrite carrying up to 5 per cent of copper and a little silver, usually less than 10 ounces per ton, and a fraction of an ounce of gold. The pyrite usually forms a rather fine-grained and sometimes crumbling aggregate, but in the Zuñi mine occurs in beautifully sharp octahedra embedded in white kaolin. Bismuthiferous ores of silver and lead occurred in some of the ore bodies, notably in the Silver Bell and Genesee-Vanderbilt. Specimens from the latter mine contain a bright lead-gray mineral having the physical properties of cosalite, which chemical tests show to be slightly argentiferous sulphobismuthite of lead containing a little copper. Cosalite occurred also in the Yankee Girl. The rare mineral, kobellite, a sulphantimonite of lead with bismuth partly replacing the antimony, was found in the Silver Bell mine,[1] carrying from 3 to 4 per cent of silver. The only other known occurrence of this mineral is at Hvena, Sweden. The rich silver ores proustite and polybasite occurred in the Yankee Girl, Genesee-Vanderbilt, and probably other mines of this district. Tennantite is reported to have occurred in the National Belle mine. Zinkenite has been described by Hillebrand[2] from the Brobdignag claim, a now abandoned prospect in the Red Mountain Range near Chattanooga. Guitermanite and zunyite occur in the Zuñi ore body and in a few prospects close to the latter, the zunyite as small sparkling, colorless tetrahedrons embedded in the massive, bluish lead-gray guitermanite. Barite is apparently always present in the ores of the stock deposits. It often occurred as isolated crystals or crystalline masses embedded in the argentiferous bornite of the Guston

[1] H. F. Keller: Zeitschrift für Krystallographie, Vol. XVII, 1890, pp. 67-72.
[2] Proc. Colo. Sci. Soc., Vol. I, 1883-84, p. 127.

and Yankee Girl mines, and these inclosed masses occasionally carried free gold. In the Zuñi it is intimately associated with pyrite and kaolin. The latter mineral is found associated with the ores of all the Red Mountain mines, sometimes occurring in the ore or serving as a matrix to pyrite, as in the Zuñi, but more abundantly in fissures in the wall rock and as a direct product of the alteration of the latter.

Although the common structure of the principal ore bodies was that of a solid and massive aggregate, yet there is evidence that vugs or cavities sometimes occurred in the unoxidized ores. Stalactites of pyrite, having the radial structure usually considered as characteristic of marcasite, were seen, which were said to have come from caves in the Genesee-Vanderbilt. Specimens of enargite from the National Belle mine, preserved in various cabinets, show a free development of clusters of radial prisms of enargite, such as could only have formed in open spaces. The well-crystallized specimens of polybasite which have come from the Red Mountain district were also probably formed in vugs.

The large caves, however, such as were a feature in the upper workings of the National Belle mine, appear to have occurred above the ground-water surface. A few of the smaller of these caves only could be seen in 1899, and we are dependent upon the description of T. E. Schwarz[1] for knowledge of the occurrence of the oxidized or secondary ores. According to him, they occurred "above a former water line, either attached to walls of caves as broken detached masses, or as a bed of clayey mud or sand, more or less completely filling the cave." The ores are said to have been mainly "carbonates of lead and iron, with iron oxides, lead sulphate, and arsenates." Kaolinite and zinc blende were common, and galena occurred as residual kernels. "Such are the ores of the National Belle, Grand Prize, and Vanderbilt mines. Other mines omit lead minerals and carry oxide and sulphide of bismuth." Several hundred pounds of this metal are said to have been reduced in Durango. These caves are stated to have ramified through the knolls of silicified andesite, such as that described at the National Belle mine, which "are cut up by cross fractures." The following general conclusions are cited from Mr. Schwarz's paper:

1. The secondary ores are richer than the sulphide ores occurring below them.

2. The ores of adjoining or connecting caves are sometimes greatly different in grade.

3. In some cases the formation of the caves along fracture or cleavage planes is evident, but in others all traces of such planes are quite obliterated.

4. The cave walls are a porous sandy quartz (see description of

---

[1] Trans. Am. Inst. Min. Eng., Vol. XVIII, 1890, pp. 139-145.

National Belle mine, p. 231), the sand from the disintegration of which forms part of the cave filling.

5. The line of change from oxidized to unoxidized ores, or the former water level, is very marked. It varies as much as 100 feet in elevation in properties within 1,000 feet of each other, rising to the south and west. The quartz outcrop rarely rises more than 200 feet above it.

6. In isolated cases may be found masses of the unoxidized ore, the enargite,[1] above the line of change, in the vicinity of the secondary ores.

Mr. Schwarz considered the caves and secondary ores to have been due to surface waters, which dissolved the sulphide ores and country rock as they moved along fracture planes. It appears unlikely, however, that the surface waters exercised much solvent action upon the country rock. Their function was probably limited to the oxidation of the ore and to the removal of some of its constituents.

CHANGES IN THE ORE WITH DEPTH.

As indicated in the preceding section, the ores of some of the stocks were partially oxidized down to the ground-water surface. This secondary alteration was particularly noticeable in the National Belle mine, and to some extent in the Vanderbilt mine, but was a much less conspicuous feature in the Yankee Girl, Guston, and other mines, the ore bodies of which outcropped nearer the bottom of the valley of Red Creek. The oxidation apparently presented no unusual features, and seems to have consisted chiefly of the transformation of galena to carbonate and sulphate of lead, with removal of more soluble constituents, reduction in the bulk of the ores, and the consequent formation of caves. Even when, as in the case of the National Belle, the ore of the deeper workings was chiefly enargite, the oxidized ores nearer the surface were evidently largely composed of lead compounds, as no reference is found to the occurrence of copper carbonates. This result is probably less a consequence of any peculiar process of oxidation than of the relative disposition of the sulphide or unoxidized ores prior to superficial weathering. It is with the changes in these latter ores relative to depth that this section has chiefly to do.

In spite of the diversity shown by the different ore bodies, there is after all remarkable uniformity to be found in the change at very moderate depths—usually less than 300 feet—from an ore consisting chiefly of argentiferous galena to highly argentiferous silver-copper ores, and then a gradual diminution of value downward through the increasing proportion of low-grade pyrite in the ore bodies. These changes are best recorded in the Yankee Girl, Guston, and Silver Bell

---

[1] It is evident that in this description Mr. Schwarz has the National Belle deposit most in mind.

mines. There are certain exceptions to what may be regarded as the normal or complete sequence. Thus the Congress mine appears to have had an ore consisting chiefly of enargite, with some bunches of galena, from the croppings downward. That the enargite at greater depth will gradually give place to iron pyrite, with diminishing amounts of chalcopyrite, can scarcely be doubted in the light of what is known of other deposits of this district. In the National Belle, enargite—in this case low grade—was encountered near the surface, but the oxidized ore in the siliceous knoll which forms the croppings appears to have been derived chiefly from galena, while below the third level (fig. 20) practically no pay ore was found. Explorations on the fourth level resulted only in the finding of small bunches of good ore and masses of crumbling iron pyrite carrying less than 0.1 ounce of gold and about 5 ounces of silver per ton, with from 1 to 3 per cent of copper.

Chalcopyrite appears to have been usually good ore. It carries, theoretically, 34.5 per cent of copper, and in the Red Mountain district ran well in silver and gold; but its appearance in great quantity was apparently a precursor of the ultimate change to ores carrying chiefly the low-grade iron pyrite.

Although there was on the whole a general change from argentiferous lead ores to argentiferous and auriferous copper ores, and finally to slightly argentiferous and auriferous iron sulphide (pyrite), yet the progression was an overlapping and irregular one in detail. Iron pyrite and chalcopyrite occurred at practically all depths, while galena in small bunches was sometimes found far below the point at which it had ceased to be the principal ore.

VALUE OF THE ORES.

The ores of the stocks, like those of the lodes, vary widely in value. In 1883, 3,000 tons of ore extracted from the Yankee Girl averaged nearly $150 per ton. A lot of 10 tons from the richest stopes of the same mine carried 3,270 ounces of silver per ton and 29 per cent of copper, corresponding to an average value per ton of about $3,000. Much richer ore than this occurred in small quantities in the Yankee Girl and Guston. The average sales value per ton of the ore of the latter mine has been given in the table on p. 222 as $91.81 for a period of eight years. The highest annual average was $363.25 per ton, and the lowest $10.70. These, it should be borne in mind, are the prices for which the ore was actually sold to the smelters. Values based directly upon metallic contents, without reference to cost of treatment, would be higher. The highest annual average represents the ore from the rich stopes above the seventh level. The lowest corresponds to the low-grade pyritic ore of the deeper levels. The richest ore recorded from this district was some taken in 1891 from the Guston at a depth of 378 feet, which carried 15,000 ounces of silver and

3 ounces of gold per ton, worth, at the then current price of silver, $14,880.

The ore of the Genesee-Vanderbilt was of much lower grade than that of the Yankee Girl, Guston, and Silver Bell. Its gross value for any one year probably never averaged more than $40 per ton. No records, however, are available for the ore taken out prior to 1893, when the price of silver stood higher. The ore of the National Belle was also low grade, but its average value is not known.

MINERAL WATERS ASSOCIATED WITH THE ORE DEPOSITS.

Probably all the waters met with in the mines of the Silverton quadrangle now accessible are meteoric waters variously modified by the materials though which they have passed in their more or less direct descent from the surface. The abundance of these waters varies much in different mines, and fluctuates with the seasons. In the Gold King mine scarcely a drop of water reaches the lower levels. Many other mines, such as the Silver Peak and Iowa, which are moderately wet in summer, become nearly dry in winter, owing to the freezing of the ground near the surface. In no case was any noticeable spring or ascending mineralized water encountered in the ore deposits now being worked. The descent of the meteoric water through masses of pyrite and other ore minerals is, however, often sufficient to give it a strong acid reaction and render it highly ferruginous.

In the Red Mountain district the troublesome acid character of the mine waters was notorious. In spite of their strong mineral nature and abundance, however, it is not improbable that these, too, were meteoric waters which had become strongly charged with sulphuric acid by their oxidizing passage through the adjacent bodies of pyrite and through the masses of altered pyritized country rock which surround the latter.

Strongly ferruginous springs are abundant within the drainage of Cement Creek and in the upper basin of Red Creek. As their waters are brought in contact with the atmosphere the iron in solution is oxidized and deposited near the spring as a mound or apron of limonite, or accumulates as bog iron in the swampy ground along these creeks. Such springs, with there accompanying deposits, may be well seen near Burro Bridge, also just below Chattanooga, and near the village of Red Mountain. A spring of the same general character issues from the hillside about 150 feet above the Guston mine, and has deposited a large apron of limonite. At present the water issues from a tunnel which was run into the limonite mass, and has cemented the dump of the tunnel into a firm, ferruginous mass. Some of this water was collected for analysis. As it issues from the tunnel it is clear, moderately cold, and noneffervescing, and has an astringent acid taste. When received in the laboratory at Washington the water gave a strongly acid reaction, and a deposit of ferric hydroxide or basic ferric sulphate, together with some silica, had settled in the bottom of the bottle. The total contents of sample were subjected to partial qualitative tests by Dr. Hillebrand, with the following results:

Partial analysis of water from spring above the Guston mine.

Constituent.	Remarks.	Constituent.	Remarks.
SO_3	Very much.	Zn	A little.
Cl	Very little.	Mn	
Al_2O_3	Much, about 214 parts per million.	Ni	Not tested for.
		Co	
Fe	Much.	Cu	0.65 parts per million.
Ca	Do.	Pb	
Na	Not tested for.	Mo	
K	Do.	As	Absent.
SiO_2	About 75-80 parts per million.	Ag	
		P_2O_5	Not tested for.

It appeared from these preliminary tests that the water of this spring carried no appreciable amounts of the important ore-forming metals, and it was not considered advisable to devote time and labor to further chemical work upon it. It was at first hoped that its analysis might throw some light upon the nature of the solutions which originally deposited the Red Mountain ores. But the preliminary tests indicate no characters that might not be expected in ordinary meteoric waters after passing for some distance through the mineralized and altered rocks of the immediate vicinity. There is, so far as known, no definite evidence connecting the ferruginous springs of this region directly with the original processes of ore deposition, nor is it necessary to assume that the waters issuing from them have come from deep-seated sources. The mine waters of the Yankee Girl and Guston are said to have carried much copper in acid solution. This is to be expected wherever surface waters carrying oxygen pass downward for any distance through pyrite and chalcopyrite. The presence of considerable alumina has an important bearing on the question of the metasomatic alteration of the country rock, as will be later shown.

METAMORPHISM OF THE COUNTRY ROCK IN CONNECTION WITH ORE DEPOSITION.

The metamorphism or change which has been effected through the agency of the mineral-bearing solutions (hydrothermal metamorphism) in the rocks adjacent to the ore bodies differs very markedly in degree, and to a less extent in character, in different portions of the quadrangle.

The mode of alteration is in general by *metasomatism* (literally change of body), by which is meant the process by which a mineral, through chemical reactions, undergoes a partial or complete change in its chemical constitution. Rocks or aggregates of minerals are "metasomatic" if any or all of the constituent minerals have undergone such changes.[1]

In Silver Lake Basin the simple fissure veins which there predominate, carrying galena, sphalerite, chalcopyrite, pyrite, and sometimes tetrahedrite, in a gangue of quartz with frequently some barite, are not accompanied by any very evident or striking alteration of the immediate wall rock.

As elsewhere pointed out, the rocks of Silver Lake Basin belong to the Silverton series. No attempt will be made in this report to give the petrographical characteristics of the series as a whole (for which the reader is referred to the forthcoming report of Mr. Whitman Cross); it is enough to say that the prevailing rock of the basin, and the one which constitutes the greater part of the country rock met with in the workings of the Silver Lake and Iowa mines, is a firmly consolidated, homogeneous, andesitic breccia. This differs from the older San Juan breccia in its greater homogeneity and in the frequent occurrence of more or less orthoclase. Indeed, it is not unlikely that the whole should be referred rather to the latites than to the andesites proper. With the breccia are associated massive andesites and latites. The breccia is in general somewhat altered, even when not close to an ore-bearing lode, a result that is to be expected when the thorough fissuring which the rocks have undergone and the extent to which they must have been permeated by ore-bearing solutions are taken into account.

A specimen of this rock taken on the trail from Arrastra Gulch up to the basin, at an elevation of about 11,800 feet, may be regarded as representative of the relatively unaltered facies. It is a tough, greenish-gray rock with pale lilac-gray mottlings, and, although showing a few feldspar phenocrysts, is generally of rather compact texture. The breccia structure, not very apparent on fresh fracture, owing to the homogeneity of the component fragments, becomes visible when the rock is breathed upon or has been slightly weathered. Under the microscope its clastic structure is evident. Fragments of general andesitic character lie in the matrix of fine andesitic detritus, which is not always easily separable from the larger fragments. This clastic matrix contains broken phenocrysts of andesine or labradorite, and an occasional somewhat rounded grain of quartz, in a fine feldspathic and obscure groundmass. There is some magnetite present, but ferromagnesian constituents are lacking. If formerly present, they have been destroyed by secondary alteration. The feldspars are largely changed to sericite (with perhaps some kaolinite) and to calcite, while bunches of calcite and flecks of chlorite are thickly disseminated through the rock. Some epidote is frequently present, although not

---

[1] Lindgren; Trans. Am. Inst. Min. Eng., Vol. XXX, 1900, p. 580.

seen in this particular slide. Such of the massive andesites and latites as were examined in thin section show a similar development of secondary chlorite, calcite, and sericite, indicating an alteration that is essentially propylitic in character.

Within the Silver Lake and Iowa mines the amount of alteration in the country rock which can properly be referred to the action of ore-bearing solutions moving along any given fissure is slight. The breccia usually preserves its greenish color and general appearance close up to the fissure walls. In the Iowa mine specimens were taken in the main crosscut on the fourth level at various distances from the Iowa vein, for the purpose of studying the alteration of the country rock dependent upon the formation of this lode. All of these specimens are pale greenish-gray rocks of one kind—a breccia of the type already described.

At points 150 feet east of the lode the country rock is fine grained and faintly mottled, showing only a few pale, minute phenocrysts of feldspar and an occasional tiny grain of quartz. Under the microscope the rock reveals the character of a much-altered andesitic tuff or fine breccia. The feldspars have been completely altered to aggregates of sericite and calcite, while areas of calcite and chlorite probably represent former phenocrysts of augite. The groundmass is a rather indistinct aggregate of secondary quartz, sericite, and chlorite, with a little apatite and rutile. The rock is thus practically wholly recrystallized into a secondary aggregate while retaining the grosser structure of the original.

At a distance of 100 feet from the vein the country rock is megascopically like that just described. Under the microscope it shows less alteration than the preceding. Chlorite and calcite are abundant, but much of the plagioclase is still recognizable. Sericite and quartz are not such prominent constituents as in the other rock.

At places 50 feet from the vein the country rock is still of the same general appearance as that already described. In thin section under the microscope it shows a general andesitic structure, but is almost wholly recrystallized. The feldspar phenocrysts have been changed to aggregates of calcite and sericite, while areas of chlorite and calcite, with sometimes rutile, are all that remain of the phenocrysts of augite or biotite. The groundmass, also, while preserving the outlines and in small part the substance of former lath-shaped feldspars, is now an aggregate consisting chiefly of quartz, chlorite, sericite, and a little rutile and apatite. The thin section does not show clastic structure, and is probably from a fragment in the breccia.

At a distance of 2 feet from the vein the only megascopical change in the country rock consists of the presence of an occasional speck of galena. Rather curiously, finely disseminated pyrite, which is a common feature of most country rock near a productive lode, does not in this case accompany the galena as a megascopical constituent. Under the microscope it is seen that the alteration has been more thorough than in the specimens previously studied. The forms of some of the phenocrysts are preserved by pseudomorphous aggregates—sericite, with some chlorite, calcite, and rutile, apparently after biotite, and calcite, quartz, sericite, and chlorite in varying proportions after augite and plagioclase. The groundmass is entirely recrystallized as a fine aggregate composed chiefly of quartz and sericite, with less chlorite. Occasional larger grains of quartz with rounded and embayed outlines are probably original phenocrysts. A little apatite and a few minute grains of pyrite occur in the groundmass. The dominant minerals of the rock are quartz and sericite. A specimen taken from the same crosscut at a distance of only 10 inches from the west wall of the vein resembles the others closely. It does not, however, show breccia structure, and is sparingly sprinkled with small crystals of pyrite. Under the microscope the rock is seen to be essentially an aggregate of sericite, quartz, calcite, and chlorite, named in the order of relative abundance. The former feldspar phenocrysts are now pseudomorphs of sericite, calcite, and quartz, and biotite has been altered to chlorite, sericite, and rutile. The groundmass consists chiefly of quartz, chlorite, calcite, and sericite. Although this facies is nearer the vein than the preceding, the character of its metamorphism is perhaps even less differentiated from the more general kind of propylitic alteration which the country rock in the mine, even at a distance of several hundred feet from the principal lodes, has undergone. Lastly, a specimen was taken from the east wall of the Iowa vein in the main stope above the fourth level. At this point the vein contained good ore, and the absence of gouge allowed the specimen to be taken immediately in contact with the ore. It differs from those thus far described in its light-gray color and more evident alteration. It is sprinkled with pyrite and a little galena, and is traversed by minute veinlets of quartz. Its breccia structure is still discernible in the hand specimen. Microscopically examined, the rock, while retaining traces of clastic origin, is wholly recrystallized. The former phenocrysts of feldspar are replaced by pseudomorphous aggregates of quartz and sericite. The quartz in such cases is often rather coarsely crystalline (grains up to 0.5 millimeter in diameter), the sericite being included as minute wisps, or gathered into bunches in the interstices between the allotriomorphic quartz grains. Crystals of pyrite are sometimes inclosed in the quartz. Of augite no trace remains, but some sericite inclosing rutile is apparently pseudomorphous after biotite. The groundmass is a finely crystalline mosaic of quartz and sericite. The notable feature of this wall rock is the absence of calcite and chlorite.

The alteration studied in the foregoing set of specimens, collected for that purpose in the Iowa mine, has been traced with similar results in collections from the neighboring Silver Lake mine. Calcite and chlorite are as a rule prominent constituents of the altered country rock to within a few inches of the productive lodes. Frequently, but not always, the rock in contact with the ore, or occurring as small horses in the veins, is free from chlorite and calcite and consists mainly of quartz and sericite, with some disseminated ore minerals. The latter, however, are not limited to the immediate walls of the fissure, but specks of galena and crystals of pyrite may occur 25 feet or more from the nearest lode, in rock carrying much calcite and chlorite. As an example of the extreme alteration which the country rock has undergone in the formation of these ore bodies may be described a specimen from the third level of the Silver Lake mine, taken from the wedge of country rock between the Stelzner and Royal veins at their junction. The usual breccia is here bleached almost white, and shows a fine, even-granular texture to the naked eye, but as a whole is shattered, full of quartz stringers, and contains flakes of galena. Under the microscope, pseudomorphs of sericite after feldspar and biotite preserve the only remnants of original structure. The groundmass is a very finely crystalline aggregate of quartz and sericite traversed by microscopic quartz veinlets. As usual, there is a little rutile present, and some crystals of pyrite.

To sum up briefly, the rocks in which the ore deposits of Silver Lake Basin occur are chiefly andesitic or latitic breccias. In the neighborhood of the mines these breccias have been generally altered to an unknown depth. The alteration involves the change of feldspar to sericite, calcite, and quartz; of augite to calcite and chlorite; and of biotite to chlorite, sericite, and rutile. Although this metamorphism is probably connected with the ore deposition, it is so generally prevalent that it can not in any case be recognized as being connected with the deposition of ore in any given fissure. It appears to have been effected through the agency of water charged with carbon dioxide or carbonates. The change involved in the rocks is propylitic in nature, and, as Lindgren[1] justly maintains, should be distinguished from ordinary weathering, with which it is often confused. Close to the veins, usually within a few inches, and in small horses of country rock within the veins, metamorphism of a different kind frequently occurs. Here calcite and chlorite have diminished in amount or are wholly absent, and quartz and sericite constitute the bulk of the rock. This alteration, which plainly emanates from the individual fissure, differs from the more general metamorphism less in kind than in the relative proportions of calcite and chlorite on the one side and of quartz and sericite on the other. This usually inconspicuous and very local alteration of the wall rock to quartz and sericite is rather common within the quadrangle, especially near lodes in the andesitic rocks of the Silverton or San Juan series. It occurs in somewhat pronounced degree in the Ridgway mine, which pro-

---

[1] Metasomatic processes in fissure veins: Trans. Am. Inst. Min. Eng., Vol. XXX, 1900, p. 645.

duces an argentite ore. It is the usual alteration seen in the mines of Savage Basin, Sultan Mountain (in monzonite), and in many other lodes where metamorphism of the country rock other than that of propylitic nature is not a conspicuous phenomenon, and where no special microscopical study of wall rock was made.

The alteration of the country rock in Silver Lake Basin may be accounted for on the assumption that the solutions which filled the fissures contained alkaline carbonates. But that the alkaline carbonates were present in small amount seems to be indicated by the uniform presence of chlorite (except in the actual walls of the vein, where it as well as calcite may be absent); for, as Lindgren[1] has pointed out, it is probable that chlorite can not exist under the action of strong solutions of alkaline carbonates. The immediate wall rock of the fissures, being more exposed to the action of solutions, has in most cases suffered the removal or further alteration of the chlorite and carbonates which were probably first formed. Connected with this later stage is the accumulation of the insoluble sericite, derived from labradorite, biotite, and probably other minerals, and very likely the direct addition of quartz in place of the removed calcite. Although no chemical analyses have been made of these wall rocks, it is probable that the abundance of sericite (derived in great part from lime-soda feldspars), which is common in the rock in contact with the vein filling, indicates a direct addition of potash to the substance of the rock, as was found by Lindgren[2] to be the case in the Grass Valley and Nevada City veins. Kaolinite was not certainly identified in the altered wall rocks of Silver Lake Basin, but in the Dives mine (on the North Star lode) it apparently occurs with sericite and quartz, as seen in an altered andesitic horse in the lode. Metamorphism in connection with the Dives ore body is similar to that just described in detail for the Silver Lake mines; but in its greater intensity, in the presence of kaolinite, and in the more evident silicification of portions of the wall rock it is intermediate in character between that metamorphism and the kind next to be described.

Of a somewhat different kind from that which has just been discussed is the metamorphism observed in connection with the ore deposits of Engineer Mountain and of the Red Mountain district. As the ore of Engineer Mountain occurs in lodes, while that of the Red Mountain district is prevailingly in stocks, and as the metamorphism presents some phases of difference in the two modes of occurrence, they will be separately treated.

The Polar Star lode, which has produced some rich silver ore carrying argentite and proustite, may be considered as a type of the Engineer Mountain deposits. Unfortunately, its workings are no longer

[1] Trans. Am. Inst. Min. Eng., Vol XXX, 1900, p. 610.
[2] The gold-quartz veins of Nevada City, etc.: Seventeenth Ann. Rept. U. S. Geol. Survey, Pt. II, 1896, p. 148.

accessible, but the metamorphic processes connected with ore deposition may be studied to considerable advantage from surface exposures and the material of the dumps. The country rock of the Polar Star is a dark-gray rock of andesitic appearance, carrying abundant glistening phenocrysts of plagioclase up to 7 or 8 millimeters in length. In thin section, under the microscope, the plagioclase is seen to be chiefly labradorite (ab$_1$ an$_1$) which is almost perfectly fresh. Phenocrysts of a pale augite or diopside are almost wholly altered to calcite and chlorite, and some bastite pseudomorphs indicate the former presence of phenocrysts of an orthorhombic pyroxene. Quartz, and probably some orthoclase, are present in the groundmass. The chemical analysis of this rock is given in Column I, page 122. This analysis differs from that of a normal andesite in its rather high alkalies and relatively high potash. It might be classed as a latite. Owing to its somewhat weathered condition and lack of knowledge as to the accurate chemical constituents of the resulting chloritic and serpentinous products, an attempt to calculate the mineralogical composition from a chemical and microscopical analysis is not entirely satisfactory.

Calculating the carbon dioxide as calcite, the soda as labradorite (ab$_1$ an$_1$), the phosphoric anhydride as apatite, the residual lime as diopside, all the potash as orthoclase, and dividing up the other constituents between chlorite, serpentine, magnetite, etc., the following approximate mineralogical composition was obtained.

Mineralogical composition of andesitic rock (latite?) from Engineer Mountain.

Constituent.	Per cent, by weight.	Constituent.	Per cent, by weight.
Labradorite (ab$_1$an$_1$)	33.9	Kaolinite	2.6
Orthoclase (?)	22.4	Magnetite	2.3
Quartz	14	Hematite	3.8
Diopside (CaMg(SiO$_3$)$_2$)	4.7	Apatite	1
Chlorite (H$_{40}$(FeMg)$_{23}$Al$_{14}$Si$_{18}$O$_{90}$)	7	Rutile and leucoxene	1.4
Calcite	3		98.9
Serpentine	2.8		

There still remains about 1 per cent of alumina not accounted for in the above calculation. Combined with part of the ferrous iron here calculated as magnetite and chlorite, and with some of the silica, reckoned as quartz, it may enter into the composition of the pyroxene, which probably does not correspond exactly to the theoretical diopside molecule. The percentage of kaolinite is rather higher than the microscopical investigation would indicate. The specific gravity as calculated from the foregoing mineralogical composition, ignoring porosity, is about 2.8. The foregoing specimen was taken a few hundred feet west of the mine, and, although fresher rock occurs in the vicinity, this was selected for chemical analysis as being undoubtedly the particular facies in which the Polar Star lode was formed.

The dumps of the Polar Star mine are made up largely of a locally

altered form of the foregoing latite. The original porphyritic structure is perfectly preserved, but the rock is bleached to a very light-
gray tint, and the feldspars are completely kaolinized. Small crystals
of pyrite are abundant, occurring both in the altered phenocrysts and
in the groundmass. This is evidently the country rock which immediately incloses the ore, although it is not known how far this pronounced alteration extended on either side of the ore-bearing fissure
zone. Under the microscope the rock appears as a fine-granular
aggregate, consisting chiefly of quartz, kaolin, and diaspore, with
considerable disseminated pyrite. The porphyritic structure is partially lost between crossed nicols, although the abundance of diaspore
in some of the altered phenocrysts differentiates them from the
groundmass. The diaspore occurs in irregular grains up to about 0.5
millimeter in diameter. The relatively high refractive index, strong
double refraction with consequent brilliant polarization colors, and
the conspicuous cleavage render the mineral very noticeable in thin
section. It is biaxial, with dispersion red less than blue.

In order to confirm the microscopical determination, a portion of
the rock was powdered and treated with hydrofluoric acid by Dr.
Hillebrand. The resulting residue proved, on microscopical examination, to consist of grains of the supposed diaspore, with a few
particles of pyrite and a little amorphous material. A rough quantitative analysis showed that the residue consisted of about 84.4 per
cent of alumina and 15.2 per cent of water, thus placing the identification of the mineral as diaspore beyond question.

A chemical analysis of this altered rock is given in Column II on
page 122. The two analyses of fresh and altered rock, as they stand
side by side, represent a comparison of unit weights, and therefore
are not, in all probability, directly comparable for the purpose of ascertaining the actual change that has taken place in a unit volume of the
original rock. It may, however, be tentatively assumed that the
alumina has remained constant. An analysis of the altered rock
recalculated on this basis gives the result shown in Column IIa. A
comparison between Column IIa and Column I shows that, under the
assumption made, the altered rock has suffered a total loss of substance
of over 14 per cent. A little silica, 3.15 per cent of water, and some
sulphur have been introduced, while magnesia and carbon dioxide
have been wholly, and lime-soda potash almost entirely, removed.
Of the original iron present in the rock, a little more than one-half
remains combined with sulphur to form pyrite. It will be noted that
in the recalculated analysis the phosphoric and titanic acids are very
nearly the same as in the fresh rock, which lends some support to the
assumption previously made that the alumina has remained practically constant in the process of metamorphism. It seems at first sight
probable that the loss of substance in the altered rock may be accompanied by an increase of porosity without actual shrinkage of the rock
as a whole. If this be so, the specific gravities of the unaltered and

altered rock, as determined on considerable fragments, should be to each other approximately as 100 is to 86 (i. e., to 100 less the loss of substance due to metamorphism). The specific gravities as actually determined are 2.764 and 2.734, and their ratio is as 100 to 99. This result does not accord with the idea that the loss of substance has resulted in corresponding increase of porosity with retention of initial volume. In so far as the laboratory determinations of specific gravity may be relied upon in drawing conclusions as to the mass as a whole, they indicate that the loss of substance has been followed by diminution of volume, presumably associated with actual cracking or fissuring and not with increased porosity.

Analyses and recalculations of rocks of Polar Star mine, Engineer Mountain.

Constituent.	I.	II.	IIa.	Loss or gain of each constituent in percentage of the whole original rock.	IIb.	Loss or gain of each constituent in percentage of the whole original rock.
	Per cent.	Per cent.	Per cent.	Per cent.	Per cent.	Per cent.
SiO_2	55.61	64.79	56.13	+0.52	64.14	+8.53
Al_2O_3	16.40	18.93	16.40	0	18.74	+2.34
Fe_2O_3	5.44	None.	None.	−5.44	None.	−5.44
FeO	2.37	None.	None.	−2.37	None.	−2.37
MgO	3.25	None.	None.	−3.25	None.	−3.25
CaO	5.85	.43	.37	−5.48	.42	−5.43
Na_2O	2.61	.15	.13	−2.48	.14	−2.47
K_2O	3.77	.24	.20	−3.57	.23	−3.54
H_2O-	.46	.50	.43	− .03	.49	− .03
H_2O+	1.51	5.39	4.66	+3.15	5.33	−3.82
TiO_2	1.10	1.21	1.04	− .06	1.19	+ .09
CO_2	1.33	None.	....	−1.33	....	−1.33
P_2O_5	.45	.51	.44	− .01	.50	− .05
S	Trace.	....	....	....	....	....
Cr_2O_3	None.	None.	....	....	....	....
NiO	None.	None.	....	....	....	....
MnO	.09	None.	....	− .09	....	− .09
BaO	.03	.06	.05	+ .02	.06	+ .03
SrO	.05	Trace.	....	− .05	....	− .05
Li_2O	Trace.	Str. tr.	....	....	....	....
FeS_2	....	7.19	....	+6.22	7.12	+7.12
Total	100.32	99.40	....	−14.25		−1.96
Sp. gr	2.764		2.734 at 19° C.	(W. F. Hillebrand.)		

I. Latite. Country rock of the Polar Star mine, Engineer Mountain. Analysis by H. N. Stokes.

II. Metasomatically altered latite. Wall rock of Polar Star mine, Engineer Mountain. Analysis by H. N. Stokes.

IIa. Recalculation of II on basis of constant alumina.

IIb. Recalculation of II on basis of constant volume.

It will be remembered that the preceding discussion rests upon the initial assumption that the alumina has remained constant during the metamorphism. This assumption may not be correct, and it will further the inquiry, as well as illustrate the complexity of the problem, to trace the results growing out of another probable premise. If it be assumed that during the metamorphic processes the volume of any given mass of the rock has remained constant, then the analysis of the altered rock may be recalculated on this basis, using for this purpose the determined specific gravities, and be then compared with that of the unaltered rock. The result of this comparison is shown in Column II*b*

It appears from this total that under the assumption of constant volume the rock has lost only about 2 per cent of its substance through metasomatism. Moreover, there has been an addition of 8.5 per cent of silica and 2.3 per cent of alumina. In other respects the alteration is not significantly different from that deduced upon the hypothesis of constant alumina.

It is, of course, possible that neither the hypothesis of constant alumina nor that of constant volume is strictly true; but it is believed that the latter is a closer approximation to the truth than the former. The altered rock is not notably porous, nor does it, in the field, show evidence of contraction in bulk. Moreover, the very obvious introduction of quartz into some facies of it and the occurrence of diaspore point to that assumption involving the introduction of silica and alumina as being nearest the truth.

In order to further test this point a special examination was made of the spring water issuing above the Guston mine (see p. 114) to determine whether it carried alumina in solution. It was found to be present to the extent of about 214 parts per million, thus proving that the alumina in the rocks can not be regarded as a constant constituent during metasomatic alteration in this region.

The mineralogical composition of the altered rock may be roughly calculated from Analysis II as follows, the soda being considered as accompanying the potash in sericite:

Mineralogical composition of metasomatically altered wall rock of Polar Star mine, Engineer Mountain.

Mineral.	Percentage by weight.	Mineral.	Percentage by weight.
Quartz	48.8	Rutile	1.2
Kaolinite	30.3	Apatite	.7
Diaspore	6.6		98.7
Pyrite	7.2		
Sericite	3.9		

The specific gravity of the rock, as derived from the foregoing mineral composition and excluding pore spaces, is about 2.9, as compared with 2.8 similarly obtained for the unaltered rock. This rough calculation is thus in agreement with the general rule that near fissure veins the replacing minerals are as a whole denser than those replaced.[1]

---

[1] Lindgren: Trans. Am. Inst. Min. Eng., Vol. XXX, 1900, p. 594.

Although the altered facies just described appears to indicate only slight silicification or actual addition of silica to the rock, this process has taken place to a greater extent in the immediate walls of the fissures and in fragments of country rock included within the vein quartz. Such facies show little or no diaspore, but more sericite. The secondary quartz in such rocks is often more coarsely granular, and in thin section can be distinguished from the quartz which fills veinlets only through its association with nests of sericite, kaolinite, and a little diaspore. All original structure of the latite disappears between crossed nicols.

Summing up briefly, the characteristic features of the Polar Star metamorphism are (1) its restriction to the immediate vicinity of the lode, (2) the removal of at least 2 per cent of the substance of the rock, including all of the magnesia and carbon dioxide, nearly all of the alkalies and lime, and much of the iron, (3) the addition of silica, water, alumina, and sulphur, and (4) the formation of a secondary aggregate chiefly of quartz, kaolinite, pyrite, diaspore, and sericite.

The depth to which the alteration of the character extends is not known. The specimens studied evidently came from the deepest workings, probably over 500 feet from the surface.

It is not probable that the metamorphism described can have been effected by solutions in which alkaline carbonates were the principal constituents. It is more likely that the effective agent was acid water. If, however, the water contained sulphuric or hydrochloric acids it is, at first thought, rather difficult to understand the formation of kaolinite and diaspore. But as these minerals themselves are not soluble in acids, they may be capable of forming, under certain conditions, even in the presence of such a solvent for alumina as free sulphuric acid. Acid waters containing sulphuric acid circulate through the rocks of the Red Mountain region at the present day and have not prevented the formation of kaolinite nor effected its removal. In fact, they have probably in some cases aided directly in its formation by serving as solvents from which the alumina has been precipitated as kaolinite or diaspore under favorable conditions. The formation of diaspore instead of kaolinite in the presence of abundant silica indicates a very moderate hydration.

The metasomatic alteration of the rocks of the Red Mountain Range has frequently been referred to in this report. It is not limited to the immediate vicinity of known ore bodies, but is so widespread and noticeable a feature as to become an important problem in the general geology of the region and a serious embarrassment to the geologist intent upon unraveling the structure and relationships of the various igneous rocks involved.

As a rule, the metamorphosed rocks, where not superficially stained by iron oxide, are nearly white. Frequently traces of original structure, such as outlines of feldspar phenocrysts or of breccia fragments.

can be recognized, but not uncommonly these, too, have vanished, and the rock has become a white granular aggregate resembling a fine-grained quartzite. Of such origin and character is most of the so-called "quartz" of the Red Mountain mines, which forms the siliceous knobs in which many of the ore bodies outcrop.

The mound at the White Cloud mine may be taken as a typical example of such a knob or knoll. It is elliptical in plan and about 50 feet long, trending N. 10° E. It rises about 30 or 40 feet above the general slope of the hill, and is composed of bleached siliceous rock which occasionally shows traces of breccia structure and which is evidently an altered form of the Silverton andesitic breccia. It contains considerable very finely disseminated pyrite.

A thin section of this rock shows a finely crystalline, rather turbid aggregate of quartz, kaolin, pyrite, rutile, and some minute indeterminable minerals. Some outlines of former phenocrysts are recognizable in ordinary light, but these disappear between crossed nicols. The quartz grains are very irregular in outline and intricately interlocked. They are full of minute inclusions of kaolinite, rutile, pyrite, and other minerals too small for determination. The kaolin occurs in bunches showing the characteristic foliated aggregation of the individual scales, and the usual low double refraction.

A chemical analysis of this rock, from a specimen taken in a short tunnel in the knoll, is here given:

Analysis of altered andesitic breccia from White Cloud mine.

[George Steiger, analyst.]

Constituent.	Per cent.	Constituent.	Per cent.
SiO_2	85.49	TiO_2	.63
Al_2O_3	5.49	CO_2	None.
Fe_2O_3	.23	P_2O_5	.07
FeO	.25	SO_3	.46
MgO	None.	MnO	None.
CaO	.27	BaO	.07
Na_2O	None.	FeS_2	3.43
K_2O	None.		99.92
H_2O-	.46		
H_2O+	3.07		

No analysis of unaltered rock is here available for comparison, and it is not possible to determine accurately the additions and subtractions to the original rock which have resulted in the metamorphosed facies. It is evident, however, that either the alumina has been very greatly reduced or the silica has been enormously increased. In all probability there has been some removal of alumina and a large addition of silica. Ferric and ferrous iron have been partly removed and the residue has been changed to pyrite. Lime has been almost all carried away and soda and potash have been wholly removed. Carbon dioxide is absent.   The little sulphuric anhydride present is probably combined with baryta as barite.

The mineralogical composition of the altered rock may be roughly calculated as follows:

Mineralogical composition of altered andesitic breccia from White Cloud mine.

Mineral.	Per cent.
Quartz	78.5
Kaolin	13.8
Pyrite	3.4
Rutile	.6
	96.3

About 1 per cent of water is not accounted for, as there are no alkalies or alumina to combine with it.   It may exist as chalcedony or opal, or as some hydrated compound of iron.   No attempt has been made to calculate the small amounts of ferric and ferrous iron, lime, phosphoric acid, sulphuric acid, and baryta, which probably correspond to a little barite, apatite, and other minerals present in minute quantities in the rock.   The total absence of sericite is rather remarkable.   The character of the alteration points to acid waters as the cause of metamorphism.

Of the same general character as the foregoing is the metamorphism which has transformed andesitic tuffs or breccias of the Silverton series into the highly siliceous rock forming the greater part of the picturesque knob in which the National Belle mine lies, at the village of Red Mountain.   The resulting rock is nearly white and decidedly porous.   Examined with a lens, it appears to consist wholly of quartz with a little pyrite, and is full of minute cavities sparkling with quartz crystals.   Locally this porous structure may be so pronounced as to give the rock a pumiceous appearance.

Under the microscope, in ordinary light, the remnants of porphyritic structure are easily recognizable.   Relatively clear areas having the shape of feldspar phenocrysts and dark areas containing much finely divided pyrite and other opaque particles, clearly pseudomorphs after some phenocrysts—probably ferromagnesian constituents—lie in a groundmass somewhat turbid with minute dark particles.   With the nicols crossed, however, this structure practically vanishes, and the whole appears as a finely crystalline granular aggregate of quartz. The space formerly occupied by the phenocrysts of feldspar is now filled with quartz, somewhat more coarsely crystalline than in the groundmass.   Frequently the quartz does not quite fill the space, but has left a microscopic vug in the center.   The quartz making up the rock is not pure, being crowded with indeterminable inclusions.   But the rock in all probability contains over 90 per cent of silica, and the evidence of thorough silicification, with removal of nearly all of the bases save a little iron to form pyrite, is unquestionable. It shows that under certain conditions even the alumina may be removed, and that the assumption as to the constancy of this constituent in unaltered and altered rock must be used with caution.

An intrusive porphyry, belonging with the monzonite-latite group of rocks, is fairly abundant in the Red Mountain region, where it forms relatively small masses of circular or elliptical plan. It has been intruded in the form of plugs or bosses into the volcanic rocks, and, like the latter, has undergone metasomatic alteration. Where not greatly altered—as, for example, at an exposure by the roadside a few hundred yards south of the Yankee Girl mine—the rock is conspicuously porphyritic, showing phenocrysts of pink orthoclase up to three-quarters of an inch in length, smaller crystals of white plagioclase, and anhedrons (crystalline grains) of quartz, all lying in a compact, greenish-gray groundmass.

Under the microscope the rock shows considerable decomposition. The orthoclase is fairly fresh, but the plagioclase (apparently in part acid oligoclase) is partly sericitized. Areas of calcite and chlorite with a little pyrite are apparently pseudomorphous after augite or diopside. The groundmass is a rather cloudy, fine-granular aggregate which seems to consist chiefly of quartz and orthoclase, but contains considerable calcite and chlorite, with some apatite, a good deal of finely disseminated pyrite, and some titanic iron altering to leucoxene. A chemical analysis of this rock is given in Column I in the following table:

Analyses of monzonitic porphyry.

[H. N. Stokes, analyst.]

Constituent.	I.	II.	Constituent.	I.	II.
	Per cent.	Per cent.		Per cent.	Per cent.
SiO_2	58.78	73.61	P_2O_5	.53	.33
Al_2O_3	13.62	13.97	SO_3	....	Not det.
Fe_2O_3	1.70	None.	Cr_2O_3	None.	None.
FeO	2.27	None.	NiO	None.	None.
MgO	3.29	None.	MnO	Trace.	Trace.
CaO	4.31	.31	BaO	.12	.04
Na_2O	3.24	.04	SrO	.05	Trace.
K_2O	4.06	.06	Li_2O	Trace.	Strong tr.
H_2O-	.25	.58	FeS_2	1.56	5.62
H_2O+	1.57	4.18		99.73	99.34
TiO_2	.99	.60			
CO_2	3.49	None.			

I. Monzonitic porphyry, from near Yankee Girl mine.
II. Metasomatically altered facies of foregoing, from near National Belle mine.

From this analysis and the microscopical study the mineralogical composition may be calculated as follows:

Mineralogical composition of monzonite-porphyry.

Constituent.	Per cent.	Constituent.	Per cent.
Albite (molecule)	27.4	Titanic iron ore	
Quartz	21.3	Magnetite and rutile	} 4.2
Orthoclase (molecule)	21.2	Pyrite	1.5
Chlorite	11	Apatite	1.2
Calcite	7.8		
Sericite	4		99.6

All the lime shown in the analysis is required to form calcite with the carbon dioxide, so that the plagioclase, whatever its original composition, must at present contain very little lime and is calculated as albite. The appearance of the thin section indicates that the chemical determination of the pyrite may be a little low and that a portion of the iron, here calculated as titanic iron ore, should really be estimated as pyrite.

On the north side of the National Belle knob a mass of this same porphyry has been altered by metasomatic processes to a very light-gray rock in which the feldspars are transformed to dull kaolin-like aggregates while the quartz phenocrysts are apparently unchanged. A few minute veinlets, filled with kaolin, occur in the hand specimen. The groundmass is gray in color, compact in texture, and abundantly sprinkled with minute crystals of pyrite. This alteration is evidently directly connected with the deposition of the National Belle ore bodies, and is merely one phase of the metamorphism which has given rise to the siliceous mass in which the ore occurs.

Under the microscope rounded and embayed phenocrysts of quartz containing minute fluid inclusions, and pseudomorphous aggregates after feldspar, are seen to lie in a finely crystalline, granular groundmass. The feldspar phenocrysts have been changed to pseudomorphs of kaolin, diaspore, and quartz. The diaspore is intimately associated with the kaolin, in which it often occurs embedded in ragged, shredlike areas. Near it there is sometimes a small amount of a colorless isotropic mineral with a fairly high refractive index, which has thus far defied identification. The groundmass is a finely crystalline mosaic of quartz and kaolin with small scattered crystals of pyrite and occasional crystals of rutile and apatite. Sericite was not noted, but is perhaps not entirely absent.

A chemical analysis of this rock is given under II, on page 127, and although the two rocks are not from the same mass, there can be little doubt that they were both originally of the same character.

It may be that the nearly identical contents in alumina shown by the two analyses is merely a coincidence, but it is probable that it denotes practical constancy of the alumina during the change which has taken place in the rock.

The notable features of the alteration are the introduction of silica, water, and sulphur, the total removal of iron (except that combined with sulphur to form pyrite), magnesia, and carbon dioxide, and the almost entire abstraction of lime and alkalies. The formation of diaspore brings this alteration close to that described at the Polar Star mine, the original rocks in both cases being of similar chemical composition, while the addition of silica and the formation of kaolin, usually without sericite, is characteristic of the Red Mountain mines.

Ignoring the small amount of the unknown isotropic mineral, the mineralogical composition of the altered rock may be roughly calculated from the chemical and microscopical data as follows:

Mineralogical composition of altered monzonitic porphyry from National Belle mine.

Constituent.	Per cent.	Constituent.	Per cent.
Quartz	60.9	Apatite	.6
Kaolinite	26.3	Rutile	.6
Pyrite	5.6		98.4
Diaspore	a 3.8		
Sericite	.6		

a The diaspore of this rock was also isolated with hydrofluoric acid. It appeared to be fully as abundant in the residue as in the similar residue from the Polar Star rock, described on page 121, and the above calculation may be low.

This is a very different alteration from that normally brought about by waters carrying alkaline carbonates, which, as Lindgren [1] has shown, tend to form sericite and carbonates in the wall rock. It is believed to be due to the action of acid waters.

It is unfortunate that the study of the metamorphism connected with the Red Mountain ore bodies is at the present time practically confined to the surface. However, as far as can be judged from the mine dumps, metamorphism of the character described accompanied the ore bodies to the greatest depth explored.

On Anvil Mountain, which forms the southern end of what has been called for convenience the Red Mountain Range, the bleached rocks, when critically examined, are found to present some variations in composition. Thus the country rock of the Brobdignag claim, probably originally andesitic, is converted into a fine-granular aggregate of quartz and barite, with a little sericite. It is impregnated with pyrite and is traversed by some veinlets of sericite or kaolin. At the Zuñi mine the country rock, which was evidently an andesite or latite, has been recrystallized near the ore body to a mass of quartz, sericite, kaolin, and barite, with a little epidote in small grains. Some rutile

[1] The gold-quartz veins of Nevada City and Grass Valley districts, California: Seventeenth Ann. Rept. U. S. Geol. Survey, Pt. II. 1896, p. 172.

and some undetermined minerals in minute grains are also present. At the Mystery claim, on the summit of Anvil Mountain, the rock (rhyolite?) has been changed to a nearly white, rather porous mass showing no traces of original igneous structure. Under the microscope it shows faint outlines of former phenocrysts, which become nearly invisible between crossed nicols. The rock consists chiefly of quartz and alunite in a fine-granular aggregate, with a little kaolin and rutile. At John Roland's tunnel, about 1½ miles southeast of the Mystery claim, an opportunity was presented for studying some of the steps of alteration. About 150 feet from its mouth the tunnel cuts through a very irregular vein-like mass of kaolin and continues beyond it for 200 feet or more. Near the breast the rock, although altered, can be recognized as an andesite, or closely related rock. It is greenish gray in color and shows outlines of small feldspar phenocrysts. Under the microscope, although the original porphyritic structure is partly preserved, the rock is seen to be completely altered. The phenocrysts of plagioclase are changed to sericite and quartz, while the augite and perhaps other dark constituents are represented by pseudomorphs of chlorite and calcite. At about 150 feet from the fissure the country rock is already bleached to a light greenish-gray tint and the original texture partly obliterated. Under the microscope it is seen to be a secondary aggregate, consisting chiefly of sericite, quartz, and calcite, retaining, however, outlines of the original porphyritic structure. About 45 feet from the fissure the rock is nearly white and shows very faint traces of original texture. In general appearance it is identical with most of the bleached and metamorphosed rock of the Red Mountain Range. Microscopically examined, the outlines of former phenocrysts are distinctly visible, but the whole rock is recrystallized to a very fine aggregate of quartz and sericite with some leucoxene (rutile). The former feldspar phenocrysts are now nearly pure sericite. Lastly, a specimen taken from the wall of the fissure is a nearly white, compact rock in which close scrutiny fails to detect any vestige of primary structure. Under the microscope the rock appears as a finely crystalline aggregate consisting of quartz and sericite, with a little brownish isotropic material occurring in minute veinlets. Recrystallization in this case has not only been complete, but there has been such thorough rearrangement of the crystallizing material as to obliterate all traces of primary structure. The alteration has plainly been accompanied by removal of iron, lime, magnesia, and probably other constituents. Whether alumina has been removed is an open question, but the fact that the fissure from whence the metamorphosing solutions penetrated the inclosing rock is itself filled with a vein-like deposit of kaolin shows that at some depth both potash and alumina have been removed from the country rock and concentrated as kaolinite within the fissure walls.

The conspicuous metamorphism of the Red Mountain region has

undoubtedly involved rocks of various kinds, reducing them all to light-colored secondary aggregates of similar appearance. Andesites, latites, monzonite-porphyry, and rhyolite have been altered to products which often give little indication of the nature of the original rock. But in many portions of the quadrangle the rhyolitic rocks show certain characteristic phases of metamorphism in connection with ore deposition which seem to merit some special notice. This rock, when acted upon by ore-bearing solutions, is particularly susceptible to recrystallization into secondary aggregates of quartz and sericite, quartz and kaolin, or quartz, sericite, and kaolin. As a general rule such metasomatic alteration is accompanied by more or less replacement of the rock by ore, as may be seen in the Tom Moore or Silver Ledge mines.

No favorable opportunity was presented for studying in detail the metasomatic processes which accompanied the deposition of ore in the monzonitic stocks, such as Sultan Mountain. But so far as observed the alteration is not conspicuous.

PARAGENESIS OF THE LODE AND STOCK ORES.

By *paragenesis* is here meant the association of the various ore and gangue minerals with special reference to the order and mode of their formation.

Beyond the common and well-known derivation of certain secondary minerals in the zone of oxidation, as, for example, anglesite or lead sulphate from galena, the directly observable paragenesis of the ores offers few points which can be embodied in any general statement of regular association. As a general rule the ore minerals which are found together in any one deposit have formed contemporaneously, and cases are rare in which a definite and constant succession of different ore minerals can be recognized. In the Tom Moore lode chalcopyrite appears to have formed later than tetrahedrite. The same relation was observed in the Dives lode, where masses of tetrahedrite are surrounded by an envelope of chalcopyrite from which radiate still younger quartz crystals. But in many other lodes these two minerals occur in such irregular relations as to point to practically contemporaneous crystallization. In the Empire-Victoria lode, on Sultan Mountain, hübnerite and fluorite have been formed since the ore was deposited. Native copper, wherever seen, was plainly of later origin than the sulphide ores with which it was associated. Native silver was seen only in detached specimens of ore, but it occurs characteristically in the upper portions of the deposits and is undoubtedly of secondary origin. Quartz of at least two generations is common, as shown by the relatively barren stringers of this mineral, which can be found traversing the ore in many of the deposits.

Free gold occurs usually embedded in quartz and associated with pyrite or chalcopyrite. In the Sunnyside, Sunnyside Extension, and

Camp Bird mines it is also associated with pale-yellow sphalerite and fluorite. Some specimens were seen from the Sunnyside Extension mine in which the gold occurred as implanted crystals on quartz and was therefore of later age than the latter.

Unfortunately there are at present few opportunities for studying the occurrence of proustite and argentite in the Silverton quadrangle. In the Ridgway mine much of the argentite occurs implanted on, or wedged between, the quartz crystals of small vugs, and was the last mineral to crystallize. It has been the universal experience in the Silverton district that argentite, proustite, and polybasite are not found below a very moderate depth. The data at hand do not permit of a definite statement as to the depth at which these rich argentiferous minerals change to low-grade ore, but the indications are that it is less than 1,000 feet, although it is well known that in other regions some of these minerals extend to greater depths. It is probable, although not at present demonstrable, that in the Silverton region proustite, argentite, and polybasite are of secondary origin and indicate a zone of enrichment of the kind to which Emmons[1] and Weed[2] have recently called attention.

The downward change from galena through richly argentiferous copper ores to chalcopyrite and pyrite, which was found to be a characteristic feature of many of the Red Mountain mines, is a case of paragenesis on a large scale, which will be discussed in the following section on the origin of ores. The origin of the rhodonite, which is so abundant in many of the lodes of the northern half of the quadrangle, and its exact relation to the quartz and ore which accompany it constitute a puzzling problem, for which no satisfactory solution has been found. Such a solution must account for the presence of large lenticular or partition-like masses of rhodonite, carrying a few specks of low-grade ore, and dividing the ore bodies longitudinally into two or more parts, and for the presence of patches and stringers of rhodonite. In the Saratoga mine rhodonite has formed by metasomatic replacement of limestone. It is possible that in such lodes as the Sunnyside the large, solid masses of rhodonite within the vein may represent metasomatically altered horses of country rock. This, however, is merely a hypothesis, which requires confirmation before acceptance. It has undoubtedly been deposited also in open fissures as a true vein mineral.

ORIGIN OF THE LODE AND STOCK ORES.

That the ores of the Silverton quadrangle were deposited from aqueous solution is a proposition which, in the light of present knowledge, needs no special demonstration. They were precipitated partly

[1] The secondary enrichment of ore deposits: Trans. Am. Inst. Min. Eng., Vol. XXX. 1900, pp. 177–217.

[2] Enrichment of mineral veins by later metallic sulphides: Bull. Geol. Soc. Am., Vol. XI. 1900. pp. 179–206.

in open spaces, to which the solutions had access, and partly as replacements of various rocks through metasomatic action. Their deposition was accompanied by chemical and mineralogical changes in the immediately adjacent country rock, producing effects which in the majority of cases diminish rapidly in intensity with increasing distance from the fissure walls. As far as is known, the deposition of ore within the fissures was not affected by differences in character of the wall rock. It is believed that the facts presented in the descriptive portions of this report all indicate an initial primary deposition by ascending mineralized waters. It is probable, as pointed out by Lindgren[1] and Van Hise,[2] that these waters were originally meteoric waters, which gained their heat and collected their mineral contents during slow downward and lateral percolation through the rocks, and were subsequently gathered as ascending currents into the main fissures. It is not known from what particular rocks they extracted their metalliferous burden, nor at what depth most of the solution took place.

That the entire process of ore deposition was directly connected with volcanism can scarcely be doubted. The most obvious aspects of this connection are threefold: (1) The mechanical formation of the fissures; (2) the accession of heat, whereby the chemical activity of underground water was intensified; and (3) the accumulation of vast masses of igneous rock from which, at some depth, the constituents of the ore minerals were probably derived. It is possible that there should be included here also the evolution of carbon dioxide, sulphydric acid, and probably other volatile substances, as active solvent and chemical agents.

Very different views are held by various investigators in regard to the extent to which pneumatolytic processes, due to the emanation of highly heated gases and vapors from solidifying igneous rocks, are responsible for ore deposition. Van Hise, in his suggestive paper just cited, practically ignores pneumatolysis as a factor in the formation of ore bodies. Kemp,[3] on the other hand, considers that meteoric waters have little or no efficiency in the original concentration and deposition of ores, and that the waters of ore-bearing solutions, together with their dissolved mineral constituents, are of intratelluric origin, and are given off at high temperatures by masses of igneous rock in process of congelation. Various intermediate views are held by many European geologists and writers on ore deposits, notably by Beck, Vogt, and Du Launay. The evident connection between ore deposition and volcanism, or at least the products of volcanism, the

[1] The gold-quartz veins of Nevada City and Grass Valley districts, California: Seventeenth Ann. Rept. U. S. Geol. Survey, Pt. II, 1896, p. 176.

[2] Some principles controlling the deposition of ores: Trans. Am. Inst. Min. Eng., Vol. XXX, 1900, p. 47.

[3] The rôle of the igneous rocks in the formation of veins: Trans. Am. Inst. Mining Eng., Richmond meeting, 1901.

presence of fluorite and hübnerite in some of the lodes of the Silverton region, and the occurrence of alunite and diaspore, in connection with metasomatic alteration of the country rock apparently brought about by acid waters and locally very pronounced in character, are all to some extent indicative of pneumatolytic processes. But, as Lindgren[1] remarks, fluorine and boron compounds are found in some deposits which are certainly not of pneumatolytic origin, nor even formed at high temperature. The Silverton lodes, as a whole, possess neither the distribution nor the mineralogical characters of those deposits to which in the light of present knowledge an essentially pneumatolytic origin can be most safely assigned. The known igneous masses had certainly solidified and probably had lost much of their heat before the lode fissures were formed. It is most probable that the transportation and concentration of the Silverton ores was effected chiefly by meteoric waters, which derived their chemical and mechanical energy mainly from the heat connected with volcanism and from pressure, but possibly in some minor part, also, to gases and vapors given off at high temperature by solidifying igneous rocks and taken into the deeper meteoric circulation.

It is difficult to determine the exact character of the solutions which deposited the ore of the Silverton quadrangle. Many natural solvents are known for the common gangue minerals, for sulphide minerals, and for gold. According to Doelter's[2] experiments, pyrite, sphalerite, galena, stibnite, chalcopyrite, arsenopyrite, and bournonite are all appreciably soluble in pure water under the influence of moderate heat (80° C.) and pressure. Galena and pyrite are soluble in water containing carbon dioxide, and gold and copper are soluble in solutions of sodium carbonate.[3]

Solutions of sulphydric acid or sodium sulphide are relatively active solvents of gold, as well as of pyrite, chalcopyrite, bournonite, arsenopyrite, galena, and sphalerite, according to the investigations of Becker[4] and Doelter.[5] The experiments of Doelter in particular indicate that heat and probably pressure increase the solvent action, although the law of this increase of the different minerals is unknown. Solution is also favored by a great preponderance in the mass of the solvent and by the long contact of solvent and solid under favorable conditions of heat and pressure. Obviously these latter conditions are those common in nature, but which can be only very imperfectly realized in the laboratory. It appears, therefore, that nearly all aqueous solutions that occur in nature may, under suitable conditions of heat and pressure, act as solvents and carriers of the heavy

---

[1] Metasomatic processes in fissure veins: Trans. Am. Inst. Min. Eng., Vol. XXX, 1900, p. 644.
[2] Chemische Mineralogie, Leipzig, 1890, p. 188.
[3] Doelter, loc. cit., pp. 190–191.
[4] Natural solutions of cinnabar, gold, and associated sulphides: Am. Jour. Sci., Vol. XXX, 1887, pp. 199–210. Also Mon. U. S. Geol. Survey, Vol. XIII, 1888, pp. 432–433.
[5] Loc. cit., p. 192.

metals and their sulphides. Some solutions are undoubtedly more efficient than others under like conditions, but where so many unknown elements, such as degree of heat, amount of pressure, relative mass of solvent and solute, and duration of the process, enter into the problem, it is rarely possible to arrive at even approximate quantitative results.

As the lode and stock ores were deposited from solutions, it follows that the chemical action of these solutions on the wall rocks offers a very important mode of attacking the problem of their chemical character. The nature of this metasomatic alteration has been discussed at some length on pages 114–131 of this report. It is concluded that the solutions producing it were chemically different in different portions of the quadrangle. The very slight development of carbonates in connection with some lodes, and their entire absence in others, indicate that carbon dioxide or alkaline carbonates were not abundant in the mineralizing waters, although probably not wholly absent in the case of deposits such as those of Silver Lake Basin. The silicification observed in connection with the Red Mountain deposits, with the removal of most of the bases, including in some cases part of the alumina and the addition of sulphur and water, indicate the action of acid waters, probably containing free sulphuric acid. But here another element, which can not be ignored, enters the problem. Water containing sulphuric acid and ferric sulphate is known in the Red Mountain region to-day—not as ascending thermal water, but as (in part at least) descending water, which owes its acidity to the oxidation of previously deposited iron pyrite. The capacity of this water to effect some changes in the country rock is beyond doubt. It is not unlikely, therefore, that much of the alteration visible in the Red Mountain region is the result of complex processes involving the action of solutions of different characters and origin and acting upon any given mass of the rock at different times.

Writers on the genesis of ore deposits very commonly speak of the "circulation," "currents," or "flow" of the ore-bearing solutions within the fissures. Although the use of these terms is convenient and can not be regarded as entirely incorrect, yet it is believed that it has tended to give artificial or false conceptions in regard to the actual processes of ore deposition in many veins, particularly in large veins. Ore deposited by solutions having obvious movement through a fissure should exhibit the phenomenon called by Posepny crustification—i. e., it should be deposited in successive layers or coatings upon the walls of the fissure.[1] But this structure, formerly so much insisted upon as evidence of deposition in open fissures, is, as a matter of fact, of comparatively rare occurrence even in veins which on other evidence are demonstrably simple fissure fillings. In most of the

[1] It is perhaps hardly needful to point out that the converse of this proposition is not necessarily true.

important lodes of the Silverton quadrangle no crustification can be detected. The fissures are filled by coarsely crystalline allotriomorphic aggregates of ore and gangue (massive structure), and there is no evidence to show that such a structure could result from successive deposition upon the walls of the fissure until the whole was finally filled. On the contrary, crystallization has proceeded simultaneously from many points within the solution. Quartz, galena, pyrite, sphalerite, chalcopyrite, and other minerals have formed practically contemporaneously about local centers of crystallization scattered irregularly through the solidifying mass. It is difficult to conceive all the details of a process which results in the formation from aqueous solution of an irregular allotriomorphic aggregate of minerals differing so widely in specific gravity. But reasoning from analogy with similar structures met with in petrology and in the arts, it is fair to assume that this structure in veins is the result of the undisturbed crystallization of a nearly motionless saturated solution. In other words, it was the crystallization of a reservoir so large, in comparison with the current or currents which circulated through it, as to have been itself free from megascopic motion. The conditions which induced crystallization were present throughout the mass of material which filled the fissure, and solidification, instead of proceeding gradually from the walls inward, took place almost simultaneously, although not necessarily with rapidity, throughout the mass. Obviously this massive structure and crustification may be present in the same vein. In such cases deposition appears to have begun by crustification next the walls, passing at a later stage into the more general mode of crystallization indicated by what has been described as massive structure.

Attention has been strongly drawn of late, through the writings of De Launay,[1] Van Hise,[2] Weed,[3] and Emmons,[4] to the important part played by secondary sulphide enrichment in the formation of many ore bodies, particularly those carrying copper and silver. As a probable example of such an enriched deposit, Mr. Emmons[5] has cited the Yankee Girl ore body. Unfortunately, at the time Mr. Emmons visited the mine, in 1888, but little attention had been given to the subject of secondary sulphide enrichment, while at the present day the state of the workings is such as to render impossible the verification by direct observation of this application of the theory. However, there are many facts which point strongly toward secondary enrichment as the true mode in which the rich ore bodies, not only of the Yankee Girl, but of the Guston, Silver Bell, and other mines, were formed. To have the important facts fresh in mind, it will be well to

[1] Contribution à l'étude des gîtes métallifères: Ann. des Mines, Vol. XII, 1897, pp. 119-228.
[2] Some principles controlling deposition of ores: Trans. Am. Inst. Min. Eng., Vol. XXX, 1900, pp. 27-177.
[3] Enrichment of mineral veins by later metallic sulphides: Bull. Geol. Soc. Am., Vol. XI, 1900, pp. 179-206.
[4] The secondary enrichment of ore deposits: Trans. Am. Inst. Min. Eng., Vol. XXX, 1900, pp. 177-217.
[5] Loc. cit., p. 19.

review briefly the characteristic features of these ore bodies which bear upon this question.

The ore first struck, in some cases actually at the surface, consisted chiefly of argentiferous galena. At a depth which varied somewhat in different mines, but which appears to have been usually less than 200 feet, this galena changed to an ore consisting chiefly of highly argentiferous stromeyerite. At a still greater depth—usually about 500 feet—the stromeyerite changed to argentiferous bornite, still deeper to chalcopyrite and pyrite, and finally to low-grade auriferous and argentiferous pyrite. These changes were more or less irregular and overlapping. Thus pyrite was found at nearly all levels, and bunches of galena were met with far below the levels at which it ceased to be the dominant ore. Small rich streaks of bornite were also found, with chalcopyrite and pyrite, below the levels where it occurred in large masses. According to Schwarz,[1] who opened and worked many of the ore bodies of the Red Mountain district, the rich portions of the ore bodies were always associated with open water-bearing fissures, and he apparently regarded the entire deposits as formed primarily by deposition from the water entering through these fractures. It seems to be a well attested fact that at a depth of about 500 feet there was encountered in the Yankee Girl and Guston mines a fault or "slip plane" with low westerly dip, and, further, that the richest ore occurred just above this seam, which was filled with clay-like gouge. Moreover, most of the water entering the mines came in at or above this fault. The acid character of this water was well known, and it carried in solution large amounts of the sulphates of iron and copper. Lastly, the mines are situated within a valley of erosion, within which the ore bodies outcrop at different levels. It is not known what thickness of material has been removed in the shaping of the present topography, but it probably exceeds 2,000 feet. It is hardly conceivable that the observed relation between the vertical variations in the ore bodies and the present topographic surface is a haphazard one. It is without much doubt genetic.

Ore deposits which show a fairly orderly vertical succession from rich sulphides near the ground-water level down to poor sulphides at greater depth are in nearly all cases regarded as the product of two concentrations—a concentration by ascending waters and a further concentration by descending waters.[2] By the former are deposited lean ores, found in the deepest workings, which are enriched to a greater or less depth by the downward-moving solutions. As Van Hise has pointed out, these two processes may go on at the same time, or they may have operated successively on any given vertical section of the ore body. The facts obtainable in the Red Mountain region point to the original deposition of bodies consisting chiefly of

[1] Trans. Am. Inst. Min. Eng., Vol. XVIII, 1890, p. 144.
[2] Van Hise, op. cit., p. 101.

pyrite carrying a little gold, silver, and copper. The lead may have originally been sparingly deposited as galena with the pyrite. It is into such low-grade auriferous and argentiferous pyrite, carrying 2 or 3 per cent of copper and containing an occasional bunch of galena, that the rich ore bodies passed in the Guston, National Belle, Silver Bell, Hudson, and Yankee Girl mines when the ore became too poor to work.

Subsequently, as the surface of the region was reduced by erosion, descending waters, which found ready passage through the fissured and mineralized rocks of the region and through the upper portions of the lean primary ore bodies, effected a second concentration, which produced the rich ore formerly mined. The penetration of this water into the primary pyritic ore must have been aided by the porous, crumbling character of much of the latter. These waters are known to have been heavily laden with the sulphates of iron and copper and other salts, resulting from oxidation of the upper portions of the ore bodies. As these solutions penetrated downward they probably acted upon the low-grade pyritic ore, replacing part of the iron sulphide by sulphides of copper and silver. Apparently much experimental work remains to be done before all the details of the chemical processes involved will be actually known, but the probability of the transformations outlined is shown by the following chemical equations taken from Van Hise's suggestive paper.[1]

By taking solutions containing merely the sulphates of copper the production of chalcopyrite from iron sulphides may be written as follows:

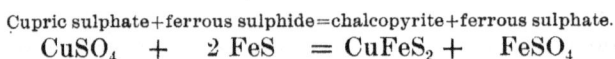

Cupric sulphate+ferrous sulphide=chalcopyrite+ferrous sulphate.
$$CuSO_4 + 2\,FeS = CuFeS_2 + FeSO_4$$

Or,

Cupric sulphate+ pyrite +oxygen = chalcopyrite+ferrous sulphate+sulphurous anhydride.
$$CuSO_4 + 2\,FeS_2 + O_4 = CuFeS_2 + FeSO_4 + 2\,SO_2$$

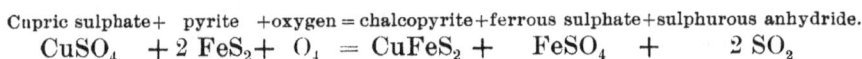

Bornite may be produced as follows:

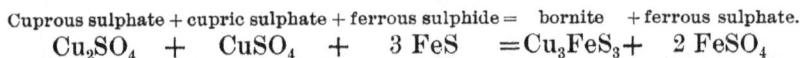

Cuprous sulphate + cupric sulphate + ferrous sulphide = bornite + ferrous sulphate.
$$Cu_2SO_4 + CuSO_4 + 3\,FeS = Cu_3FeS_3 + 2\,FeSO_4$$

Or,

Cuprous sulphate + cupric sulphate + pyrite + oxygen = bornite + ferrous sulphate +
$$Cu_2SO_4 + CuSO_4 + 3\,FeS_2 + 6\,O = Cu_3FeS_3 + 2\,FeSO_4 +$$
sulphurous anhydride.
$$3\,SO_2$$

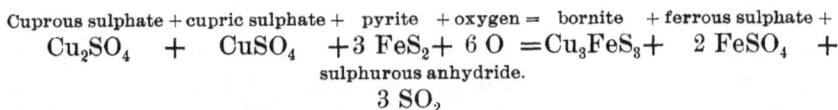

But bornite may also be formed by direct action of the copper sulphates on chalcopyrite, as follows:

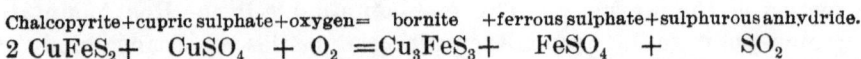

Chalcopyrite+cupric sulphate+oxygen= bornite +ferrous sulphate+sulphurous anhydride.
$$2\,CuFeS_2 + CuSO_4 + O_2 = Cu_3FeS_3 + FeSO_4 + SO_2$$

---

[1] Some principles controlling the deposition of ores: Trans. Am. Inst. Min. Eng., Vol. XXX, 1900, pp. 111-112.

Chalcocite may be produced from pyrite, chalcopyrite, or bornite, as follows:

Cupr.us sulphate + pyrite + oxygen = chalcocite + cuprous sulphate + sulphurous anhydride.
$$Cu_2SO_4 \quad + FeS_2 + \quad O_2 \quad = \quad Cu_2S \quad + \quad FeSO_4 \quad + \quad SO_2$$

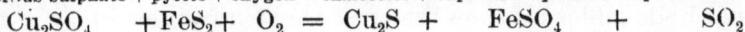

Cupric sulphate + chalcopyrite + oxygen = chalcocite + cuprous sulphate + sulphurous anhydride.
$$CuSo_4 \quad + CuFeS_2 + \quad O_2 \quad = \quad Cu_2S \quad + \quad FeSO_4 \quad + \quad SO_2$$

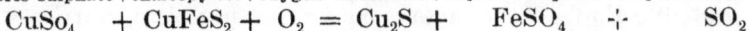

Cupric sulphate +   bornite   + oxygen = chalcocite + ferrous sulphate + sulphurous anhydride.
$$CuSO_4 \quad + Cu_3FeS_3 + \quad O_2 \quad = 2Cu_2S \quad + \quad FeSO_4 \quad + \quad SO_2$$

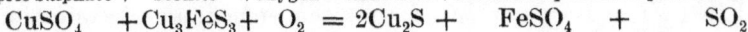

Of course it is understood that these sulphides overlap one another. Before the iron sulphide has wholly been replaced by chalcopyrite bornite may appear. At the place where bornite has become reasonably abundant chalcocite may be found. However, certain general statements may be made. If the dominating material is iron sulphide, the copper mineral which is present is likely to be chalcopyrite rather than the richer sulphurets. Chalcopyrite, on the one hand, is likely to be associated with the pyrites, and on the other hand with bornite, or even chalcocite. Bornite and chalcocite are likely to be associated with each other and with chalcopyrite, but with the first two compounds iron sulphide is likely to be subordinate or absent.[1]

When the sequence of formation, deduced largely on theoretical grounds by Van Hise, is compared with the facts of occurrence in the Yankee Girl and Guston mines, the close correspondence affords striking confirmation, not only of the probable truth of the theory of secondary sulphide enrichment, but also of its applicability to these deposits. As far as the cupriferous sulphides and pyrite are concerned, the relations are exactly those which might be expected as a result of the downward percolation of solutions of copper sulphates through a body of low-grade pyritic ores.

The iron sulphates, of course, can not act directly, as do the copper sulphates, in enriching the ore. Their rôle is limited to an attack upon the previously formed sulphides, which converts them into sulphates and redeposits them in part as sulphides below the zone of oxidation.[2] They thus aid, although in a different manner, the copper sulphates in effecting a secondary concentration of the ores. It is probable that, owing to the slight solubility of its sulphate, the lead is deposited as sulphide at an early stage in the downward movement of the solutions, and that, through a constant and prolonged continuance of this process, as the outcrop of the ore body is lowered by erosion, there have resulted the masses of nearly pure galena ore which occurred above the chalcocite or stromeyerite.

The application of the theory of secondary sulphide enrichment to any ore body naturally presupposes the existence of small amounts of the valuable metals in the low-grade ore of primary deposition. But this fact is not always easily determined, as mining operations are not often pushed very far beyond the depth at which the ore ceases to

---

[1] Van Hise, op. cit., p. 112.
[2] The secondary enrichment of ore deposits, by S. F. Emmons: Trans. Am. Inst. Min. Eng., Vol. XXX, 1900, p. 210.

pay, and it is not always possible to determine the actual downward limit of the enrichment. But the numerous assays made of the unworkable pyrite in the deepest levels of the Guston, Silver Bell, and National Belle mines all show the presence of a fraction of an ounce of gold and a few ounces of silver. Lead has not been determined. It is probable that in the assays made for practical purposes no attempt was made to determine its presence in small amounts. It should further be borne in mind that the enrichment in question is not of necessity due solely to waters which have descended from the upper portion of the selfsame ore deposit. The descriptions of the mines indicate that some of the second concentration was effected by waters which may have traveled laterally, from some distance, through rocks which are themselves mineralized and may contain considerable bodies of ore as yet undiscovered.

The theory of secondary enrichment, which is regarded as giving the most probable and satisfactory explanation of some of the peculiar features of the Red Mountain deposits, has, if its application is correct, certain very important and practical consequences. It indicates that the rich ores formerly mined have a working lower limit which is probably less than 1,000 feet in depth. While small bodies of rich ore may occur deeper than this, it is probably not worth while prospecting for them when once the depth has been reached at which the ore is practically all low-grade pyrite. On the other hand, as some pyrite occurs at all depths, and as the vertical limits of the zones of enrichment are necessarily irregular and overlapping, it is unwise to suspend operations too quickly, merely because the ore falls off locally in value and becomes pyritic.

Passing from the Red Mountain stock deposits to the lodes occurring in other portions of the quadrangle, the evidence of secondary enrichment is less easily read. It seems very probable that the rich silver minerals, such as polybasite, proustite, and argentite, are the result of secondary enrichment similar to that described by Weed[1] in the Neihart district in Montana. In the San Juan region, however, the actual steps of the alteration of lower-grade ores into those rich in silver can not be traced. The evidence for a secondary origin consists in the fact that these rich minerals occur largely in vugs, and that in several cases they pass at moderate depths into low-grade ores. These features, however, do not in themselves preclude the idea that the minerals are primary, and resulted from original deposition in a zone where ascending and descending waters mingled.

In the Tomboy mine the undoubted association of the richest gold ore with shattering and movement in the lode, accompanied by some oxidation far below the level of general weathering, is suggestive of secondary enrichment, but it is hardly safe to assume that this process

---

[1] Geology of the Little Belt Mountains, Montana: Twentieth Ann. Rept. U. S. Geol. Survey, Pt. III, 1900, pp. 421-422.

has actually been effective. Both in this mine and in the Silver Lake mine it is stated by those in charge that rich bunches of ore are commonly associated with secondary or "post-mineral" fissuring, which, if true, is significant of some secondary enrichment below the zone of oxidation.

THE GROUND WATER.

In many mining regions the level at which the ground water stands, or, more accurately, the shape of its surface and the relation of the latter to the topography, is a matter of much importance, as by it are frequently determined questions of mine drainage and the depths at which oxidized ore changes to unoxidized ore. In the Silverton region, however, erosion has proceeded with such celerity relative to processes of oxidation and weathering that the change in question is largely independent of the depth of the ground water, and the latter is accordingly of less moment to the miner.

In most regions of considerable relief the upper surface of the ground water corresponds to a very subdued copy of the topographic relief. In the exceedingly rugged and much fissured Silverton area, however, the ground-water surface is very much less accentuated than the topography. Not only is the water tapped and drained away by the deeply cut ravines, but its level is often modified by the tunnels run into the mountains, sometimes several thousand feet below the crests of the ridges. Under these conditions it is often practically impossible to predict at what depth the ground-water surface will be encountered in mining operations.

Tunnels driven in from the floors of the high basin do not, as a rule, encounter permanent ground water. Those run below the basins, such as the Unity tunnel, or from the bottoms of the deeper ravines, such as the Highland Mary tunnel in Cunningham Gulch, the Bonanza tunnel near Animas Forks, the Old Lout tunnel in Poughkeepsie Gulch, the Revenue tunnel, the North Star tunnel on Sultan Mountain, and the Empire tunnel near Silverton, usually tap the ground water and artificially lower its surface within the overlying rock masses. In the shafts of the Red Mountain district permanent water is encountered at depths varying from 50 to 200 feet below the surface, dependent upon the location of the shaft and the seasonal fluctuations of the ground-water surface, which sometimes exceed 50 feet. In the Red Mountain district the ground-water surface is nearest the topographic surface in spring or early summer.

REPLACEMENT DEPOSITS.

Between certain replacement deposits and the metasomatic impregnations frequently accompanying fissure veins or ore stocks no sharp distinction can be made. Whether the deposit shall be classed as a vein or as a replacement deposit may depend simply upon the quantitative relation between the ore found within the walls of the fissure and that occurring in disseminated particles or in larger masses in the metasomatic country rock.   The Red Mountain stocks undoubtedly owe their form partly to metasomatic replacement, and might almost as well be described under this head.   Ore-bearing fissures traversing rhyolite rocks are in this region frequently accompanied by notable replacement of the wall rock by ore.   But there are a few deposits where replacement is so conspicuous a feature and where so small a part of the ore occurs within distinct fissure walls that but little doubt need arise as to their proper classification.   They are, however, neither abundant nor relatively important in the Silverton quadrangle.

Direct replacement of limestone by ore occurs on the eastern side of Ironton Park, where, for some distance, a thick bed of white crystalline limestone, probably of Devonian age, is exposed beneath Tertiary volcanic rocks.   At the Saratoga mine the ore, consisting in its unoxidized state of pyrite, chalcopyrite, and galena, with perhaps a little argentite, has replaced the upper portions of this limestone to a varying depth.   The ore-bearing solutions have been chiefly active along the contact between the limestone and an overlying andesitic breccia or tuff belonging to the Silverton series.   A little Telluride conglomerate, consisting largely of limestone pebbles, occurs near the adit of the mine, between the limestone and the overlying volcanic series.   It has been silicified and probably partly replaced by ore, but was not recognized as present in the main workings.   The ore is sometimes associated with rhodonite, which has also replaced the limestone and frequently incloses residual kernels of the latter.   The ore was probably deposited by solutions rising through one or more northeast-southwest fissures, although no direct connection of the ore with any of the fissures could be made out.   In the Baltic mine a similar but lower-grade ore occurs, also as a replacement of the upper portion of the same bed of limestone.   This ore, however, is directly connected with an ore body filling a fault fissure (the Mono vein).   In the Maud S. claim, which is part of the Baltic group, some argentiferous copper ore occurred in bunches in the same limestone alongside of a fault fissure.   The direct connection between the replacement ore bodies of the Baltic group and a system of northeast-southwest fault fissures strongly suggests that a similar connection does or did formerly exist in the case of the Saratoga ore body.

In the southern portion of the quadrangle the Devonian limestone at the east base of Sultan Mountain carries disseminated particles of native silver on the Fairview claim, while in an unknown prospect near the King mine a body of chalcopyrite occurs partly in a fissure and partly as a replacement of the limestone near the fissure.   The ore in this case makes largely at the underside of the limestone, where it rests upon quartzite.

The facility with which the rhyolitic rocks of this region become to a greater or less extent metasomatically replaced by ore has already been noted. In most cases such replacement is a minor accompaniment of the deposition of ore in fissures. But at the Silver Ledge mine replacement is the dominant mode of occurrence. The ore, consisting of galena and sphalerite, occurs metasomatically, replacing rhyolite in the vicinity of a zone of fracturing and faulting trending about N. 20° E. These fractures contain crushed country rock and ore, and there has evidently been movement along them since the ore was deposited. The principal ore bodies, however, are not found within the fissures, but as irregular replacements of the rhyolite which forms their walls. In some instances bodies of ore have been followed into the country rock for 30 or 40 feet from the main fissures. The replacement is sometimes complete, resulting in solid masses of ore. More often it takes the form commonly described as impregnation; i. e., the ore is more or less thickly scattered through the rhyolite in specks and small bunches. In this case there is no definite boundary to the ore body, and only so much of it is removed as can be profitably worked.

The deposition of the ore has been accompanied by metasomatic change and a complete recrystallization of the rhyolite. The resulting material, which forms the gangue of the ore, is a nearly white, minutely crystalline aggregate of sericite, kaolin, and quartz, with generally a little rutile and occasionally a small amount of calcite. The proportions of the three principal constituents vary, some facies of the rock consisting chiefly of sericite and kaolin, while others are largely sericite and quartz.

MISCELLANEOUS MINERAL RESOURCES.

Compared with the extraction of gold, silver, copper, and lead, all other mineral productions of the quadrangle are insignificant. At present no deposit is being worked for zinc, but sphalerite ores are abundant. As a rule, ore bodies showing much sphalerite with little galena have been little prospected, on account of the smelter penalty attaching to an excess of zinc in ores smelted for other metals. But with the increasing demand for zinc ores it is not unlikely that some of the sphalerite lodes may ultimately be worked for this metal. This, however, may be some years hence, owing to the natural disadvantages under which mining is carried on in so elevated a region.

The occurrence of hübnerite on Cement Creek has aroused expectations of the profitable exploitation of the deposits. The mineral is easily concentrated from its gangue, but it is very doubtful whether it occurs in deposits of sufficient size or constancy to justify extensive operations.

Bog iron ore occurs in the swampy ground along Mineral Creek and in Ironton Park as a deposit from iron-bearing springs. It was formerly used to some extent in Silverton as a flux for smelting, but has at present no market.

Limestone suitable for making lime occurs at several points in the peripheral portions of the quadrangle, and has been burned for lime just below Silverton to supply local demand.

Building stone is not in much request. Good stone might undoubtedly be quarried from the monzonite of Sultan Mountain were there any demand for a building material involving so much labor in its production.

PART II. DETAILED DESCRIPTIONS OF SPECIAL AREAS AND INDIVIDUAL MINES.

INTRODUCTION.

The following descriptions are arranged in general conformity to a geographical plan. The mines in the Silver Lake Basin, in the southeastern part of the quadrangle, are first described. Proceeding northward, other groups are taken up, in the eastern and northeastern portions of the area. Then follow, successively, accounts of mines in the northern, northwestern, western, and southern districts, and, finally, descriptions of the Cement Creek ore deposits in the southwest-central part of the quadrangle.

LODES OF SILVER LAKE BASIN.

General features.—The basin in which Silver Lake lies is a typical high cirque of a kind common in the San Juan Mountains. Its rocky, hummocky floor, with an altitude between 12,000 and 12,500 feet, is inclosed on three sides by cliffs and precipitous slopes culminating in Kendall and Little Giant peaks, which attain altitudes of over 13,400 feet. The basin opens to the north-northwest, and its sloping floor is terminated, at an altitude of about 12,000 feet, by precipices which plunge down to the talus-covered bottom of Arrastra Gulch. Silver Lake,[1] a characteristic mountain tarn, occupies a depression of considerable depth, excavated by glacial erosion in the softer rock behind the sheet of massive andesite at the northern end of the lake, over which the present drainage escapes to Arrastra Gulch. The marks of glacial scoring are still plainly visible on this resistant sill, through which the clear water escaping from the post-Glacial lake has been unable to cut. The original beauty of this little sheet of water has been marred by mining operations, particularly by a partial filling with tailings from the Silver Lake mill. The basin is well above timber line, and the mine buildings are sometimes seriously damaged by snow slides. Mine timbers are packed in from Deer Park by burros, over the col or saddle at the southern end of the basin. The main communication with the mines, however, is by the trail and the wire-rope tramways running up Arrastra Gulch.

The rocks in which the basin has been carved, and which form the

---

[1] Formerly known as Arrastra Lake, and still so designated on the maps of the General Land Office. This is also the name of the post-office at the Silver Lake mine.

country rock of the mines to the depth now worked, belong to the Silverton series and illustrate its typical development. They are prevailingly breccias, in which the angular fragments and the interstitial matrix are composed of the same materials, with practically no admixture of foreign particles. The induration of these breccias has produced rocks which can with difficulty be distinguished from massive lavas. A careful inspection of weathered surfaces, or a microscopical examination, will, however, usually reveal their true clastic nature. The general character of these breccias is andesitic. But they differ from normal andesite, such as occurs in the San Juan formation, by the presence of occasional porphyritic crystals or phenocrysts of orthoclase and quartz, and in the occurrence of orthoclase in the groundmass. The abundance of the phenocrysts of orthoclase feldspar and quartz varies in different varieties or facies of the breccia. As a whole these rocks appear to be most closely related to the latitic group of intermediate rocks. Associated with the breccia are massive rocks of the same general type as the breccias, which have been intruded as sheets and dikes, and possibly in part as contemporaneous lava flows. As examples of these may be cited the masses forming the summit of Round Mountain and the hard sill that holds up the waters of the lake at its northern end. The latter, a dark-gray rock of splintery fracture, shows abundant fresh-looking phenocrysts up to about a centimeter in length, and smaller crystals of hornblende, lying in a compact groundmass. A thin section shows, under the microscope, phenocrysts of labradorite feldspar, brown hornblende, and pale-greenish augite, in a feldspathic groundmass. The augite is partly decomposed to calcite and chlorite, and certain areas of chlorite appear to represent former phenocrysts of biotite. The groundmass, while somewhat obscured by calcite, chlorite, and other products of alteration, appears to consist chiefly of plagioclase, orthoclase, and glass. The accessory minerals are magnetite and apatite. The rock from the summit of Round Mountain is very similar, but the presence of orthoclase in the groundmass is more clearly shown and the augite is fresher and more abundant.

The lodes of Silver Lake Basin are large and prominent, and on this account were discovered and located at an early stage of the mining development of the region. But it was not until the erection of the Silver Lake mill, about 1890, that they were worked on the scale necessary for the profitable and permanent extraction of low-grade ore. They are nearly all fissure veins of comparatively simple form, and have clearly defined walls. The wall rock has been chemically altered by the solutions traversing the fissures and, to a slight degree, impregnated with ore minerals, but it is not regarded as ore and is always readily distinguished from the vein filling.

The lodes wholly or partly within the basin may be divided into two groups. The first group comprises at least two very strong and per-

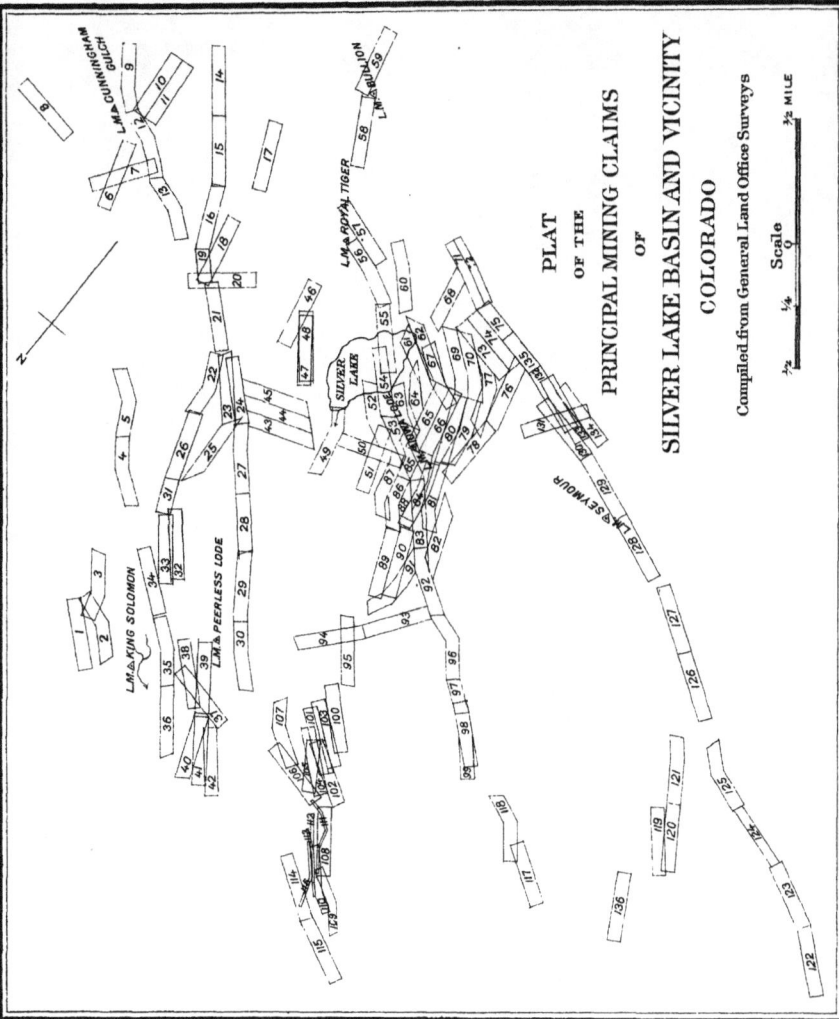

NAMES OF MINING CLAIMS

1 MOUNTAIN QUEEN	47 LAKE VIEW	93 UNITY
2 KING SOLOMON	48 SINA LOA	94 PROFESSOR WINKLER
3 JURA	49 MARTHA	95 EL PASO
4 NIGHT BIRD	50 ARABIAN BOY	96 NEVADA
5 WILHELM	51 SILVER LAKE SPUR	97 NEVADA EXTENSION
6 HIGHLAND CHIEF	52 MAXWELL	98 HILL
7 LONG TOM	53 SILVER LAKE	99 HAPPY JACK
8 BLACK CROOK	54 GRETCHEN	100 PEACOCK
9 WM. H. NICHOLS	55 ROYAL TIGER NO. 1	101 LITTLE BELLE
10 NUMBER 1	56 ROYAL TIGER	102 LUCY
11 NUMBER 3	57 HOMESTAKE	103 ORIENTAL
12 HIGHLAND MARY	58 ANNIE	104 TOM PAINE
13 ROBERT BRUCE	59 BULLION	105 TOLEDO
14 LOOKOUT	60 DIAMOND L	106 GREY EAGLE
15 MOUNTAINEER	61 DUMP	107 EZRA R
16 SHENANDOAH	62 GRIP	108 MAMMOTH
17 BIG PINE	63 BLACK DIAMOND	109 PATHFINDER
18 QUEEN	64 J. W. COLLINS	110 PROSPECTOR
19 SHENANDOAH NO. 3	65 IOWA	111 ASPEN
20 - - - -	66 STAG	112 SUSQUEHANNA
21 DIVES	67 CABO	113 MAC GREGOR
22 CREMORNE	68 DANON	114 LEGAL TENDER
23 YELLOW JACKET	69 AMERICAN BOY	115 SILVER STAR
24 NORTH STAR	70 PROFESSOR WEDDING	116 CHLORIDE
25 NORTH STAR EXTENSION	71 BUCKEYE NO. 2	117 FREE COINAGE
26 POTOMAC	72 BUCKEYE	118 LAST CHANCE
27 TERRIBLE	73 PROFESSOR ZEUNER	119 CONTRA COSTA
28 SLIDE	74 PROFESSOR WEISBACH	120 DOLLY VARDEN NO. 2
29 MAYFLOWER	75 LAST CHANCE	121 ETHAN ALLEN
30 ARGENTINE	76 CAIRO	122 IDAHO NO. 1
31 GENERAL GARFIELD	77 JAFFA	123 IDAHO NO. 2
32 MOUNTAIN QUAIL	78 ALGIERS	124 BOWERY
33 BIG GIANT	79 SMYRNA	125 SAL CUSTER
34 BLACK PRINCE	80 WHALE	126 POMPEII
35 ESMERELDA	81 ROCHESTER	127 PYRAMID
36 SILVER CROWN	82 LENA ALLEN	128 EMMA
37 JEBS	83 ROUND MOUNTAIN	129 SEYMOUR
38 EMMA S	84 NEW YORK CITY	130 CAROLINA
39 PEERLESS	85 ROYAL	131 LEUCOCITE
40 GOLDEN FLEECE	86 RIVINGTON ST.	132 BERA
41 LITTLE GIANT	87 ESSEX ST.	133 TITUSVILLE
42 IRON MASK	88 PROFESSOR STELZNER	134 SOLSTICE
43 LITTLE SHAVER	89 GALENA	135 PANUCO
44 CROSS CUT	90 LAST CHANCE	136 J. B. SMITH
45 GOOD FORTUNE	91 BADGER	
46 ECLIPSE	92 LOWVILLE	

PLAT

OF THE

PRINCIPAL MINING CLAIMS

OF

SILVER LAKE BASIN AND VICINITY

COLORADO

Compiled from General Land Office Surveys

Scale

A. HOEN & CO. BALTIMORE.

sistent lodes, which may be referred to as the Silver Lake and Titusville lodes. The general course of the fissures of this group is nearly northwest and southeast. The second group embraces a number of shorter lodes, having a general average course about N. 25° W., and lying between the two chief lodes of the first group. The members of the second group are locally spoken of as "north-and-south veins"—an obviously inaccurate expression.

The general dip of all the productive lodes in the basin is northeasterly, although some of them are nearly vertical and often dip locally to the southwest. Those of the first group frequently have a much flatter dip than those of the second group, and the ore is generally of lower grade and more irregular in its occurrence. Large bunches of ore are separated by relatively barren portions of the lode. As a rule the product is a lead-silver ore, consisting chiefly of galena accompanied by sphalerite, chalcopyrite, pyrite, and sometimes tetrahedrite. In some cases the silver may be so low as to be subordinate in importance to the lead. The lodes of the second group have a much steeper dip, the average being probably near 80°. Their ore bodies are usually more regular and of higher grade. Galena is still the most prominent ore mineral, but chalcopyrite and pyrite are generally more abundant, and the ore may carry several ounces of gold per ton. These lodes have been best explored toward the north, where some of them form junctions with the Silver Lake lode. Several of them pinch and die out in a few small, irregular stringers before reaching this fissure. Others, such as the New York City and Royal veins, reach and connect with the master lode. The general relations of the lodes of the two groups can perhaps be best understood from fig. 2, page 51, which is a generalized and diagrammatic sketch of the plan of the principal fissures exploited in the Silver Lake and Iowa mines, on the west side of Silver Lake. The Silver Lake lode is fully as long again as shown. It has been exploited on the northwest through the old Nevada mine and is known to extend into the head of Blair Gulch. Toward the southeast it passes under the lake and is probably continuous with the lode of the Royal Tiger mine. Just before crossing the southeast rim of the basin, the Tiger lode divides into two branches, neither of which has been traced far beyond the point of divergence. A plat of the principal claims of Silver Lake Basin and vicinity is shown in Pl. XIII.

As the lodes of the second group approach the junction with the Silver Lake vein they become poor. Their courses are also deflected toward the west, as indicated in the sketch. The country rock in the V near the junction is usually shattered and traversed by numerous quartz stringers. Some of these are irregular, but there is often noticeable a radiation of the stringers outward from the angle of junction. In some instances a breccia has resulted, the fragments being cemented by quartz and themselves altered into aggregates of quartz, sericite, and other secondary products by the action of the vein solutions. A close study of all exposures of such conjugations of the veins of the two groups failed to show any evidence of different age. The filling of both sets of veins is continuous, as it would be in synchronously filled branches of a single fissure, or is separated by a clay gouge due to subsequent movement. As far as is known, there is no continuation of the lodes of the second group north of the Silver Lake lode. A single small vein (the Jim Owen), shown in fig. 2, has been prospected for a short distance northward from the Silver Lake lode, but it is very improbable that this represents the continuation of any of the productive lodes south of the latter. If the Silver Lake lode were, as is supposed by the miners, younger than the lodes of the second group, the absence of the latter on the northern side of the intersecting lode could be accounted for only on the assumption that the Silver Lake fissure is a fault of at least 2,500 feet throw. The entire evidence of the region, however, as is elsewhere shown, indicates that the fissures now occupied by productive ore bodies were not formed by large faults, but on the contrary often reveal no appreciable displacement of their walls.

The relation of the lodes of the second group to the great Titusville lode on the south are not so well known as in the case of the Silver Lake lode on the north. The lodes in this portion of the basin are imperfectly exposed and have received very little underground exploration.

The ore bodies of the lodes of Group II have not been worked in such a manner as to give any very definite data as to their dimensions and shapes. In nearly every slope considerable masses remain which are regarded as pay ore. The pay shoots are supposed to pitch northward, but it is by no means certain that there is any regularity in their pitch.

In addition to the main lodes a great number of smaller fissures traverse the country rock, particularly west of the lake. The strike of these minor veins ranges from 20° to 40° west of north. They are nearly vertical, but one group of them sometimes exhibits a steep northeasterly dip, while another group dips southwesterly. The latter case appears to be more common. They are often so closely spaced, a few feet apart, as to produce a conspicuous sheeting of the country rock. This spacing does not exhibit any regularity or recognizable rhythm in its occurrence. Many of these fissures are ore bearing, and it is not uncommon to see on a clean exposure of rock a stringer, an inch or so wide, consisting of quartz next the walls and solid galena in the medial portion. Such mineralized stringers can be seen to branch or to wedge out completely in the rock and be succeeded by an overlapping stringer a short distance to one side. On a small scale they repeat the phenomena to be observed underground in the larger veins of the second group. They are believed to belong to the same system and to have been formed at the same time as the latter.

Although the lodes of the Silver Lake Basin have been classified in two groups, it should be borne in mind that the distinctions upon which the grouping is based are quantitative rather than qualitative. The difference in direction, sufficiently marked and definite on the west side of the lake, becomes less so on the east side, where the Silver Lake lode appears to begin splitting up into its ultimate ramifications. Moreover, as will appear when the separate mines are described, the general difference in the ores of the two groups is a difference in value rather than in mineralogical type, and is neither constant nor without exceptions.

It is fairly certain that the original formation of the fissures of Silver Lake Basin was unaccompanied by any considerable faulting. A careful study of all the available underground workings and of the surface geology and topography of the basin has failed to reveal the existence of any extensive tangential movement (faulting) along the vein fissures. Moreover, if the mutual relationship of the lodes has been correctly interpreted, it would seem that the conception, very commonly held, that vein-filled fissures are produced essentially by faulting involving considerable tangential movement, and usually by thrust faults, is inapplicable as an explanation in the present case. The more general aspects of this subject are discussed under the head of "The lode fissures," on pages 43–63.

Silver Lake mine.—This, the most extensively developed mine within the quadrangle, is situated at an altitude of about 12,300 feet, with its principal adit near the western edge of Silver Lake. As is frequently the case, it includes a number of individual claims, some of which, such as the Whale, were separately worked in former days and produced considerable ore. Prior to 1881, the Whale, Silver Lake, Round Mountain, and other claims now embraced in the Silver Lake mine had attracted but slight attention. In 1882 a little ore was produced, which is said to have carried 50 ounces of silver and 60 per cent of lead. In 1883, 72 tons of ore, containing 28 ounces of silver per ton and 55 per cent of lead, were shipped to Sweet's Sampling Works in Silverton, and the Silver Lake claim, with its 55 feet of drifting, was regarded as a promising prospect. Since then its development has progressed with rapid strides. In 1891 the mine was credited in the Mint reports with a product of $254,908.

At least six distinct lodes have been worked, and the vertical distance from the uppermost to the lowermost stopes is somewhat over 1,000 feet. The main adit is a crosscut for about 250 feet. It then follows the Silver Lake lode northwesterly for about 1,000 feet. Thence a crosscut to the southwest, some 900 feet in length, gives access to the various lodes of the second group, viz, the L. A. S., E. G. S., Royal, Stelzner, and New York City lodes. A northwesterly drift along the New York City lode for about 1,250 feet again comes into the Silver Lake lode, and from this junction a drift of about 400 feet runs out to the surface in Whale Basin, on the west side of

Round Mountain. In addition, the Unity tunnel, run in from the southwest side of Arrastra Gulch, cuts the Silver Lake and New York City lodes 700 feet below the present main adit level (No. 1). Still a third tunnel is being run in a southeasterly direction from a point near the Animas River, at the mouth of Blair Gulch, which will, if carried through, ultimately tap the Silver Lake lodes at the lowest practicable point, over 2,500 feet below the No. 1 level.

There are no abandoned workings, and the character and relationships of the lodes may be studied under more than usually advantageous circumstances.

The Silver Lake lode as exposed in the mine has a general strike of N. 53° W. Its dip is to the northeast and varies from 50° to 70°—the flatter dip being more common in the upper workings. The average dip is probably a little over 60°. Its croppings are conspicuous as a rust-colored zone of oxidation running up the declivity west of the lake and passing through the col just south of Round Mountain. At this point they have an apparent width of about 40 feet of stained and cellular quartz carrying galena, as well as the carbonate and sulphate of lead. Northwestward from this saddle the lode, different portions of which are known as the Round Mountain, Lowville, and Nevada lodes, crosses Whale Basin just above the cliffs of Arrastra Gulch, and passes through another saddle into the head of Blair Gulch. In its productive portions the Silver Lake lode is from 6 to 20 feet in width. Elsewhere it may be represented by a strong clay seam or gouge, or less commonly a small tight vein. The filling of the fissure is quartz, carrying variable amounts of galena, sphalerite, chalcopyrite, and pyrite. No regular arrangement of the ore minerals within the quartz could be detected. They occur in irregular bunches in all portions of the vein, and appear to be the result of a single uninterrupted process of deposition. The ore bodies, also, although often of considerable size, are without any known regularity in their distribution through the lode. The best ore is usually found in the more solid portions of the lode, a soft lode with abundant gouge being regarded as less favorable. The ore is nearly always separated from the walls by more or less gouge. A sheeting of the walls parallel with the main fissure is not infrequent, and where it occurs careful prospecting is necessary to avoid overlooking considerable masses of ore which may be lying behind what appears at the time to be the outer wall of the lode. On levels 1 and A the most productive portion of the Silver Lake lode is divided into two branches by a large horse of country rock traversed by stringers of quartz.

In the levels above No. 1, particularly in level 4, the Silver Lake vein is frequently a soft yellowish mass of more or less broken quartz, decomposed country rock, clay, cerussite, and other products of alteration. This portion of the lode thus shows considerable oxidation, particularly where there is evidence, in the form of clay seams and crushing of the quartz, of movement since the deposition of the ore.

Anglesite and cerussite were found associated with the galena in these upper levels, but never formed a very large proportion of the ore, of which the important bodies, even in these upper workings, were chiefly galena.

Movements within the Silver Lake lode subsequent to the deposition of the ore may be conveniently divided into four classes, (1) faults cutting the lode obliquely and at a considerable angle with the horizontal; (2) strike faults essentially parallel with the lode; (3) strike faults making a considerable angle with the plane of the lode, and (4) irregular fracturing. Those of class 1 will first be described.

(1) A close inspection of the southwest or foot wall of the Silver Lake vein shows that it is cut at frequent although irregular intervals by fissures, usually of small width. It is seldom possible in such cases to determine accurately the general course of these fissures. Fissures are never geometrical planes, and an observation made on a fissure exposed in the walls of a drift may give figures differing very much from the general or average course of the dislocation. The observed courses of these fissures give directions varying from north and south to N. 55° W. In general, they appear to form a system essentially parallel with the productive lodes of Group II, such as the New York City and Royal. Their dip is usually to the southwest at a high angle (over 70°), but there are a few notable exceptions. The fissures vary from narrow cracks up to openings a foot wide. In a few cases they are filled with solid quartz, carrying a little ore. These last are not known to cut the Silver Lake lode. They are probably of the same age as the latter and are comparable, in all respects save size, with such veins as the E. G. S. and L. A. S. of Group II. Most of these fissures, however, are filled with a clay seam or gouge of varying thickness, sometimes containing broken vein quartz. These fissures cut through the Silver Lake lode and are therefore younger. They record disturbance subsequent to its formation. They have not influenced the productiveness of the latter to any known extent, but are well known and carefully watched by the miners on account of their misleading tendency. Experience has shown that a miner, on running upon one of these cross seams, is very likely to follow it into the hanging wall, leaving, unknowingly, the main lode on his left. The presence of broken quartz mixed with the gouge of these seams shows that in some cases they record fresh movement along preexisting fissures which had become filled with quartz. The actual result of movement along these post-mineral fractures is present as gouge, but the net displacement seems to have been always small. It is usually impossible to detect any actual faulting of the main lode where crossed by these later fissures.

On Level A, southeast drift, a set of fissures occurs which, while apparently belonging to this system, are of unusually flat dip, in one case as low as 20°. The direction of dip is easterly. The vein itself is very irregular southeast of the first of these faults and is chiefly a breccia of country rock cemented by quartz and low-grade ore. It is such an irregularity as might be expected near the lower edge of the large horse which divides the lode on this level and the one above.

(2) Strike faults, essentially parallel with the lode, are more conspicuous in the case of the Silver Lake than in the other lodes of the mine. They are indicated by gouges, often of considerable thickness, on foot or hanging-wall, and by sheeting of the country rock when accompanied by soft gouge. Their effect upon the lode is well seen near the northwest end of level 2, where there is a pronounced strike fault on the hanging-wall side of the vein associated with approximately parallel slips within the country rock of the hanging wall. The lode here is crushed and contains much clayey gouge. The character and amount of the movement could not be determined in any case. It is believed, however, to have been in the main oscillatory, with but little net displacement. Had it been otherwise it would be very difficult to account for the present relation of the Silver Lake

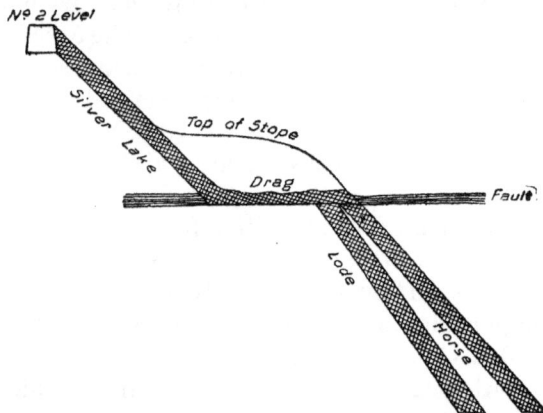

FIG. 11.—Sketch section across the Silver Lake lode between levels 1 and 2, showing probable faulting of the lode by a flat strike fault.

lode to the lodes of Group II and for the absence of gouge or other evidence of recent movement in certain portions of the former lode.

(3) Strike faults making a considerable angle with the plane of the lode are not common on the Silver Lake lode. The best example of this form of dislocation was noted in a large stope between levels 1 and 2. The fault here is nearly horizontal and cuts both branches of the Silver Lake lode near the apex of the horse which divides them. The dislocation is a thrust fault with about 12 feet throw. The character of the fault and its result upon the ore body are shown by the accompanying sketch section across the stope (fig. 11). This fault appears to die out within a comparatively short distance to the southwest, as it was not noted on the stopes of the Royal and Stelzner lodes.

(4) Irregular fracturing of the lode includes those minor dislocations which it has not been found possible to group according to any prevailing regularity of direction. They are often directly connected with the more regular fissures and are of small importance.

The ore of the Silver Lake lode is low grade, carrying very little gold. Although it has been extensively stoped in the past, the main output of the mine is from the lodes of the second group.

The New York City vein (or Whale-Rochester vein) is the most important of these, from the extent, regularity, and comparatively high grade of its ore body. Its general strike is about N. 20° W., and its average dip is about 80° to the northeast. As to the Whale-Rochester vein, its upper portions were formerly extensively worked through an adit in Whale Basin. In the Silver Lake mine it has been worked through levels 1 and C, 300 feet apart, and by means of an intermediate level halfway between them. This lode is generally smaller than the Silver Lake, and has been less disturbed by post-mineral movement. The ore is generally fast, or frozen, to the walls, and is usually solid and unoxidized. In common with the other lodes of Group II the New York sometimes pinches to a very narrow and tight seam. This is well illustrated at the point where it is intersected by the crosscut on level C. It is here a stringer less than an inch in width, showing little indication of being part of a great productive vein. The chief value of the New York City vein has been found to lie northwest of the points where it is pierced by the crosscuts on levels 1 and C. In this direction it is stoped practically continuously up to the Silver Lake lode. A typical section of the lode from the stope above level 1 is shown in fig. 3, page 67. As it approaches the Silver Lake lode it becomes smaller and more irregular, and bends westward. It is also somewhat oxidized and stained. At the junction it passes into the Silver Lake with no other lines of separation than those due to later movement, and is usually defined by seams of gouge. From the point of junction westward the ore of the Silver Lake vein contains more chalcopyrite than is usual elsewhere, and thus resembles somewhat the ore of the New York City vein. The productive portion of the New York City vein is cut by a fault about 750 feet south of its junction with the Silver Lake vein. The strike of this fault is apparently about N. 73° E., and it hades northerly about 15°. It has thrown the northern portion of the lode from 12 to 15 feet to the west. This fault was best seen in the stope above the midway level. Here the vein is cut off abruptly by a breccia zone about 6 feet wide which crosses the stope. The breccia is composed of sharply angular fragments of country rock cemented by quartz, which carries a little ore, but is of too low grade to pay. Within the breccia zone are two or three fairly regular quartz stringers. This occurrence is of special interest as showing that the deposition of quartz and ore was not confined to the period within which the important lodes were filled, but also took place at a later time. On the other hand, it seems very probable that this fault is older than the faults usually filled with soft, moist gouge, which have been described in connection with the Silver Lake vein. South of the crosscuts the New York City lode has proved of little value. This portion of it has obviously been subjected to strong disturbance connected with a pronounced zone of post-mineral faulting. This fault appears to have a general north-and-south strike, and hades to the east about 20°. Where it is cut by the crosscut on level 1, about 140 feet east of the New York vein, it shows a zone of broken and disturbed country rock about 12 feet wide, with a strong gouge on the hanging wall. This zone of disturbance appears to intersect the New York City lode about 200 feet south of the crosscut. The lode and the adjacent country rock have been broken up, and the value of the former has been destroyed. The country rock is traversed by seams of soft, moist gouge, in general parallel with the line of the fault, but often showing considerable irregularity. Small bunches and stringers of ore are found, but they are not continuous. On level C what is supposed to be the same fault crosses the crosscut about 330 feet east of the New York lode. It is here less than 6 feet wide, and is composed of broken, decomposed country rock and wet clay gouge. As on level 1, the New York lode has so far proved worthless south of the crosscut. The New York lode, probably originally somewhat irregular south of the crosscut, has been dislocated by a fault cutting it at a small angle and thus producing somewhat the effect of a strike fault parallel with the vein. The direction and amount of movement along this fault are not known. Some light may be thrown on it when the Silver Lake lode, which it should cut, is more thoroughly opened up. If any undisturbed ore bodies exist in the southern part of the New York lode they probably lie east of the present south drift on level 1.

In the autumn of 1899 the crosscut driven to cut the New York lode from the Unity tunnel, 400 feet below level C, had overshot its objective, and upon searching back the New York was found at the beginning of the crosscut near the Silver Lake lode. This was opened by a short drift and found to carry a body of excellent ore similar to that of the upper levels.

The ore of the New York City lode consists of galena, sphalerite, chalcopyrite, and pyrite in a quartz gangue. Some of the richest ore occurs in a rather friable stringer near the foot wall, which contains black oxide of copper (melaconite), with considerable chalcopyrite and pyrite. This may carry several ounces in gold, although free gold is very rarely seen. It is worth noting that the melaconite is found in the deepest workings, level C, at least 700 feet below the surface.

The Stelzner and Royal veins resemble in their general characters the New York City. The course of the Stelzner is rather curving and irregular, but may be generalized as about N. 20° W. That of the Royal is about N. 35° W. It thus nearly bisects the angle between the Stelzner and Silver Lake and illustrates the essentially radial character of the whole group of lodes. This feature is still further brought out by the fact that the Stelzner and the Royal with its two branches, called the right-hand and left-hand Royals, are, in the lower levels, all ramifications of a single vein, which in turn is a branch of the Silver Lake lode. Both the Royal and the Stelzner veins dip to the northeast. The Royal is very nearly vertical, while the Stelzner has an average dip of 75° to 80°. Consequently the line in which the two lodes join pitches southeastward. On the other hand, the much flatter dip of the Silver Lake lode causes the lines of junction between it and the veins of Group II to pitch to the northwest. As a result of these conditions the Stelzner and Royal join the Silver Lake on level 4 as separate veins. On level 3 the Stelzner, Royal, and Silver Lake lodes all come together at about the same point. On level 1, however, the two veins unite about 200 feet before they join the Silver Lake lode as a single vein. The junction of the Royal-Stelzner vein with the Silver Lake is similar to that described in the case of the New York lode. On level 2, at the point where the two veins come together, no clay seam could be seen separating them, although a little farther northwest the Silver Lake lode shows well-marked seams of soft, wet gouge on foot and hanging walls. At the junction there is much brecciated country rock, cemented by quartz which is full of small vugs.

On level B, north drift, the combined Royal-Stelzner vein has been stoped for about 400 feet, to a point where the lode diminishes to a small stringer. A short crosscut to the west, however, discovered the continuation of the lode. There is apparently no fault at this point. The lode has merely pinched out and is succeeded by a slightly overlapping parallel fissure in the manner already described.

The productive portions of the Stelzner vein are sometimes 10 or 12 feet in width. The Royal is usually somewhat narrower. The ore in both veins is usually solid and not accompanied by any considerable gouge. Post-mineral movement along the fissures is not, however, wholly lacking, and is best seen where the veins pinch to very narrow dimensions. As usual, in this mine the country rock near the veins has not undergone any alteration which can be recognized by the eye as evident and characteristic. The ore of the Stelzner and Royal veins consists of abundant galena, with sphalerite, chalcopyrite, and pyrite in a gangue of quartz. A little dolomite occurs in the vein, usually associated with included fragments of country rock. The quartz also shows greenish patches of chlorite, which are regarded as an indication of good ore. Chalcopyrite is less abundant than in the New York vein. The ore may carry 2 or 3 ounces of gold and is of higher grade than that of the Silver Lake lode, although somewhat inferior as a whole to the ore of the New York vein.

The E. G. S. is a small vein approximately parallel with the Royal and lying about 60 feet to the northeast of it. In its general character it resembles the Stelzner and Royal, but is smaller. Its junction with the Silver Lake can be seen on level 2. At the junction the quartz of the two lodes is separated by a narrow seam of gouge. Together they constitute a solid vein of low-grade ore about 10 feet wide. The E. G. S. has contained some of the richest pockets of ore encountered in the mine, but its value falls off, like the other veins of Group II, as it approaches the Silver Lake. The L. A. S. is another similar small vein, which was not worth working until level B was reached. It illustrates the fact, thus far exemplified in this mine, that the small veins of Group II may prove more valuable and have longer ore bodies as they are worked downward and as their junctions with the Silver Lake are carried farther and farther to the northeast on account of the relatively flat dip of the latter lode.

The Silver Lake mine is worked by overhand stoping, the ore being broken down onto canvas, and the stopes are filled with waste as the work is carried up. Siemens and Halske electric drills are used in most of the stopes. They are portable and convenient, but somewhat too lightly constructed for the work required of them. The ore from the various levels is all dumped into bins at the central station on level 1, that from the lower levels being raised through a vertical shaft by a cage and an electric hoist. From the central station the ore is drawn out in trains of cars by mules to the mill at the mouth of the adit.

At the mill it is sorted, the larger and heavier pieces of nearly solid ore being sent down the tramway as crude, or shipping, ore. The equipment of the mill is 2 Blake crushers, 50 stamps, 4 sets of rolls, 10 Wilfley tables, 8 Hartz jigs, 8 Woodbury tables, and 18 "end-shake" slime tables. Power is furnished by three 100-horsepower boilers and a 150-horsepower engine, and the mill has a daily capacity of about 200 tons. The water for milling is pumped from the lake, into which the tailings are allowed to run at another point. A new mill is now in progress of erection on the Animas River, at the lower terminal of the tramway.

As already stated, the ore of the mine is low grade. Ore running $10 a ton can be worked, but even $8 ore is often run through the mill with the better grades. About half the total value of the output is in gold, the rest being in silver, lead, and copper. Free gold is very rarely seen, and no attempt is made at amalgamation. The concentrates are said to average about $45 per ton. The crude ore and concentrates are carried down from the mill on a substantially constructed Bleichert tramway to a large terminal station near the mouth of Arrastra Gulch, furnished with capacious bins. From these bins the ore is loaded directly onto the railroad cars. A second tramway from the Unity tunnel was completed in October, 1900, and will soon handle the bulk of the ore.

Power for hoisting, pumping, and lighting the mine and buildings is generated by dynamos run by two Pelton waterwheels located on the Animas opposite the mouth of Boulder Gulch, the water for this purpose being taken from the Animas River just below Howardsville. In 1900 a substantial plant was also being installed at the same place for generating power by steam.

Iowa mine.—The Iowa mine also lies on the west side of Silver Lake and joins the Silver Lake property on the south. A prospect in 1881, its growth has, in a general way, followed that of its larger neighbor. The principal development has been on the Iowa, East Iowa, Stelzner, and Melville lodes. The last is probably continuous with the Black Diamond-Royal lode, but has not proved profitable in the Iowa mine, as it has in the Silver Lake ground. Practically all of the stoping has been on the Iowa, East Iowa, and Stelzner lodes. The relation of these three lodes, as shown in fig. 2 (p. 51), illustrates admirably the arrangement of veins en echelon, as observed in the small ore-bearing stringers on the surface. The Stelzner lode becomes small and poor at its southern end and is succeeded on the west by the overlapping East Iowa lode, which in turn is similarly succeeded by the Iowa lode. The Iowa and East Iowa lodes can be seen to die out in a few small stringers toward the north. That they actually terminate as individual lodes and do not open out again still farther northwest is shown by the fact that no lodes corresponding to them have been found in the two long crosscuts to the New York vein in the Silver Lake mine. Similarly, any southeasterly extension of the Stelzner and East Iowa should have been noted in the crosscuts of the Iowa workings. A lode plan such as is here described is not consistent with any extensive faulting along the lines of the fissures.

The Iowa has been worked on four levels, connecting with the surface by adit tunnels. They are numbered from above down. Work at present is confined to levels 3 and 4. The distance between levels is not uniform. Thus from level 1 to level 2 is 100 feet; from level 2 to level 3, 200 feet; and from level 3 to level 4, 130 feet.

The best ore in the mine occurs in the Stelzner vein and is identical with that already described in the Silver Lake workings. It may carry 2 or 3 ounces of gold. The ore of the Iowa and East Iowa lodes is a heavy lead ore carrying little gold, and resembling the ore of the Silver Lake lode more than is usual in the veins of the second group.

The main Iowa vein varies in width in its productive portions up to 15 feet. It is commonly between 2 and 5 feet wide. It dips steeply to the northeast, the average angle being near 80°. It is a generally simple and regular fissure vein, with well-defined walls. At its best it consists of nearly solid galena, with some quartz, chalcopyrite, sphalerite, pyrite, and chlorite. Comb structure is occasionally shown in the quartzose portions of the vein, and irregular vugs occur lined with crystals of quartz and galena. The galena sometimes occurs chiefly in the middle portion of the vein, as if it were of later crystallization than the quartz. In many cases, however, it is scattered through the vein in bunches without any apparent regularity.

The vein filling is solid and shows very little trace of movement subsequent to the deposition of the ore. It generally lies close against the walls of the fissure, with little or no gouge. As in the Silver Lake mine, the wall rock does not, as a rule, exhibit any very marked alteration close to the vein. The microscope shows, however, in the rock alongside the vein a thorough recrystallization of the original rock constituents into aggregates of quartz and sericite, with some calcite and specks of pyrite. A specimen 10 inches from the vein shows similar alterations, but with more calcite and abundant chlorite. The same general character of alteration is found 2 feet from the vein. At 50 feet from the vein quartz, chlorite, and calcite make up most of the rock, in which traces of the original structure remain, and the sericite is confined chiefly to the altered feldspar phenocrysts.

The Stelzner lode presents the same characteristics in the Iowa mine as in the Silver Lake. It is a nearly solid, simple vein, showing as much as 8 feet of excellent ore in the stopes. There is usually no gouge between the vein filling and the walls.

The East Iowa vein has not proved of very great importance. It has been stoped for a distance of over 100 feet, but presents no features of special interest.

. The Melville lode is chiefly notable from the fact that in its southern portion it dips to the southwest at from 70° to 75°. It carries galena, sphalerite, chalcopyrite, and pyrite in a gangue of quartz, with a little barite. It varies in width, being in some places as much as 4 feet wide. The ore is too poor to work. Northward this vein becomes irregular and in places is a nearly vertical fissure containing some gouge, but no quartz. It finally dies out in a few barren stringers. Although in general line with the Black Diamond-Royal lode, it is by no means certain that it occupies the same fissure. The low grade of its ore and its westerly dip, contrasted with the good ore and steep easterly dip of the Royal, seem rather to indicate that it is an overlapping fissure, bearing a similar relation to the Royal that the East Iowa does to the Stelzner.

Regular parallel fissuring of the country rock, the fissures being usually filled by stringers of quartz and ore, is even more conspicuous in the Iowa than in the Silver Lake mine. As in the latter mine, these fissures are apparently minor results of the same forces that produced the productive veins of Group II, seeming, in fact, to belong to that group. In the Iowa these fissures usually dip very steeply to the southwest.

Generally the lodes in the Iowa are very regular in their productive portions, having comparatively few horses. The ore is confined within the fissure walls. The disturbance and movement which has taken place since the veins were formed is scarcely appreciable. Tetrahedrite is reported as occurring in the Iowa workings, but none was seen at the time of visit.

The Iowa ore is carried on a Bleichert tramway, built in 1897, down to a mill situated in Arrastra Gulch, about a mile from its mouth. In 1899 an additional tramway was built from the mill to the railroad, on the Animas River. As a whole the ore is of low grade, averaging from $10 to $14 a ton, and carrying a large proportion of lead. The mill is equipped with 1 9 by 15 Blake crusher, 2 sets 14 by 27 rolls, 7 3-compartment jigs, 2 5-foot Huntingtons, 1 5-foot Chilian mill, and 5 Wilfley and 8 Cammett tables. The power is steam, and the daily capacity of the mill is given as from 175 to 200 tons of crude ore.

Royal Tiger mine.—This property, which is in the early stages of development, is on the east side of Silver Lake, and is supposed to be on the continuation of the Silver Lake lode. This, indeed, seems probable, although it is by no means certain, as there are too many nearly parallel veins in this vicinity to permit offhand correlation across the width of the lake. The Tiger lode, moreover, presents several characters foreign to the Silver Lake lode as exploited on the west side of the lake, and may fill an overlapping fissure nearly in line with the latter. The main adit is a crosscut at an elevation of 135 feet above the lake, and the levels are numbered from this upward. A little stoping has been done on level 4, but the bulk of the work has been on levels 2 and 3.

The lode dips steeply northeast, the average being nearly 80°. Locally, however, it may sometimes have a steep southwesterly dip, as on level 2, about 400 feet from the entrance of the tunnel. The width of the lode varies up to about 10 feet. It is usually fairly regular, but sometimes pinches for considerable distances, showing crushed rock accompanied by gouge, but very little ore. Near the entrance to level 2 the lode is divided by a horse 30 or 40 feet in width. Both branches of the lode come together in the stope above this level, where the lode shows about 19 feet of low-grade milling ore. It here takes on the character of a stringer lead, consisting of a network of stringers with much shattered country rock. The lode in this stope shows no gouge and no regular walls, the stringers running out irregularly into the country rock. Some of the ore-bearing stringers were faulted by later stringers of barren quartz. Small vugs, lined with quartz crystals, are frequent. The Tiger lode is frequently free from gouge and often without well-defined walls. In such cases the country rock near the lode is intersected by stringers and may carry bunches of ore. In other places there may be clay gouge on foot or hanging wall, or both, and even within the lode itself. Thus, movement subsequent to deposition of the ore has affected some portions of the lode, but not others.

The mine produces essentially a lead ore carrying from 20 to 60 per cent lead, rarely more than 12 ounces of silver, and occasionally a little gold—up to half an ounce. One specimen of free gold has been found. The gold usually occurs associated with sphalerite and chalcopyrite. The chief ore mineral is galena, which is accompanied by sphalerite and a little tetrahedrite and chalcopyrite. The tetrahedrite was most abundant in level 6, near the top of the ridge. It occurs in the lower levels, but less abundantly. The gangue of the ore is quartz with some barite. The quartz and galena are crystallized together irregularly in the larger stringers and in the main lode. A little native copper has been found on level 3. The alteration of the wall rock adjacent to the lode is somewhat more noticeable in the Tiger than in the other mines of the Silver Lake Basin. A specimen from the south wall of level 3, about 3 feet from the main ore body, is a compact, light greenish-gray rock, showing indistinct traces of former porphyritic structure. It is traversed by numerous veinlets. The largest of these, about one-fourth inch wide, carries quartz, galena, and sphalerite. This veinlet shows comb structure, and the ore minerals tend to crystallize along the medial line. Smaller veinlets consist of galena without quartz. The rock itself is thickly speckled with minute crystals of pyrite. Under the microscope the rock shows the general structure of an andesite, but no remnant of the original minerals remains. The feldspars are changed to aggregates of quartz and sericite, and the same minerals, very finely crystalline, seem to compose the entire mass of the rock, exclusive of the thickly sprinkled pyrite.

In the breast of the drift, level 4, the lode, which is about 4 feet wide, contains a dark horse-like mass. This is said to have been 4 feet wide in places and to carry 1 to 2 ounces in gold. This material shows numerous white spots, which suggest altered feldspathic phenocrysts, lying in a dark-gray groundmass. The whole is mineralized, and contains specks of galena, sphalerite, chalcopyrite, and finely crystalline pyrite. Under the microscope fragments showing the structure of an andesitic rock are seen embedded in a matrix of quartz and a very finely crystalline material, which is apparently kaolin, although there seems to be some sericite present also. The andesite fragments are altered to aggregates of these same minerals, but preserve their original porphyritic structure. The material is a much altered and mineralized breccia, but whether an original tuff-breccia or a breccia formed during the formation of the fissure it is impossible to say. The lode at this point shows indications of some brecciation of the ore and subsequent cementation by barren white quartz.

The Royal Tiger mine has not yet a separate mill. It is connected with the Iowa mine by a wire tram across the lake, and the ore is sent down to the Iowa mill. Although extensive work had barely begun in 1899, in 1900 the Royal Tiger was producing more ore than the Iowa.

Buckeye mine.—This mine, situated about one-third of a mile south of the lake, is on the Titusville lode. Work on it was abandoned some years ago, but the old tunnels were being cleaned out and reopened in 1899. The ore is chiefly galena, with some tetrahedrite, the latter being more abundant in the upper tunnel. There is no mill at this mine, and the ore was formerly packed out on burros.

Titusville mine, at the head of Kendall Gulch, although not in Silver Lake Basin, may be most appropriately mentioned here as being on the same great lode as the Buckeye. This mine was first worked about 1886. A small mill was subsequently erected in Kendall Gulch, and the ore was carried down on a lightly constructed Huson tramway about 8,000 feet in length. The mine ceased operating in 1893. It was developed on four levels, aggregating about 1,150 feet of drifts. The lode is about 30 feet wide. Its general dip is to the northeast at about 80°, although it is locally vertical. It consists of chalcopyrite, galena, sphalerite, and pyrite, in a gangue of quartz and rhodochrosite. Chalcopyrite is the most abundant ore mineral. The ore is generally of low grade, that containing abundant chalcopyrite running about $6 per ton. There is a richer streak, however, usually about 14 inches wide, which may run as high as $30 per ton. Some of this richer ore contains very finely crystalline galena scattered through the quartz in minute particles. Where this occurs the ore may carry from 1 to 3 ounces of gold and 60 to 100 ounces of silver per ton, the value being associated with the presence of the galena. The product of the Titusville mine in 1888 was $51,300.

From this mine the lode can be traced over the slope to the summit of Kendall Mountain, where it appears to divide. It has been prospected along this distance on the Seymour, Emma, Pyramid, Pompeii, and Sal Custer claims, and some ore has been taken out, chiefly galena. On the west slope of Kendall Mountain, the Idaho prospect, in Silverton Gulch, is apparently on the same fissure or fissure zone, although it is here very different in character. The country rock at this point is monzonite, part of the great stock of Sultan Mountain. The vein here is usually without gouge and narrow—18 inches or less. It dips southwesterly about 80°. The ore is mostly chalcopyrite and pyrite, with pockets of galena in a quartz gangue. The galena usually occurs in the middle of the vein, with quartz, chalcopyrite, and pyrite on the sides. This prospect has been worked intermittently for twenty years and has produced some ore. The development consists of two tunnels and some small stopes.

LODES OF HAZELTON MOUNTAIN.

General.—Hazelton Mountain, as the spur separating Blair and Arrastra gulches is called, was the scene of some of the earliest mining in the quadrangle. In 1874 work was already in progress on the Aspen, Prospector, Susquehanna, Mammoth, and other claims, most of which were afterwards consolidated as the Aspen mine.

Aspen mine.—The workings of this mine, which are extensive, are shown in plan and longitudinal section in fig. 12. As early as 1875, the Susquehanna and Aspen claims were being worked at a profit, while in 1876 the Prospector lode was developed by two shafts, 100 and 130 feet deep, connected by a drift 100 feet long. The product of the group up to 1880 has been estimated at over $200,000. In 1882 the Aspen group, embracing the Aspen, Susquehanna, Mammoth, Legal Tender, Matchless, and McGregor claims, was one of the best developed properties in the district and had over 5,000 feet of workings. During this year its operators contracted to furnish Sweet & Co., in Silverton, with 3,000 tons of ore, which were crushed, sampled, and sent to the new smelter in Durango. In 1893 the Mint reports credit the mine with a product of $75,000, and in 1897, $78,000. In 1892 the output, according to the same authority, was $50,916. In all, 11,423 tons of ore have been produced since January 1, 1884.

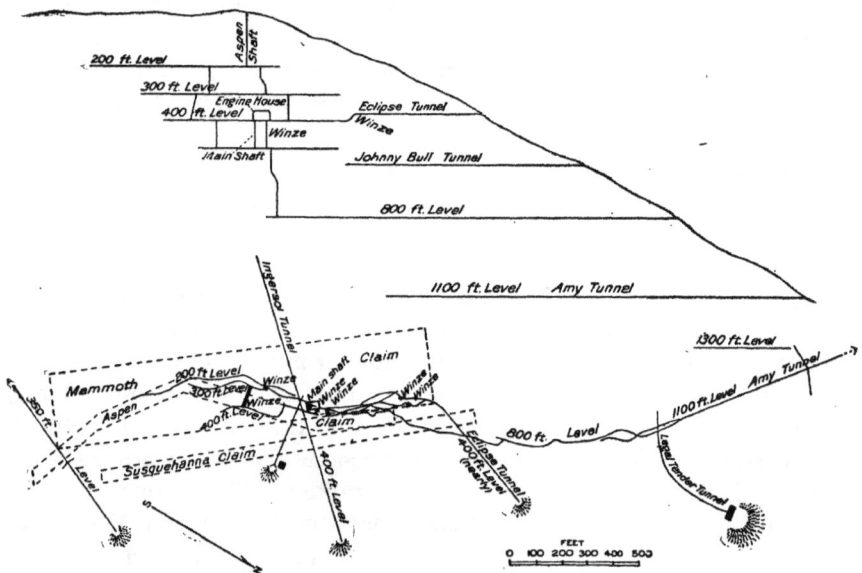

FIG. 12.—Plan and longitudinal section of the workings of the Aspen mine, Hazelton Mountain.

As the mine has been idle for several years, it was impracticable in 1899 to make a satisfactory study of its underground workings. In 1900, however, some progress had been made toward opening or retimbering the Amy tunnel, connecting with the 800-foot level. Access to a small portion of the lode was obtained through the Legal Tender tunnel. The vein at the point cut by the tunnel strikes N. 50° W. and dips southwest at about 80°. It is small at this place, accompanied by soft gouge and disturbed by some cross faulting. The Legal Tender tunnel shows the country rock to be traversed by several other veins running generally parallel to the Aspen.

The Aspen workings are in rocks belonging to the Silverton series, exhibiting chiefly the characteristic breccia structure already described and containing much rock of rhyolitic character. The lowest tunnel, which, judging from the dump, has not cut the lode, is in massive andesite.

According to information received from W. H. Thomas, there were several small veins near the surface, which came together in the deeper workings. The plan of the workings (fig. 12) indicates that the lode is of rather irregular character and that the general dip is northeasterly, at angles varying from 40° to vertical. It is often assumed that the Aspen lode is the northwestern continuation of that branch of the North Star lode upon which are located the Slide and Mayflower claims. But, as elsewhere pointed out, it is more likely that they are distinct.

The ore of the Aspen is chiefly a heavy lead-silver ore, consisting of galena, sphalerite, chalcopyrite, pyrite, tetrahedrite, and a little native silver in the upper workings. The gangue is quartz, with considerable green fluorite. The tetrahedrite is said to have occurred in pockets or bunches in the other ore minerals. One trial lot of ore, consisting of 1,000 tons, is reported to have run 110 ounces of silver and 60 per cent of lead, no returns being made for gold. The Aspen ore was smelted in the old Greene smelter, north of Silverton, as long as that pioneer plant was in operation.

It is commonly stated that the Aspen ceased working on account of the poverty of the ore in the lowest workings. It is probable, however, that good bodies of ore will yet be found, and that the mine will again become productive.

NORTH STAR[1] (KING SOLOMON) GROUP OF LODES.

General.—The prominent North Star lode crosses Little Giant Peak, northeast of Silver Lake Basin, very near its summit. Its general course is a little more northerly than that of the Silver Lake vein, or about N. 48° W. The dip is northeasterly, but nearly vertical. Northwest of the peak the lode continues through the Terrible, Slide, Mayflower, and Argentine claims, down into Arrastra Gulch. It is often supposed that the Grey Eagle-Aspen lode, on the south side of Arrastra Gulch, is part of the North Star lode. But the connection, if it exists, which is doubtful, is concealed by the talus in the gulch. It seems more probable that the Aspen is a distinct, essentially parallel, overlapping lode. Southeast of Little Giant Peak the North Star lode continues through the Dives, Shenandoah, Mountaineer, and Lookout claims into Mountaineer Gulch.

North Star (King Solomon) mine.—This mine was first regularly worked in 1883, although it had been exploited intermittently for three or four years previous to that time. Since then it has been producing steadily up to a year or two ago. The total product since 1883

---

[1] There are two North Star mines within the quadrangle—one on Little Giant Peak, commonly spoken of as "The North Star of King Solomon," and the other on the north base of Sultan Mountain. The former will be referred to in this report as "The North Star (King Solomon) mine."

has been $1,200,000. The lode on the North Star property is a strong stringer lead, consisting of a zone of country rock traversed irregularly by ore-bearing stringers. At the croppings, near the summit of Little Giant Peak, this zone is about 100 feet wide. The vein as a whole has no walls, and the stringers are usually not accompanied by gouge. The country rock is that found in the Silver Lake Basin, and belongs to the Silverton series. The hard, massive sheets of andesite are found to be less favorable for the deposition of ore than the softer breccias. The character of the lode makes it very difficult to work. Ore-bearing stringers may be encountered anywhere within the zone of fissured country rock, and frequent crosscutting is a necessity. The reduced horizontal projection of the levels, shown in fig. 13, will give some idea of the irregular character of the workings. The pay

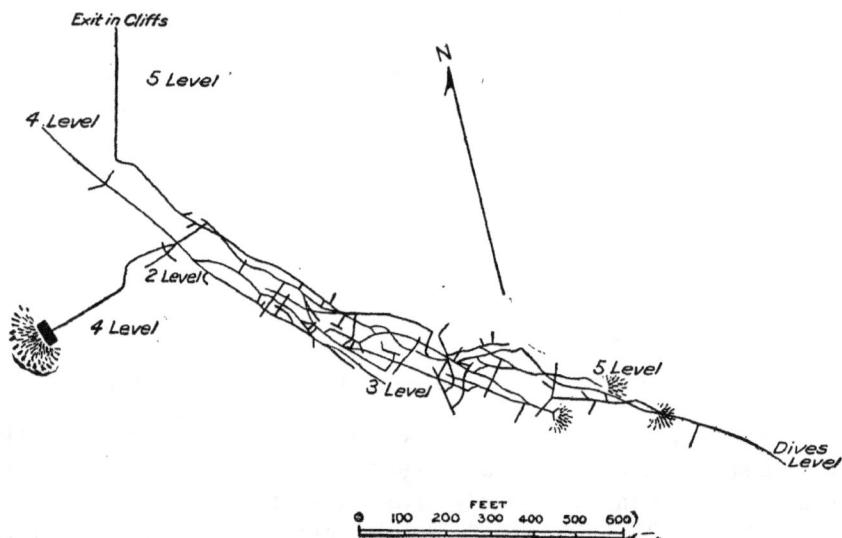

FIG. 13.—Sketch plan of underground workings of the North Star (King Solomon) mine.

shoot as a whole pitches to the southeast, apparently at an angle of about 45°. The upper 200 feet, near the surface, was largely sulphate of lead (anglesite) not carrying much silver. This changed in depth to galena, which was associated with increasing amounts of tetrahedrite as the ore was followed downward. In the lower levels the galena was subordinate to the gray copper. This was the most valuable ore mineral in the mine, being a highly argentiferous tetrahedrite, or the variety of gray copper known as freibergite. Below the fifth level the pay shoot failed, although the fissure continues. No serious prospecting has yet been undertaken to determine whether there may not be other ore bodies beneath the pay shoot. That there are such ore bodies is rendered probable by the fact that ore similar in character to that of the North Star mine occurs at a lower level in the Dives mine, which adjoins the former on the southeast. The gangue of the North Star ore was quartz. Some gold occurred occasionally in the ore of the lower levels, assays of 6 ounces to the ton having been obtained. This metal, however, formed an exceedingly small proportion of the whole output, which was chiefly in silver.

Owing to the cessation of mining and the accumulation of ice and mud in the abandoned workings, it was impossible to make any satisfactory study of the detailed structure of the lode. Small stringers and gouge seams are said to come in from the foot wall (southwest), as in the Silver Lake lode.

The ore from the North Star was sorted and packed on burros down to the road in Big Giant Gulch. From this point wagons carried it down to the Crooke mill, near the mouth of Boulder Gulch. Built to treat ore from the Polar Star mine, on Engineer Mountain, this mill was originally equipped with a modification of the Augustin leaching process. Its present plant consists of 1 7 by 10 Blake crusher; 10 1,000-pound stamps, dropping 5 inches at 90 to the minute; 5 Harz jigs, 1 Wilfley table, 3 Frue vanners, and 6 canvas tables. Power is furnished by a 60-horsepower engine and boiler. The capacity of the mill is about 50 tons. It is evident that ore so irregularly mined, and milled under such disadvantages, must have been high grade to have paid for extraction and treatment. Much ore that might have been profitably milled in a better-equipped mine has undoubtedly gone on the dump or remained in the stopes.

Dives mine.—This claim joins the North Star (King Solomon) on the southeast and is on the same lode. The workings here are not extensive and the mine has not yet produced much ore. It contains two stopes, one on the Dives and the other on the Shenandoah claim. The general character of the lode resembles that described in the North Star. It is an irregular stringer lead, without gouge and without regular walls. The ore is argentiferous gray copper (perhaps not a normal tetrahedrite, as it carries some lead), galena, chalcopyrite, and pyrite, in a gangue of quartz and barite. Small vugs lined with quartz crystals are common. Much of the vein matter is merely altered country rock, in which the outlines of former porphyritic crystals of feldspar are still recognizable. The ore-bearing stringers are distinctly cut by one, and possibly two, sets of later stringers of quartz, generally parallel to the course of the lode. These later stringers, as far as observed, are small, and are usually barren, though some carry a little ore. Comb structure is particularly common in them. A well-marked spherulitic structure was noted in some of the ore. More or less rounded, irregular lumps of tetrahedrite were surrounded by a skin of chalcopyrite, from which, as a foundation, the quartz had crystallized radially outward (see Pl. XI, C). The high-grade ore may run about 200 ounces of silver. On the Dives claim there is a solid vein of barite, 6 to 7 feet wide, on the hanging wall of the productive lode. Its relation to the latter was not determined. Its presence and the occurrence of barite within the ore body are perhaps connected with the Potomac lode, which is known to carry barite, and should join the North Star lode from the north at about this point. Barite was not recognized in the ore of the North Star mine.

The country rock of the Dives is that of the North Star. It shows some alteration close to the ore bodies, being bleached in color, traversed by veinlets of quartz, and full of minute cubical crystals of pyrite and occasionally specks of galena. The microscope shows that it has undergone entire metasomatic recrystallization and consists chiefly of quartz, sericite, and pyrite.

The Dives was worked on a lease in 1899, the ore being treated in the North Star (Crooke) mill. In 1900 it was idle.

The Mountaineer and Lookout mines have not proved steadily productive claims, although in the Mint report for 1891 the Lookout is credited with a production of $27,152. Both are now idle. The tunnel of the latter is run in Algonkian schists, just beneath their contact with the overlying volcanic rocks of the San Juan series. The schists are shattered and decomposed, and the deposit appears to be very bunchy and irregular. The ore shows galena, chalcopyrite, and pyrite. The Mountaineer is said to have been located by prospectors from New Mexico in 1870. At least three veins can be seen cropping in the steep bluffs of San Juan breccia above the Lookout tunnel, and apparently converging just south of the tunnel mouth in what is locally termed a "blow-out." This is a mass of shattered limestone (probably the Ouray limestone (Devonian), which overlies unconformably the Algonkian schists south of the Highland Mary mine) and fragments of schist and andesite, cemented by comminuted andesitic material. Some of the limestone fragments are surrounded by envelopes of hematite. The whole mass is somewhat mineralized and is traversed by quartz stringers. The origin of this area of local disturbance is not certainly known. The original shattering may have been due to the movements which formed the veins that converge at this point. There has, however, been some recent movement of a landslip character, which has tended to obscure the original relationships.

Big Giant lode and others.—In Big Giant Gulch are several lodes which may be conveniently described in connection with the North Star lode. The most important of these is the one on which the Potomac claim is located. Some work has been done on this claim, and also on the Big Giant, which appears to be on the same lode. On the latter claim the lode is apparently a stringer lead, without any considerable gouge and carrying much chalcopyrite in its ore. It dips southwest at from 75° to 80°. A small mill was erected on this claim, but the mine was not successful, and was idle in 1899. Toward the southeast this lode passes over the northern shoulder of Big Giant Peak and is supposed to join the North Star vein in the Dives claim. There are other lodes in this gulch, two of which are shown on the map, but they have not proved to possess any value.

Little Giant lode.—This lode, which lies between Big Giant and Arrastra gulches, has not been worked for many years, and its chief interest at present is a historical one. Discovered by Miles T. Johnson in 1871, it was the first mine to be opened in the district, and, in contrast with the present mines in the vicinity, produced a gold ore. Such portions of this ore as were not shipped to Pueblo in the crude state were originally treated in arrastras. About 27 tons were treated in this way and yielded $150 per ton. In 1872 the Little Giant Company was formed in Chicago to work the mine, and in the following year the company erected a mill equipped with a Dodge crusher, ball pulverizer, and 5 stamps. Power was supplied by a 12-horsepower engine. The mill was constructed 1,000 feet below the mine opening and the ore was brought down by the pioneer wire-rope tramway of the region. About 100 tons of ore were milled this year, producing about $14,500. The saving is said to have been about 65 per cent of the assay value. The mine became involved in litigation, and this, coupled with the diminution of the pay shoot, led to its abandonment.

The strike of the vein is N. 40° W., and its dip southwest. The pay streak is said to have been about 8 inches wide.

LODES OF CUNNINGHAM GULCH.

General.—Cunningham Gulch, named from one of the pioneers of Baker's company, is the largest of the various side gulches which open on the Animas River above Silverton. At an early date it was the scene of mining activity, and then gave promise of a development which has not yet been fulfilled. In 1899 and 1900 a little work was in progress on the Pride of the West, but otherwise the gulch above its junction with Stony Gulch contained no working mines. Above the Green Mountain mine the bottom of the gulch is excavated in the Algonkian schists, which immediately underlie considerable areas of the volcanic rocks of the quadrangle, but are seldom exposed in its central portion. The San Juan and Silverton formations lie upon the characteristically uneven erosion surface of the older schists. The lodes traverse both the schists and the overlying volcanic rocks, and thus afford opportunity for studying the changes, if any, that take place in the lodes when they pass from the older to the much younger formations. This investigation, however, is unfortunately seriously handicapped by the abandonment of most of the old workings, which has rendered many of them inaccessible. The general course of most of the lodes in Cunningham Gulch is between N. 30° W. and N. 40° W. The Highland Mary vein, however, is about N. 62° W., and still other lodes run more nearly north and south.

Highland Mary mine.—This mine has it workings in the ancient schists at the point where Cunningham Gulch, rising by an abrupt ascent to a higher level, loses its narrow linear character. The schists which compose the country rock vary somewhat in mineralogical composition, but the most common facies is an ordinary hornblende-schist consisting essentially of green hornblende and quartz. The strike of the schistosity is here about N. 48° E. The vein has a strike of N. 62° W., and is practically vertical. Its average width is from 2 to 3 feet, and it is a simple quartz-filled fissure, the vein being adherent, or "frozen," to the schist walls. Several tunnels have been run into the vein, but none of them were accessible in 1899. The only ore minerals seen on the dump were chalcopyrite and pyrite. According to Endlich,[1] of the Hayden Survey, who visited the mine in 1874, the width of the vein is from 4 to 5 feet, and galena and tetrahedrite formed the main ore, accompanied by pyrite, chalcopyrite, and sphalerite. The lower tunnel has a somewhat devious westerly course, and was designed by Edward Innis, its projector, to tap the Royal Tiger and other lodes of the Silver Lake Basin. It was started about 1875 and had reached a length of about 3,000 feet when work was abandoned in 1884. Several veins are said to have been cut in this tunnel, but none were prospected. An inspection of the dump shows that the whole of the tunnel is in schist. The total product of the mine was about $50,000.

Toward the northwest the Highland Mary vein passes upward from the schists into the volcanics of the San Juan and Silverton series. As examined at a point 1,500 feet above the mine buildings, the lode in the volcanics is somewhat larger than in the schists below, having a width of about 12 feet. It is here composed of apparently barren quartz. The lode probably joins the North Star vein on the Shenandoah claim.

Green Mountain mine.—This mine, which is credited with a production of about $100,000, was being opened when visited by the officers of the Hayden Survey in 1874. In 1883 it is said to have produced about 300 tons of ore, averaging 38 ounces of silver and 45 per cent of lead per ton, valued at about $25,000. It was idle in 1899, and evidently had not been worked for several years. The mine was opened through four tunnels, entering on or near the course of the lode. The country rock of the three lower tunnels is a rhyolitic flow-breccia. The upper tunnel is in Algonkian schist. The schist of course underlies the rhyolite, and the fact that it is encountered in the upper instead of the lower workings of the mine is due to the irregularity of the contact between it and the younger volcanic rocks, and to the course of the fissure, which makes an angle with the contact. The strike of the lode is N. 38° W., and the general dip nearly vertical, or at a high angle to the northeast. In the rhyolite the vein is generally about 3 feet wide and is destitute of gouge. It is a simple fissure filling, containing a few small horses of altered country rock. The ore contains abundant sphalerite and galena, with some chalcopyrite and pyrite, in a quartz gangue. In the schists the vein is from 3 to 4 feet wide and fills a clean-cut fissure. The quartz and ore are "frozen"

---

[1] Report on the San Juan region: Hayden Survey, 1874, p. 234.

to the cleanly fractured edges of the schists. The latter are slightly impregnated with pyrite, but show less alteration than the rhyolite breccia. The differences in the country rocks have not affected the vein or ore in any way recognizable by the eye. Pay ore was evidently stoped in both. From a scientific point of view it is much to be regretted that this mine is not now in operation and open to inspection.

Pride of the West mine.—This was one of the earliest mines in the district to ship out ore. The first lot was taken out by pack animals to Del Norte, on the Rio Grande, and there sold, in 1874. It was from the upper workings, and is said to have contained handsome masses of wire silver. The total product of the mine has probably been near $100,000. In the Mint report for 1884 the mine is credited with a total product from the upper tunnel (level 3) of 1,500 tons, with an average value of $65. This would give a total of $97,500 up to the end of 1884. The product since that time has probably not been large. Work of a desultory character only was in progress in 1899. The general course of the Pride of the West lode is N. 32° W. The general dip appears to be to the southwest at about 80°. Local variations, however, are common. The country rock is a greenish, indurated tuff, belonging to the Silverton series, and probably latitic in character. The upper and older workings are situated some 500 feet above Cunningham Creek, near the croppings of the vein, which are here from 20 to 25 feet wide. A crosscut tunnel of about 80 feet gives access to the lode, which has a threefold character. Next the hanging wall, where cut by the crosscut, lies about 3 feet of apparently fair milling ore, upon which no stoping has been done. This is succeeded on the northeast by 15 feet of quartz, carrying abundant pyrite and chalcopyrite. A division plane, carrying a little gouge, separates these two portions of the lode. Northeast of this wide mass of quartz, and next the foot wall, is the main " pay streak " of the mine. This ore body is a stringer lead. Its thickness, where stoped, was apparently about 15 feet at the most, and usually considerably less. There are no true walls. Stringers from the lode penetrate the foot wall irregularly. For a distance of several feet the latter is traversed by stringers generally parallel to the lode. The lode has been stoped from the main level, and from a level 60 feet below, reached through a winze.

The ore is chiefly galena, occurring in stringers and bunches, with more or less quartz gangue. Pyrite, chalcopyrite, and sphalerite are also present. The richer pockets, however, carried rich tetrahedrite (freibergite), and sometimes wire silver.

A tunnel has been run at a point on the road about 100 feet above Cunningham Creek, for the purpose of cutting the Pride of the West lode at a greater depth and at a more convenient point for working. Many hundred feet of crosscut and drift were run, without, however, finding any pay ore. Much of this rock was on the Silver

Fountain lode, which runs southwest of and approximately parallel with the Pride of the West lode. Its general course is N. 42° W The crosscuts show that between this and the Pride of the West lode, which appears never to have been reached in these lower workings, are several other generally parallel lodes, none of which carry pay ore, as far as prospected.

The ore from the upper workings of the Pride of the West is brought down to the road by a small wire-rope tramway, and is thence hauled to the railroad at Howardsville in wagons.

Philadelphia mine.—This is a small mine higher up the slope and apparently on the same lode as the Pride of the West. Work on it began in 1875, and it is said to have contained a single pocket of ore, chiefly freibergite with a little galena, which afforded about 100,000 ounces of silver. It has been idle for years.

Other lodes.—There are several other lodes on Green Mountain between Stony and Cunningham gulches. Several of these have produced small pockets of ore, but have never been extensively mined. The stronger lodes run approximately northwest and southeast. One of these is prominent along the crest of the mountains. Some large masses of barren-looking white quartz outcrop along the course of the lode, but often the rock (andesite) is bleached and full of small quartz stringers for a width of about 30 feet. About a quarter of a mile northwest of the summit of Green Mountain this lode cuts an older vein of solid white quartz, about 10 feet wide, accompanied by some parallel stringers. This older vein strikes N. 60° E. and dips southeast at 80°–85°. The lode along the ridge at this point dips southwest. On the northern spur of Green Mountain the veins form a complex network. A few only of the better-defined ones are shown on the map. As far as could be determined they are all of the same age. As a rule they show small indications of mineralization. I was informed, however, by the owner of one of the claims that ore occurred in scattered pockets, from one of which he had taken out a carload running 3 ounces of gold per ton.

North of the Pride of the West is a lode running generally parallel to the road. There are several prospects upon it, but it has never produced pay ore. In one of these prospects, the Osceola, just north of the Pride of the West, occurs a very curious banded ore, described more fully in the general section on the ores. Owing to the caved state of the tunnel this banded ore could not be studied in place.

LODES OF GALENA MOUNTAIN AND VICINITY.

General.—Under this heading there will be described not only the fissure deposits on Galena Mountain itself, but those of Maggie, Porcupine, and Stony gulches. Galena Mountain is an exceedingly precipitous peak, the upper part of which is composed chiefly of andesite, both massive and fragmental (tuff-breccias), belonging to the Silverton series. In its lower portions occur some rhyolite and intrusive masses of monzonite. The mass of the mountain is traversed by very numerous lodes, in great part true veins, many of which are prominent on the surface. Only a few of the more conspicuous ones could be platted on the map, and those with merely approximate accuracy. The number, prominence, and rough parallelism of certain sets of these lodes suggest their division into at least four groups.

(1) N. 45° E. lodes.—Lodes having a nearly northeast and southwest strike are not so numerous on Galena Mountain as they become farther north. The best example in this portion of the quadrangle is the Ridgway lode, the general course of which is about N. 40° E., with local variations. Thus, in the Ridgway mine the strike, as observed in the drifts, is about N. 65° W. The lode dips southeast at from 75° to 80°. The Ridgway lode is faulted by the Alaska lode, with a strike of N. 40° W.

(2) N.–S. lodes.—In this group are included a number of rather short lodes, of which the Veta Madre may be taken as an example. They usually dip to the east at an angle of about 80°. No important ore bodies have been found on any of them, nor was any work on them in progress in 1899. The larger of these N.–S. lodes are often accompanied by parallel fissuring and veining of the country rock.

(3) N. 25° W. lodes.—Lodes of this group are very well seen on the south declivity of the mountain in the vicinity of the Veta Madre mine. The massive andesite is here traversed by a very prominent set of fissures running N. 25° W. and generally nearly vertical. The fissures are closely spaced, a foot or less apart, and are usually occupied by rather small stringers, which are sometimes mineralized. In the Veta Madre tunnel the N.–S. lode is cut by a vein belonging to this group (N. 25° W.), which was followed for about 300 feet, and carried galena with chalcopyrite and probably sphalerite and pyrite. The veins of this group are thus, in part at least, younger than the N.–S. veins of group 2.

(4) N. 65° W. lodes.—Lodes whose courses are within a few degrees of N. 65° W. are generally larger and more prominent than the other lodes of Galena Mountain. Just southwest of the summit of the peak a prominent lode of this group has been extensively prospected. The lode at this point is a large one, striking N. 60° W. and dipping 75° to the northeast. The ore is chiefly chalcopyrite and galena in a quartz gangue. The Neigold mine is also working one of these strong lodes (N. 75° W.), in this case practically vertical. The ore shows galena, chalcopyrite, a finely disseminated sulphobismuthite of lead, specularite, and pyrite, in a quartz gangue. A gray prismatic mineral of brilliant metallic luster, determined as probably bismuthinite, is also present. The veins of this group cut the N. 25° W. veins of group 3, and in one case were observed to cut a vein striking about N. 30° E. They are thus, as far as is known, the youngest veins on the mountain.

Although Galena Mountain has been prospected at many points and has produced some ore, no important ore bodies have yet been developed. The veins in general contain milk-white quartz and some barite and are apparently not heavily mineralized.

Antiperiodic mine.—This is a small property on the east side of Rocky Gulch. The lode has a general strike of about N. 38° W. and dips southwest about 75°. It is irregular and the ore occurs in pockets. The country rock is andesite. The ore consists of galena, with cerussite and wire silver near the surface. Tetrahedrite, chalcopyrite, pyrite, and sphalerite are also present. The total product of this little mine during the twenty years it has been worked amounts to about $12,000. The ore is often of high grade, carrying sometimes 1,100 ounces of silver and 5 ounces of gold, but it is difficult to find and variable in tenor.

Ridgway mine.—This mine is situated on a spur which runs out northeast from Galena Mountain to Maggie Gulch. It was first worked in the summer of 1896 and has paid continuously. The strike of the lode is N. 40° E. and the dip southeast at 75°. The workings consist of two tunnels, with drifts and stope on the upper level. The lower tunnel has just cut the lode, and a winze is being sunk from the upper level to meet it. The country rock of the two main levels is massive andesite. The stope, however, is partly in well-bedded andesitic tuff about 200 feet thick, resting upon an uneven surface of the massive andesite. The tuff is in turn capped by an andesitic flow. The lode in the stope shows a maximum width of 7 feet, nearly all ore. The vein is "frozen" to the walls, the rock of which exhibits considerable alteration. The microscope shows that the andesite near the ore has been changed into an aggregate of quartz and sericite. The gangue of the ore is chiefly quartz, with a little calcite. Vugs are common, usually lined with crystals of quartz, which are sometimes faintly amethystine. Toward the northeast the vein, as followed in the stope, becomes a barren stringer lode. Toward the southwest the ore body is sharply terminated by the Alaska vein, which faults the main lode, apparently offsetting its southwestern continuation to the southeast. No ore has yet been found south of the fault, although the vein continues beyond the offset.

The valuable constituent of the ore consists chiefly of a dark-gray or black, sectile, argentiferous mineral, which contains neither arsenic nor antimony and is, without much doubt, argentite. Curiously, this mineral, with sectility as its most characteristic physical property, is called "brittle silver" by the miners. It occurs finely disseminated through the quartz (Pl. X, *A*) and as small irregular crystals, with quartz, in the vugs. A little ruby silver is reported. There is very little galena in the ore, and the product of the mine is wholly in gold and silver. The ore shipped averages $110 per ton and carries about 3 ounces of gold. Some of it runs much higher. In 1898, 300 tons of ore were shipped, and in 1899 about 150 tons.   Ore under $20 in value is thrown on the dump.   The crude ore is taken down on pack animals to Middleton and shipped to Durango by rail.   It costs $6 per ton to transport it to the smelter.   The total product of the mine has been about $60,000.

Gold Nugget claim.—In Maggie Gulch are numerous lodes and veins, nearly all of which have been superficially prospected.   The Gold Nugget is one of the oldest of these prospects and has produced about $10,000, chiefly from a single rich pocket near the surface.   The value is in gold and silver, and the ore is the so-called "brittle silver," probably in this case argentite.   The strike of this vein is N. 10° E. and its dip about 60° to the east.   It is a strong, solid, white vein, containing some calcite and barite with the quartz.   The country rock is andesite.

Little Maud claim.—This is on a strong vein about 12 feet wide, striking N. 20° E. and dipping southeast about 77°.   The vein is frozen to the walls.   The country rock is andesite.   The ore is similar in character to that of the Ridgway, consisting of the so-called "brittle silver" with pyrite in a quartz gangue.   It has thus far been found near the hanging wall, but the pay streak appeared to be changing to the footwall.   About 40 sacks of high-grade ore had been taken out in 1899, but none shipped.

As a whole the veins in Maggie Gulch, including the Ridgway, differ markedly in the character of their ores from those hitherto described in Silver Lake Basin, Cunningham Gulch, and on Galena Mountain.   They do not contain noticeable quantities of galena, and are mined for gold and silver only.   The important ore mineral is the so-called "brittle silver," which usually is finely disseminated through the quartz.   As usual, it is not brittle silver (stephanite) at all, but either an argentiferous galena or a mixture of galena and argentite.   Common pyrite is also usually an abundant constituent.   As far as observed, veins carrying ore of this character all strike east of north.   The value of the ore is about equally distributed between gold and silver.   The ore is usually of a high grade, but has not been found in large bodies.   It is generally in irregular bunches and streaks in prominent, generally barren lodes.   Except the Ridgway mine, there are no workings in the gulch which can be regarded as anything more than prospects.

Hamlet mine.—This mine, about three-fourths of a mile south of Middleton, although hardly in Maggie Gulch, may be conveniently referred to in this place.   This is a silver-lead-copper mine, which was worked intermittently from 1883 to 1887.   Most of the work was done during the last two years, when the mine shipped several hundred tons of chalcopyrite and galena ore.   The strike of the vein is N. 50° W. and the dip 80° to 85° to the northeast.   The ore is galena, sphalerite, and chalcopyrite in quartz.   The country rock is a fine-grained monzonite, and the vein is a simple fissure filling, with little or no gouge. The workings are rather extensive, embracing three adit tunnels, with drifts, stopes, and crosscuts. The ore was taken down to the railroad on pack animals, and it is remarkable that so much work was done on an ore deposit of this character without more effect-ive means of handling and concentrating the ore.

Other lodes.—Just north of Howardsville the precipitous slopes of Dome Peak are traversed by several prominent lodes. There are two sets of these, one with a course of about N. 63° W., and the other about N. 25° E. They are not distinct fissures, such as would result from decisive faulting, but are veins showing considerable regularity for some distance, then branching, anastomosing with other veins or dying out in networks of irregular stringers. There is a tunnel, the name of which was not ascertained, about half a mile north of How-ardsville, at the junction of two of the most prominent veins. In it much fruitless work was done, but no extensive ore body was ever discovered and the mine was abandoned. The lodes, so prominent in the cliffs above, seemed to be pinched, barren, and irregular when followed under ground.

LODES OF BURNS GULCH.

General.—Although the lodes in this gulch were prospected at an early date in the history of the region and have produced some ore, none of them have proved permanently productive.

Tom Moore lode.—This lode, reported to have been located in 1870, was being worked in 1881, and then produced a massive galena said to carry about 70 ounces of silver. But its development has been slow and intermittent. Its strike is N. 53° E. and its dip NW. 68°. The country rock is a rhyolitic flow breccia, part of the Silverton series. This, when examined at some distance from the lode, is a light-gray rock showing very conspicuous mottling, due to the brecciation which its material has undergone. Both the fragments and the matrix are essentially of rhyolitic character, although an occasional andesitic particle occurs. The rock is usually streaked and spotted with little nests of epidote. The microscope shows that the original glassy base of the rock has been largely devitrified and is now a fine-grained granular aggregate in which quartz is the only recognizable constitu-ent. Phenocrysts of orthoclase and plagioclase are abundant and are usually partly altered to sericite. Other phenocrysts, presumably ferromagnesian silicates, have been changed to aggregates of quartz and epidote. Areas of chlorite represent what were probably former phenocrysts of biotite. There is also a little calcite present. The accessory constituents noted are allanite and titanite. The lower tunnel, near the Animas River, shows an irregular stringer lead with a well-defined gouge slip along the foot wall. The ore shows chiefly galena, with sphalerite, chalcopyrite, pyrite, and sometimes tetrahedrite, in a quartz gangue. A little rhodochrosite occurs in vugs. The country rock alongside the lode is impregnated with pyrite and specks of galena to a greater extent than is usually observed in this region. When examined in thin section under the microscope, it is seen to be altered to an aggregate of quartz, sericite, and ore, traversed by veinlets of quartz. The original brecciated structure and the outlines of many of the feldspar phenocrysts have been preserved in spite of complete recrystallization. The feldspars have usually been altered to aggregates of sericite.

A second tunnel, several hundred feet above the lowest opening, was not accessible. The dump, however, showed that this portion of the lode was originally a breccia zone in the rhyolite which became filled with ore, accompanied by considerable impregnation and replacement of the country rock. This ore deposit was itself shattered and veined with white quartz, carrying little or no ore.

The upper tunnel, 200 feet above the last, was being worked on a small scale in 1898. The ore streak here is frozen to the walls and carries no quartz save in some small later stringers, often accompanied by kaolin, which cut the ore, and in a few drusy vugs, where it is sometimes accompanied by a little pale-lilac fluorite. The ore is chiefly tetrahedrite, with some chalcopyrite and galena. Sphalerite was not seen. There is said to be some bismuth present, probably as a sulphobismuthite of lead, and the occurrence of hübnerite was noted in small crystals in the later quartz stringers. The width of the lode is usually from 3 to 6 feet, and it is in part a replacement of the rhyolitic walls. The chalcopyrite occurs chiefly in the middle portions of the ore stringers. The tetrahedrite on each side of it is solid for a short distance and then fades off into the altered country rock as minute veinlets and specks associated with chalcopyrite or pyrite. Such gangue as is present with the ore is altered rhyolitic flow breccia, consisting chiefly of finely crystalline quartz and sericite. The lode is thus not sharply differentiated from the country rock on each side of it. The tetrahedrite is not highly argentiferous. The first-class ore runs about $60 to the ton and the second-class about $48. The chief value is in copper, but there is some silver and a little gold, the last sometimes in visible particles. Free copper is not uncommon in these upper workings, but is confined, as far as is known, to the small stringers of younger white quartz which cut the ore. These sometimes show comb structure, and the copper occurs along with the medial suture of the stringers.

The middle tunnel is said to have furnished ore carrying more galena and higher in silver, with subordinate amounts of tetrahedrite, while in the lowest tunnel tetrahedrite was rare.

Silver Wing mine.—This property comprises several lodes, of which one is the northeast extension of the Tom Moore. Some development of these lodes was in progress as early as 1880, but they have never been successfully worked.   An extensive tunnel has been run in from the Animas River near the mouth of Burns Gulch, and a costly mill has been erected to treat the ore when it should be found.   The property became involved in legal difficulties, and in 1899 all work had ceased before ore had been discovered in paying quantities.

LODES OF SUNNYSIDE BASIN.

General.—Although the Sunnyside mine is the only one in the basin now producing, there are several other large lodes which have in the past produced ore from superficial workings.   In the general character of their ores they show much similarity.   They are all of prevailingly low grade, although the Sunnyside contains masses of richer ore carrying free gold.   The lodes are unusually massive, in some cases over 50 feet wide, and all carry abundant rhodonite.   The ore minerals are galena, sphalerite, chalcopyrite, pyrite, and sometimes tetrahedrite and free gold.   The ore minerals are usually rather evenly scattered through a quartz gangue, and the ores are more siliceous and the galena is more finely crystalline than in those of the Silver Lake Basin.   The large veins at the head of Lake Emma are curved and show abrupt changes in their courses.   The cause of these irregularities is not apparent.

Sunnyside mine.—Situated in the basin at the head of Eureka Gulch, this mine enjoys the distinction of having been an almost continuous producer from the first discovery of the region.   It is to-day one of the few mines in the quadrangle that are actively working on a considerable scale.   The deposit, like many another prominent lode in the district, was located in 1873, while the land was still held by the Indians, and afterwards relocated when the San Juan country was thrown open to settlement in 1874.   It was worked as a prospect in 1875, and about four years later produced considerable ore.   Thompson Brothers worked the mine for several years, during one of which they are said to have taken out $385,000.   The rich gold ore which made the early exploitation profitable finally gave out.   A 10-stamp mill was erected at the outlet of the lake and an attempt was made to concentrate the lower-grade ore.   The effort resulted in failure. Finally, about 1890, J. H. Terry put up the present mill in Eureka Gulch and operated it successfully.   About 1896 this mill was connected with the mine by a wire-rope tramway.   In 1899 a second mill was completed at Eureka, designed to relieve the work of the upper mill by concentrating the lower grades of ore which do not require amalgamation.   A tramway was also constructed connecting the two mills.

The Sunnyside lode is very large—30 to 50 feet in width.   Its general strike is about N. 50° E. and it dips to the southeast at 65° or 70°. Toward the northeast it passes through a saddle, where it is known as the Sunnyside Extension lode, and continues down Placer Gulch.

Southwest of the Sunnyside the lode becomes the Belle Creole, curves sharply westward, and is concealed by alluvium at the head of Lake Emma. A short distance northeast of the Sunnyside lode, and nearly parallel to it, is the No Name lode, dipping 75° to the southeast, and joining the Sunnyside in the Sunnyside Extension ground. Between this lode and the Sunnyside are at least three other veins, making in all a very large body of low-grade ore. The Sunnyside vein is frozen to the walls, and gouge is usually absent. The ore minerals of the vein are galena, sphalerite, chalcopyrite, pyrite, tetrahedrite, and free gold, in a gangue of quartz and rhodonite. The ore streaks occur irregularly in the vein, between partitions of rhodonite. As a rule the ore occurs near the walls, and when an ore shoot pinches on one wall it usually thickens on the other. The best ore, carrying free gold, occurs in irregular lenses parallel with the plane of the lode and sometimes 30 or 40 feet in diameter. It is usually associated with rosin-colored sphalerite and small amounts of lilac fluorite. These are regarded as useful indications, but are not invariably accompanied by free gold. The ore has been brecciated to some extent since its original deposition and is sometimes traversed by a network of veinlets. These are usually filled with quartz carrying a little chalcopyrite. But where the veinlets traverse rhodonite they are often filled with that mineral instead of quartz. The exact relation of the rhodonite "partitions" to the ore is not always clear. In some cases the rhodonite appears to form the mass of the vein, in which the ore occurs as lenticular bodies, and sometimes lenses of rhodonite are surrounded by ore. The large masses of rhodonite, or "pink," as the miners term it, are never regarded as ore, although they are not destitute of the ore minerals. The material is rose-pink in color on fresh fracture, very compact, and exceedingly tough to work. It always contains a little quartz, calcite, and rhodochrosite. The rhodonite sometimes occurs as distinct stringers cutting the quartz and ore, while in other places it may be seen intimately and irregularly intercrystallized with them. As far as observed, neither the rhodonite nor the ore shows any regular megascopic structure, such as banding, crustification, or comb structure, other than the general relationship already indicated.

The No Name lode has not yet been commercially worked. Where seen it showed about 6 feet of quartz and ore on the foot wall and about 4 feet on the hanging wall, the two streaks being separated by 10 feet of rhodonite (fig. 7, p. 88). The ore is chiefly galena.

The Sunnyside is worked through a main adit tunnel about 100 feet above the lake. There are two other important levels above the present adit. Underhand stoping is employed, resulting in empty stopes of great height. Fortunately the nature of the country rock is such that the men working in the bottoms of these stopes are fairly safe from falling material. There are no stopes as yet below the main adit level.   As is too often the case in the mines of this region, adequate maps, showing the underground workings and the shapes of the ore bodies as developed up to date, are lacking.

The chief product of the mine is in gold and lead, the silver averaging not much more than 11 ounces per ton.   Occasional streaks of tetrahedrite may, however, carry as much as 400 ounces.   The ore is carried on a Finlayson wire-rope tramway down to the mill in Eureka Gulch.   This mill is equipped with 1 Blake crusher, 15 stamps of 650 to 700 pounds weight, dropping 55 to 60 times a minute, 3 Bartlett tables, 2 Wilfleys, and 2 canvas tables.   The power is water.   The batteries are provided with one amalgamated copper plate each.   The capacity of the mill is 30 to 35 tons daily.   In 1900 this mill was not in use, and all the ore was sent down on a Finlayson tramway to the new mill at Eureka, originally designed for a concentrating mill and furnished with rolls.   The equipment of this mill is at present one 11 by 15 Blake crusher, 20 stamps, 3 Wilfley tables, and 5 Bartlett tables. Power is supplied by 2 Pelton water wheels.

La Belle Creole.—This claim is on the main Sunnyside lode, which has here turned so as to strike east and west, dipping south 75°.   The croppings are very prominent and have been worked by open cuts and through a small tunnel.   Several pockets of good ore have been taken out near the surface.

The Lake claim is on a prominent lode just north of Lake Emma. This has also produced some good ore and has been worked by an open cut and small shaft.   The strike of the fissure is N. 74° W., and the dip practically vertical.   The continuation of this lode toward the east is lost beneath the talus.   On the west it appears to be connected with the George Washington lode, although the two are not in line.

The George Washington is a strong lode, striking east and west and dipping south 70°.   It also has been superficially worked by an open cut and a shallow shaft.   In 1900 a new prospecting shaft had been sunk south of the vein and a crosscut driven across the latter, here over 100 feet wide.   About 20 feet of very compact rhodonite was cut through about in the middle of the vein, and alongside the rhodonite, on the north, is about 5 feet of ore, carrying in places as much as 9 ounces of gold.   Most of the vein, however, is of low grade. The relation of the rhodonite to the ore is similar to that described in the Sunnyside mine.   The George Washington vein shows no gouge and no regular walls.   The ore resembles that of the Sunnyside in general mineralogical character, but perhaps carries more galena. The lode apparently curves and passes through the saddle into Ross Basin, down which it is easily followed by its remarkably prominent croppings.

Ben Franklin mine.—This is an enormous lode of low-grade ore, in places over 100 feet in width, which crosses Eureka Gulch about half a mile below Lake Emma. Its strike is N. 43° E., and it dips northwest from 45° to 60°. This mine caused considerable excitement in 1883. Two tunnels, 20 and 80 feet in length, had been run and some 300 tons of ore averaging $36 per ton had been taken from the longer tunnel. An inclined shaft was also sunk for a short distance on the lode. The ore consists of sphalerite, galena, chalcopyrite, and pyrite, in a gangue of quartz and rhodonite. Masses of richer ore, carrying tetrahedrite and native silver, are said to have occurred in caves within the lode. A road was built up Eureka Gulch to the mine, but the latter did not fulfill expectations and was abandoned.

PLACER GULCH AND TREASURE MOUNTAIN GROUP OF LODES.

General.—As may be seen by reference to the map, on which only a few of the most prominent viens are shown, the veins of this areal group form a complex network in which certain dominant directions of fissuring are plainly recognizable. There is a series of strong, continuous lodes, represented by the Sunnyside and Scotia veins, having a general course of about N. 40° E. The dip of these is southeasterly at a high angle. The Scotia lode appears to be a very persistent one. There is good reason to suppose it to be continuous with a lode which passes up Cinnamon Creek and over Wood Mountain to the Lake Fork of the Gunnison. A second system of shorter, less conspicuous veins crosses the first series almost at right angles, with a course approximately N. 55° W. These are nearly vertical, but the majority of them dip southwest at high angles. Whether they are older or younger than the dominant northeast veins could not be satisfactorily determined with the exposures at hand, although some of them appear to cut the Scotia lode. A third system of transverse veins, represented by the Golden Fleece, has a course about N. 75° E., with usually steep southerly dip. Besides the three systems already noted, there is a very prominent set exposed at the head of Placer Gulch, which have an average trend of about N. 25° E. Their dip is usually steep and easterly. They resemble in character the great Sunnyside lode, and in most cases are branches of the latter, forming with it a fan-like group somewhat similar to that previously described in Silver Lake Basin.

There are several other lodes in this area which do not strictly belong with any of the systems noted. Several of them have a nearly east-and-west strike and might perhaps be grouped together, or possibly included with veins of the Golden Fleece system.

The country rock of all these lodes is in the main the Silverton series, with a few masses of intrusive rhyolite.

The Sunnyside Extension lode and those just described as its branches resemble very closely those in the Sunnyside Basin. They are for the most part strong, heavy lodes with bold outcrops, to which the abundant rhodonite, by oxidation, generally gives a black color. Underground it is seen that these lodes are as a rule composed of alternating bands or streaks of ore and barren pink rhodonite. The ore bodies may vary from a few inches up to 30 feet in width. They are sometimes continuous for long distances, but are elsewhere plainly lenticular. The miner has sometimes followed a single ore streak for a considerable distance without suspecting that another ore body lay close alongside, separated from him by a partition of rhodonite. The ore streaks frequently show irregular banding, by deposition, and often contain small vugs. The ore minerals are sphalerite, galena, pyrite, chalcopyrite, free gold, sulphobismuthites of lead and silver, and sometimes small amounts of tetrahedrite, molybdenite, native silver, and native copper. The sphalerite varies from dark brown to rosin-yellow in color, and usually about equals the galena in amount, and exceeds the other ore constituents. The light-colored zinc blende often accompanies ore rich in free gold. The gangue is quartz, rhodonite, and locally a little fluorite and calcite. The rhodonite, which forms the "partitions of pink" between the ore bodies, is tough and compact, and varies from light pink to rose-red in color. It is always poor in ore minerals when it occurs in large masses, but small amounts of rhodonite may be embedded in the richest ore.

In spite of the banding on a large scale shown by the veins, they are as a rule simple, thick plates of vein filling, frozen to the walls and containing very few horses of country rock. Of low grade as a whole, they have hitherto been mined for the richer portions, carrying free gold.

Sunnyside Extension mine.—This property joins the Sunnyside on the northeast, and is on the same lode. The main lode dips southeast about 75°. Just northwest of it is a second lode, nearly parallel and bearing to it a relation similar to that which the No Name bears to the Sunnyside. This second lode also dips southeasterly, but in places it is as flat as 40°. It probably joins the Sunnyside vein in depth.

The Sunnyside Extension mine was first extensively worked in 1886, when $3,965 worth of ore was shipped. It was worked steadily until 1892, producing in all $337,687. After a short intermission it was again worked up to about 1897, but the output for this period is not known. The mine is now (1900) being again reopened. The ore resembles in general character and occurrence that of the Sunnyside mine, but is more irregular in its distribution. There are three main levels, all connecting with the surface by adit tunnels. Above level 2 is a stope about 30 feet wide, 200 feet long and 125 feet high, which is said to have been practically all ore. Some very rich ore has been taken from this mine, carload lots running as high as 74 ounces of gold per ton, together with silver and lead. The gold occurred very largely free, and, as in the Sunnyside, was limited to certain pockets or bunches within the mass of the otherwise low-grade vein. Specimens of this ore in the possession of Mr. Rasmus Hanson, of Eureka, showed the gold scattered through masses of friable spongy quartz, and also as implanted crystals on the quartz crystals of drusy cavities or vugs. It also occurs embedded in a dark-gray mineral locally termed "graphite," but which in reality is molybdenite,[1] in a sulphobismuthite of lead and silver not specifically determined, and also in light-yellow zinc blende. A little native silver sometimes occurs with the gold, and some native copper was in the upper workings. The general mineralogy of the vein is identical with that of the Sunnyside.

The Sunnyside Extension (Hanson's) mill, erected in 1889 near the head of Placer Gulch, is equipped with a Blake crusher, 20 750-pound stamps, dropping 9 inches at 64 to the minute, 4 Gilpin County shaking tables, and 2 Woodbury tables. The mill was driven by steam and has a capacity of 47 tons per day. Ore was brought down to the mill on a Huson tramway 2,300 feet in length, with 61 buckets and a daily capacity of 175 tons.

The Mastodon is on the same lode as the Sunnyside Extension mine, and lies just northeast of the latter. This claim has never been extensively developed. It is known to contain some rich silver ore, but this is irregularly distributed through the large low-grade lode. A mass of this ore, consisting chiefly of an argentiferous sulphobismuthite of lead in a fine aggregate of quartz and barite, was assayed by Mr. Hanson and found to carry 400 ounces of silver. The sulphobismuthite occurs as a bright lead-gray mineral, finely disseminated through the quartz and barite, or in minute ragged prismatic crystals. It is not practicable to obtain the mineral in sufficient purity and quantity to warrant quantitative chemical analysis, by which alone the mineral species could be determined.

The Mastodon mill was one of the earliest to be constructed in the district. In 1899 it had been repaired and was working on ore from the Silver Queen mine. Its present equipment is a Blake crusher, 2 sets of rolls, 1 Huntington, 1 double compartment jig, 2 sets of revolving screens, 1 large Bartlett table, 2 "double-deck" Wilfleys, and 4 slime tables. Its capacity was said to be 35 tons.

Silver Queen.—This is a small mine at the head of Placer Gulch and about one-fourth mile east of the Sunnyside Extension. The lode is somewhat irregular, but appears to strike about N. 10° W. Its dip is to the west and unusually flat for this region, being near 40°. Considerable work has been done, but in a most unsystematic manner. The ore seen showed galena, pyrite, chalcopyrite, tetrahedrite, and sphalerite, in a gangue of quartz and rhodonite, with a little rhodochrosite. It is also said to carry some free gold. This mine has produced some good ore, but how much is not known. The silver is said to

---

[1] T. A. Rickard has erroneously referred to this as graphite, supposing that it aided in precipitating the ore. See The Enterprise mine, Rico, Colorado: Trans. Am. Inst. Min. Eng., Vol. XXVI, 1897, pp. 976–977.

have occurred largely in combination with bismuth, probably as
a sulphobismuthite of lead and silver. In 1899 a crosscut was
being driven to cut the extension of the Sound Democrat lode to the
eastward.

Hidden Treasure, Pride of the Rockies, etc.—These claims are
located on lodes lying radially in the angle between the Silver Queen
and the Sunnyside Extension. They have been prospected in a small
way and have produced inconsiderable amounts of ore. According
to Mr. Rasmus Hanson, ore from the Hidden Treasure vein has assayed
300 ounces of silver and 4 ounces of gold per ton. In their general
character these lodes resemble the others of this group already
described. The lodes at the head of Placer Gulch are exceedingly
numerous, and those named or those indicated on the map comprise
merely a few of the more prominent ones.

Sound Democrat.—This also is a small mine which was being developed in 1899. The principal lode strikes N. 20° E. and dips northwest
about 68°. The portion of the lode now being worked is very irregular, the ore occurring in bunches, as if the lode had been broken up
and the parts displaced. Much of the irregularity is apparently due
to the fact that two or more transverse or intersecting lodes occur in
the workings. Owing to the absence of maps and the small scale of
the present workings, it was impossible to make a satisfactory study
of the ore bodies. In its general mineralogical character the lode
resembles the Sunnyside. The ore shows pyrite, sphalerite, galena,
chalcopyrite, and free gold, in a gangue of quartz and rhodonite, and
a little rhodochrosite. A gray metallic mineral occurring in minute
particles in some of the ore is probably one of the sulphobismuthites
of lead. Gold is the most important constituent, and occurs free in
fine particles in the quartz and in the pyrite. Chalcopyrite is not
abundant, and when present is regarded as an indication of the presence of gold. The galena in the best ore occurs in such fine particles
as to make its recognition difficult. Quartz is the principal gangue
mineral, and often shows small vugs. In 1899 the ore of the Sound
Democrat was being packed on burros to the upper Sunnyside mill
in Eureka Gulch and there treated.

Scotia lode.—This is a very prominent lode, which outcrops along
the southeast slope of Treasure Mountain. It is probable that the
same fissure or fissure zone crosses the Animas River and extends up
Cinnamon Gulch and over the southeast shoulder of Wood Mountain.
Its strike on Treasure Mountain is N. 43° E. and its dip about 75° to
the southeast. It is occupied by a large low-grade lode, with a more
or less black outcrop, owing to the oxidation of the rhodonite of which
it is partly composed. Although prospected at many points, it has
not been profitably worked. The Scotia claim has produced some
high-grade ore, but it has come from a small transverse vein, the
Golden Fleece.

Golden Fleece vein.—This is a very small and tight vein which has produced some rich free-gold ore. The strike of the vein is N. 75° E. and its dip about 75° to the south. Its width varies from a fraction of an inch to 6 inches, and it is unaccompanied by gouge. Although so small, the vein is remarkably regular and has been extensively stoped. As it approaches the Scotia lode it breaks up into stringers and grows poor. This splitting up of the vein into stringers as it approaches the Scotia lode is at least an indication that the latter is older.[1] Toward the east it bends southward and joins the Golden Fleece extension, a low-grade lode approximately parallel with the Scotia. The ore of the Golden Fleece is free gold, which occurs in beautiful dendritic forms made up of thin branching plates of gold embedded in a gangue of quartz with some pale-pink rhodochrosite. The quartz near the gold is usually dark colored, and this darker shade is apparently due to the presence of extremely minute particles of pyrite, galena, and sphalerite, and possibly other ore minerals. The gold occurs both in the quartz and in the rhodochrosite. Some beautiful specimens of rich gold ore from this little vein were seen in the possession of Mr. Hanson, of Eureka, who worked the vein for years. The product was all sorted and shipped as crude ore. There was no work in progress in 1899 when I visited the mine, but a resumption of mining was reported in 1900.

San Juan Queen.—This is a prospect in Picayune Gulch just southeast of the Scotia and Golden Fleece mines. The lode strikes N. 83° E. and dips south about 80°. Its average width is from 3 to 4 feet. The ore consists of pyrite, galena, sphalerite, and a little free silver, in a gangue of quartz with some rhodonite. Much of the galena is very finely crystalline, giving a dark-gray tint to the quartz. The prospect has been developing for about seven years, and has shipped about 10 tons of ore. The best-grade ore is said to carry about 5½ ounces of gold and 180 ounces of silver.

Toltec lode.—This is a prominent lode on the ridge east of the upper part of Picayune Gulch. Its course, N. 40° E., is nearly parallel with that of the Sunnyside and the Scotia. It dips northwest at from 50° to 60°. It has not been worked on any extensive scale and is low grade. In the Toltec claim the vein is in places about 60 feet wide, of solid quartz carrying finely disseminated pyrite, with a more heavily mineralized streak carrying galena, sphalerite, and chalcopyrite near the hanging wall. An upper tunnel on this claim has afforded some coarsely crystalline galena resembling more the ores of the Silver Lake Basin than those usually found in this vicinity. The Toltec was idle in 1899, and has never emerged from the condition of a prospect. Farther southwest the vein carries abundant tetrahedrite, probably in bunches.

Mountain Queen mine.—Situated at the head of California Gulch,

[1] See von Cotta: Erzlagerstätten, Freiberg, 1859, p. 115.

under Hurricane Peak, this mine may be conveniently described with those of Placer Gulch. It is one of the oldest workings in the district, and in 1877 shipped 370 tons of ore to Lake City. This ore was carried by pack animals to Rose's Cabin at $3 per ton, and is said to have contained 30 ounces of silver and 64 per cent of lead. The mine was most active from 1878 to 1880, and during one of these years is reported to have shipped $60,000 worth of ore. It was still shipping ore in 1882. It was worked by an open cut and through a vertical shaft. There are at least two large lodes, and the shaft is sunk at or near their junction. The largest strikes N. 80° E. and dips south 75° to 80°. Its croppings are very prominent and in places over 200 feet wide. It is quite likely the same lode as that on which the Custer claim is located, but is not worked in the Mountain Queen. The lode exploited in this mine appears to strike about N. 42° E. and to dip southeast at nearly 85°. It apparently curves somewhat and passes through a saddle east of Hurricane Peak, but the abundant talus and numerous intersecting veins obscure the course of the main fissure. As seen in one of the old levels, 60 feet below the surface, the vein is wide and adherent to its walls. The galena occurs as rather irregular bunches in the quartz near the footwall. At the present time work is carried on by a few leasers in a shallow shaft south of the old works. The ore as sorted for shipping is nearly solid galena. It is associated in the vein with chalcopyrite, pyrite, sphalerite, and sometimes tetrahedrite, in a quartz gangue. No rhodonite or rhodochrosite was seen. Fluorite was observed with some galena ore in a prospect about 100 feet north of the old workings, but may possibly not be on the same vein. The country rock of the Mountain Queen belongs to the Silverton series.

Other lodes.—Several other lodes than those mentioned have been worked to a small extent in Placer Gulch and on California Mountain. The Evening Star, on the west side of Placer Gulch, has shipped some ore, consisting of galena and tetrahedrite with pyrite, sphalerite, and chalcopyrite, in a quartz gangue. The Custer is a large lode which crosses California Mountain just north of the northern peak. Its strike is N. 81° E., and it is supposed to dip south. It is possibly the same lode as that upon which the Mountain Queen mine is located in California Gulch. The ore of this prospect resembles the Ridgway ore, save in the presence of rhodonite. It carries chiefly an argentiferous lead sulphobismuthite, wrongly called "brittle silver" by the miners, and pyrite. The workings are of a most superficial character. The Black Diamond prospect is on a strong lode striking N. 56° E. and dipping 80° NW., on the eastern slope of California Mountain. It is crossed by another lode striking N. 81° E. and dipping 80° NW., which passes through the saddle between the two summits of the mountain. The ore in this lode is chiefly a gray copper, carrying both antimony and arsenic, with lead and a little silver replacing part of the copper.

LODES OF MINERAL POINT.

General.—There is probably no area within the quadrangle where outcrop so numerous and such conspicuous lodes and veins as in that immediately surrounding Mineral Point, at the headwaters of the Animas and Uncompahgre rivers. The general trend of these lodes is roughly northeast and southwest, but they are not all parallel. A radial disposition of the fissures is plainly discernible south of Mineral Point, recalling the similar arrangement at the head of Placer Gulch and in Silver Lake Basin. From the knob south of the deserted town (from which the latter takes its name) the croppings of many of the lodes can be followed with the eye for long distances to the northeast and southwest. Unlike large quartz veins in some other regions, the prominent veins here do not stand up uniformly as topographic ridges, but they are, nevertheless, often easily traceable as long white lines stretching away over the uneven country. The summit of Mineral Point knob is formed by a heavy vein of nearly solid white quartz at least 60 feet wide, which dips about 75° to the southeast. Toward the northeast this vein divides into three. The middle branch is that upon which the San Juan Chief is located, while the Bill Young mine is on the eastern branch. Toward the southwest this vein passes over the eastern spur of Tuttle Mountain. The triangle between Houghton and Tuttle mountains and Mineral Point is occupied by a most complicated network of fissure veins. Only the more prominent ones can be shown on the map, and their representation on a map of this scale and character is necessarily somewhat diagrammatic. Some of them follow curved courses, as is well shown by a large vein which crosses the road just west of Mineral Point town and extends southwest toward Lake Como.

Although so numerous and persistent, the lodes in the vicinity of Mineral Point have been productive of more disappointment than profit. Excluding the Old Lout, which will be described with the Poughkeepsie Gulch mines, there has not been a mine in this vicinity which can be said to have yielded a substantial profit over and above its total cost. The Polar Star, on Engineer Mountain, was formerly extensively worked, and produced some rich silver ore; but the ore shoots apparently had little depth, and the net profit of this mine was probably not large when it was finally compelled to close. In 1899 and 1900 the work in this vicinity was practically limited to prospecting or the extraction of small amounts of ore from shallow workings.

The prevailing country rock of the Mineral Point area is andesite, belonging in the Silverton series. There are also considerable bodies of rhyolite, which are intrusive in the andesite, and some latite.

Red Cloud mine.—This claim was first worked about 1874, and in 1881 is reported to have had about 1,200 feet of drifting and to have taken out some 800 tons of gray copper and galena ore in the progress of development. The present workings consist of three shafts, a lower adit tunnel, and an unknown extent of drifts. The mine never proved successful, and has been idle for ten or fifteen years. No examination of the underground workings was possible in 1899.

The Red Cloud lode is a very strong one. It passes through the saddle just east of Tuttle Mountain, where its individuality is somewhat obscured by a multitude of intersecting veins. It appears, however, to continue along the southeast slope of Tuttle Mountain and to form one member of the remarkable plexus of veins about Lake Como. Toward the northeast it passes through the Ben Butler claim and over the dividing ridge between the Animas River and Henson Creek drainage. At the Red Cloud mine the lode is split into several members, of which the three most prominent are indicated on the map. The mine is located on the western or main fissure.

Inspection of the dumps shows that much of the ore originally filled a brecciated zone in rhyolite. The ore contains galena, often disseminated in minute particles through the quartz, sphalerite, chalcopyrite, and pyrite, in a quartz and rhodonite gangue. Some of the pyrite has the radial structure usually taken as indicative of marcasite. (See p. 77; also Pl. XI C.) Ruby silver (proustite) is also reported. The crude ore is said to have contained about $40 and the concentrates about $15 per ton. The process of ore deposition was attended by much alteration, impregnation, and possibly some replacement of the country rock. Subsequently the ore was brecciated by renewed movement along the fissure and recemented by quartz carrying small amounts of sphalerite and galena.

Ben Butler mine.—This is a prospect on the same lode as the Red Cloud. The lode dips southeast about 75°. It is less evidently brecciated than in the Red Cloud. The country rock is andesite. The ore shows much sphalerite, with galena, pyrite, and ruby silver (proustite), in a quartz gangue. The different ore minerals are not uniformly distributed through the lode, but are more or less concentrated into zones or bands. The best ore is the very finely crystallized galena and ruby silver in a quartz gangue. This galena, minutely disseminated through the quartz, is called "brittle silver" by the miners, but chemical examination shows that this name is erroneously applied. These two minerals are practically confined to two streaks 4 to 8 inches wide. The lode is a large one, and the walls are not exposed in the workings. The latter are 100 feet in depth and opened through an inclined shaft. The shipping ore is said to run about 40 ounces of silver, 40 per cent of lead, and a little gold, which sometimes amounts to 4 ounces per ton. Work on this property was in progress in a small way in 1899 and 1900, and a tunnel was being run to cut the lode at greater depth.

Uncompahgre Chief claim.—This prospect, about a quarter of a mile southwest of the Ben Butler, on an approximately parallel lode, had a shaft 50 feet deep in 1899. The vein is in two main branches,

25 feet apart, and dips southwesterly at about 75°. The ore near the surface consisted of tetrahedrite, presumably argentiferous, and a little galena, in a quartz gangue. The tetrahedrite is erroneously termed "brittle silver" by the prospectors. Another mineral of bright metallic luster and gray color occurs near the surface. It is usually in small, more or less radially grouped prisms in the quartz and is locally termed "star silver." It proves to be a sulphobismuthite of lead and copper, probably argentiferous. The ore in the bottom of the shaft is pyrite in a quartz gangue. A little chalcopyrite and rhodocrosite in small vugs were also observed.

Bill Young mine.—This mine, just east of the dismantled town of Mineral Point, is on the easternmost of the three lodes which join on the summit of Mineral Point knob. It produced some ore about 1876 and was worked to a considerable extent, but it has been idle for many years and few data concerning it could be secured.

San Juan Chief mine.—This mine is on a very prominent lode striking N. 40° E. and dipping southeast at about 75°. It is crossed by a second lode striking about N. 60° E., also dipping steeply southeast. The present developments consist of a shaft 300 feet deep, sunk at the intersection of the two lodes, and a crosscut and drift at the 300-foot level. The work has been unsystematically done and without accurate surveys or maps. No ore has yet been found in these workings, and it is highly probable that the drifting has all been done on the barren intersecting lode and not on the San Juan Chief lode at all.[1] The latter has produced in all nearly $75,000 from open cuts and shallow shafts. The ore from these is very siliceous and consists of quartz and pyrite with a little tetrahedrite. The British Queen claim, which joins the San Juan Chief on the northeast, has produced some gold ore from an open cut, the gold occurring in finely disseminated pyrite in quartz.

In spite of its small output and the fact that no ore body has yet been systematically developed in depth, this mine has been equipped with substantial buildings and expensive machinery. The present mill is the second one that has been erected and has never worked on pay ore from the San Juan Chief. In 1899 it was supplied with two roasting furnaces, a Blake crusher, rolls, 15 stamps, and lixiviation vats. In 1900 the vats had been removed and 3 Cammett tables and a canvas plant substituted. These were running on ore from the Ben Butler. The power is steam.

Polar Star mine.—This property, situated on Engineer Mountain, has been more thoroughly exploited than any other in the vicinity. The average strike of the lode is about N. 25° E., with a dip of about 75° to the southeast. The croppings on the Polar Star claim show

---

[1] The mine was closed in 1900, but I was informed that the main lode had been found by crosscutting on the 300-foot level. The ore, however, ran only $8 or $10 per ton. It is reported to be 30 feet wide.

very little solid quartz, and the ore occurs in a breccia zone in a latitic andesite.   The mine was worked in the late seventies and early eighties, and was idle in 1899.   The workings were then blocked with ice and inaccessible.

From the material on the dump it is evident that the ore, after filling the interstices in the brecciated zone of country rock, was itself shattered by later movement and cemented by white quartz.   The lode in the lower workings was apparently not accompanied by much gouge and did not possess regular walls.   The ore is said to have consisted chiefly of ruby silver and argentite in the upper workings.   It was practically free from lead, although a little galena occurs in the vein and some specimens of galena ore were seen from near the southwest end of the claim, next to the Mammoth.   The ruby silver occurred in pockets in the lode.   In the lower workings the high-grade silver ore is said to have changed abruptly to an ore consisting chiefly of pyrite in quartz, carrying about 12 ounces of silver per ton, with no gold.   A similar change is said to have been experienced in all the mines on Engineer Mountain.

The country rock immediately alongside of and included in the Polar Star lode has been bleached and altered by the ore-bearing solutions and transformed to a mixture of quartz, kaolin, and pyrite. The nature of this alteration is discussed on pages 120–124.

The ore from the Polar Star was hauled in wagons down to the Crooke mill on the Animas, near the mouth of Boulder Gulch, and there treated by the Augustine process.   It seems to have been well adapted to this treatment, a saving of about 95 per cent being reported.

Other mines.—The Mammoth, Syracuse Pride, and Annie Wood are all mines on the southern slope of Engineer Mountain upon which considerable work has been expended but which are now idle.   The Mammoth is on a branch of the Polar Star lode, which here shows croppings of solid quartz nearly 12 feet in width, dipping southeast at 70°.   The lode material as seen on the dump is generally similar to that of the Polar Star.   This mine is reported to have yielded about $30,000 in silver and gold.   The Mint reports give its total product for 1891 and 1892 as $25,655.   According to the reports of the Tenth Census, stephanite (true brittle silver) occurred in the Annie Wood and Mammoth mines, together with freibergite, ruby silver, and sulphide of bismuth.   It is quite possible, however, that this is an error, due to the very loose use of the term "brittle silver" by the miners. The Mammoth and Polar Star are often asserted to be on the San Juan Chief lode.   As may be seen by reference to the map, where the lodes are plotted as accurately as the scale and topography will allow, this assertion is improbable.   They are probably overlapping fissures. The Syracuse Pride shipped ore to Crooke & Co., in Lake City, during 1876, and is credited with a product of $10,000 in 1890, but the mine was never successful.   Some ore was found in the upper workings and a shaft sunk to cut the lode at depth, but no ore was found, and there is some doubt whether the ore-bearing vein was ever cut in the lower workings.

The Black Silver, on the crest of Engineer Mountain, east of the summit, was developed by a shaft and several tunnels run in on the north slope of the mountain. The lode strikes N. 38° E., and dips northwest at about 75°. It is 18 inches to 2 feet in width, and in general character closely resembles the Polar Star. Some very rich silver ore was formerly extracted near the surface.

The Mohawk, on the north side of Engineer Mountain, was not visited. It was not working in 1899 and 1900.

The Sunset, on the southwest slope of Engineer Mountain, is a small mine which has shipped a few tons of ore. The lode strikes N. 25° W., and dips southwest at an angle of 75° to 80°. It is 4 to 5 feet wide and consists of gouge, crushed decomposed rock, and bunches of quartz and ore. The country rock is a rather coarsely porphyritic altered andesite, like that of the Polar Star. The ore consists chiefly of tetrahedrite and pyrite in a gangue of quartz, with some kaolin.

The Wewissa is a small mine or prospect on the western spur of Engineer Mountain, and is credited with a production of about $13,000 in 1890. It was shipping a galena and tetrahedrite ore, carrying silver, in 1899, the ore being taken by burros and wagons to Ouray. The lode is irregular and is not extensively worked.

The Early Bird, on the north slope of Houghton Mountain, is on a small and irregular vein striking N. 12° E. and dipping 75° to the east. The ore is tetrahedrite and a little galena in a gangue of quartz, which is in part a replacement of the andesitic country rock. The tetrahedrite carries as much as 300 ounces of silver, and gold up to 7 ounces per ton is occasionally met with in small amounts. The prospect has produced 11 carloads of ore ranging in value from $19 to $100 a ton. East of the Early Bird are two prominent lodes striking about northeast, one of which, the Denver, has produced a little ore.

The London, formerly worked through a shaft near the road, north of Houghton Mountain, is now abandoned. It is reported to have produced some ore carrying 3 ounces of gold per ton.

LODES OF HENSON CREEK.

General.—In the northeast corner of the quadrangle are several old locations, some of which have produced ore in considerable quantity, but at the present time active work is limited to prospecting on a small scale.

Frank Hough mine.—This mine, situated in American Flats, under the northern slope of Engineer Mountain, began work early in 1882 on large and irregular bodies of argentiferous and auriferous copper ore. During 1882 the mine was worked through a shaft 250 feet deep, with drifts at 50, 80, 140, and 200 feet below the surface. The lower drift was run nearly 60 feet in solid ore and 100 tons were extracted in running it. During this year some 60 tons of ore were shipped, carrying from 50 to 60 ounces of silver and 25 to 27 per cent copper. The gold varied from a trace up to about one-fourth ounce per ton. A road was built to the mine and a great production was looked for. When the shaft had reached a depth of 350 feet the mine buildings were destroyed by fire and the mine has since lain idle. In all about 2,000 tons of high-grade copper and silver ore were shipped.[1] From the accounts of Mr. J. S. Hough, who operated the mine, the ore body seems to have been an irregular impregnation or replacement deposit alongside a fissure. The country rock is a fine-grained, soft, tuffaceous sandstone, probably of andesitic origin. Near the ore body it is hardened and impregnated with pyrite. This material, when thrown on the dump, weathers to a soft, crumbling mass, stained with the carbonates of copper. The strike of the fissure is N. 23° E., and the dip, according to Mr. Hough, about 45° to the west.

Judging from specimens collected on the dump, the ore of the Frank Hough consists chiefly of chalcocite, often intimately intergrown with quartz, associated with a little chalcopyrite. There is possibly some tetrahedrite present with the chalcocite.

Palmetto mine.—This mine, about 1 mile northeast of Engineer Mountain, is on the continuation of the Polar Star lode, which here has a strike of N. 23° E. and a dip of 75° to the southeast. It was originally owned by J. S. Hough,' who stated that the ore in the old workings, near the surface, resembled that in the Polar Star, and consisted chiefly of argentite and ruby silver. It sometimes ran as high as $500 per ton. Several carloads of this rich ore were shipped about 1878. He subsequently sold the mine, and a vertical three-compartment shaft was sunk east of the vein to a depth of 400 or 500 feet, with the intention of cutting the ore body at a depth of a few hundred feet. The ore from this shaft contained much iron pyrite and was low grade. In 1882 a 15-stamp mill was erected on Henson Creek, a mile below the mine, and the ore treated by chlorination. The mine during this year was working five levels, and produced $14,840. It shortly afterwards closed. In the Mint report for 1891 it is credited with a product of $9,566, which, if true, indicates a temporary resumption of activity in that year. The total output is not known. The mine has been idle several years and the mill is dismantled.

Specimens of the ore and vein matter obtained on the dump resemble some of those from the Polar Star. It has been brecciated at least once and healed with comparatively barren white quartz. Small vugs lined with quartz crystals are common. The country rock is andesite, and is evidently considerably impregnated with pyrite and chalcopyrite near the vein.

Dolly Varden lode.—This prospect, situated in Dolly Varden Gulch,

---

[1] Oral communication from Mr. Jas. W. Abbott, a pioneer mineral surveyor of Lake City.

was producing copper and silver ore as early as 1878.   The lode, .
revealed on the surface, is a sheeted zone in an amygdaloidal and
ite.   The fissures comprising this zone contain very little quartz, a
the only ore seen on the surface consisted of the carbonates of cop!
disseminated through the amygdaloid.   Good ore, consisting of tet:
hedrite with some "brittle silver," is said to have been taken from
shaft sunk on the lode.   A crosscut tunnel 600 to 700 feet in lengt
was run lower down in the gulch, but did not reach paying ore.   T' .
dump of this tunnel shows some chalcopyrite deposited in a brecc!
zone in the andesite.   The average strike of the Dolly Varden lode
nearly north and south, with an easterly dip.

Other prospects.—On Copper Mountain the John J. Crooke, B'
Horn, and other prospects have been worked intermittently for ma.
years and have produced some ore, chiefly argentiferous bornite at.
chalcocite (stromeyerite), with a little galena and some native silv:
in a gangue often containing considerable calcite.   The strike of thes·
lodes is usually a little east of north, with steep easterly dips.   The
country rock is andesite.

Other prospects have been opened in a small way in the gulches and
basins south of Henson Creek.   These are chiefly galena ores, carrying some silver and a little gold, in andesite country rock.

LODES OF POUGHKEEPSIE GULCH.

General.—There is no reason, other than that of convenience, for
separating these veins as a whole from those about Mineral Point, of
which they are in many cases the direct continuation.   Formerly the
scene of busy activity, the gulch, with the exception of two or three
prospectors, was deserted when visited in 1899.   The first location was
the Poughkeepsie claim, staked by R. J. McNutt in 1874.   In 1879 mining was actively proceeding in several claims.   A road was built from
Cement Creek to the head of the gulch, and lixiviation works we ·
erected at Gladstone to treat the ore.   Since 1890, however, the only
work done has been desultory prospecting on a small scale.

In general the longer fissures have trends varying from N. 30° E.
to N. 50° E., with southeasterly dips.   But they are sometimes curved
and irregular, and near Lake Como occurs the most remarkable group
of large, closely spaced lodes to be found in the quadrangle. ˙ An
observer standing on the north side of the lake may count at least five
prominent veins running down the steep slopes to the south and west
and passing beneath the deep blue water of the tarn.   At the point
now occupied by the latter the fissures were deflected sharply eastward, and from the eastern side of the lake four prominent lodes and
many smaller ones converge at a point on the ridge separating the
head of Poughkeepsie from California Gulch.   In addition to these,
numerous other lodes come in through the saddle connecting with
Ross Basin and converge toward the same point on the ridge.   The

result is a very interesting complex, which is represented as faithfully as possible on the map. But in addition to the lodes shown, the rocks in this vicinity are traversed by innumerable small veins and fissures which can not be indicated on so small a scale. Those shown are chiefly strong massive veins of white quartz, sometimes including considerable country rock, usually carrying a little rhodonite or rhodochrosite, and not showing much evidence of mineral wealth. They frequently branch and are linked together by smaller fissures.

In the saddle between Lake Como and Ross Basin the remarkably thorough manner in which the rock between the main fissures has been sheeted is beautifully shown. This minor sheeting is generally parallel in strike with the larger lodes. At the particular point where the structure was best shown, these sheeting planes dip westerly at about 60°. As many as 50 of them may frequently be counted within a width of a foot. There has been some infiltration of quartz along these minute planes of fracture, which causes them to stand out as little parallel ribs on weathered surfaces. The appearance in the western part of the saddle is as if a series of finely laminated cherty sediments had been tilted up at a high angle. The local strike of this sheeting is about N. 35° E., but it is sometimes disturbed by later cross fractures. No regularity or rhythm could be detected in the spacing of these minor sheeting planes. The country rock is apparently andesite, but too altered and decomposed for certain identification. It may be rhyolitic.

Faulting was not detected along any of the fissures about Lake Como. Study of them was, however, necessarily confined to the surface.

Old Lout and Forest mines.—These properties, situated a mile and a quarter due west of Mineral Point, being closely connected, will be described together. The Old Lout was located about 1876. Shortly afterwards it sold for $10,000 and became, in the years of its activity, the most important mine in the gulch. It has produced between $300,000 and $400,000 worth of rich silver ore, containing bismuth. According to the Mint reports, about $200,000 of this amount was produced in 1884, in which year the mine was the largest producer in the county. It employed about 30 men, raising some 5 tons of ore per day. In 1886 it was sold to a London company for $200,000, and in 1887 it is credited with a product of $86,654. The new owners ran a tunnel some 1,800 feet in length, from the bottom of the gulch, but never found the rich ore which had rendered the upper workings profitable, and about the year 1890 operations were abandoned. The tunnel house and the machinery were subsequently destroyed by fire.

The upper workings were exploited by means of a shaft of unknown depth, which in 1899 was no longer safely accessible. The lower tunnel, being partly caved and filled with a strongly flowing stream of water, can not now be entered. The ore which came from it indicates a strong and well-mineralized lode carrying galena, sphalerite, chalcopyrite, pyrite, and a little marcasite, rather equally distributed in a gangue of quartz with a little barite. This ore is, presumably, low grade. Fragments on the dump show that a part of the lode consists of brecciated and altered country rock (rhyolite flow-breccia) cemented by quartz. The richest ore of the Old Lout is said to have been bismuthiferous, in which case it is probable that an argentiferous sulphobismuthite of lead occurred in this, as in other lodes in Poughkeepsie Gulch.

The Forest mine was worked through a vertical shaft only a few hundred feet east of the Old Lout shaft. Through this it was possible in 1899 to gain access to an upper level 200 feet below the surface. The Forest lode here strikes N. 70° E. and dips south at about 70°. It consists of stringers of ore, without much quartz, traversing altered country rock. These stringers are irregular, sometimes expanding to 2 or 3 feet of nearly solid ore with included fragments of country rock, and again contracting to narrow veinlets. The ore is adherent to the country rock and the lode is without regular walls.

The ore, as seen in this drift and in the ore house, consists of tetrahedrite, pyrite, galena, sphalerite, and chalcopyrite, in a gangue of quartz and barite. The Forest mine was never a profitable producer.

The exact course of the Old Lout lode could not be determined on the ground, but it is more nearly northeast and southwest than the Forest. The two lodes come together southwest of the shafts, and at this junction occurred the rich bismuthiferous ore which made the Old Lout profitable. The lodes are said to have dipped toward each other, the Forest at a lower angle than the Old Lout. If this is true, the dip observed at the 200-foot level of the Forest is not the general dip of the lode.

The country rock of the Old Lout and Forest is a rhyolite, but so much altered in proximity to the ore bodies as to be scarcely recognizable. A specimen taken 7 feet from the Forest lode on the 200-foot level is completely recrystallized and consists of a finely crystalline aggregate of quartz and sericite impregnated with minute crystals of pyrite and galena. As seen in thin section under the microscope, the sericite (and possibly some kaolin) occurs in irregular areas without definite boundaries, and also scattered through the rock amid the quartz grains.

Maid of the Mist mine.—This mine, which has apparently been idle for many years, lies a few hundred yards northeast of the Old Lout and Forest shafts. The workings comprise an inclined shaft of unknown depth and two tunnels, one, a crosscut, cutting the lode about 150 feet below the surface. The strike of the vein is about N. 25° E., and it dips southeast about 65°. It has been drifted upon and stoped from several levels. In the northeast portion of the workings it is a solid vein of quartz 18 inches to 2 feet in width, containing fragments of altered rhyolitic country rock and ore which is apparently of low grade. Toward the southwest the vein has been disturbed by post-mineral movement and is accompanied by crushed wall rock and soft gouge. The course of the vein would carry it into the Old Lout shaft, and it is probable that both mines are on the same fissure.

Alabama mine.—This mine, situated on the east side of Poughkeepsie Gulch, about half a mile south of the Old Lout tunnel, is commonly supposed to be on the same lode as the Old Lout, the strike of the Alabama lode being N. 35° E. The dip is southeast at 70°, although locally variable. In general character the lode resembles that seen in the accessible portion of the Forest workings, but it is more irregular and shows less ore. It is essentially an irregular stringer lead, passing sometimes into a breccia of country rock cemented by quartz and sometimes dwindling to a few narrow and tight stringers. The country rock is rhyolite flow-breccia. Considerable prospecting has been done by tunnels, drifts, and winzes, but apparently no ore was stoped. The mine has been long idle. On the other side of the gulch some work has been carried on upon the same vein, which here appears to dip about 60° to the southeast and shows similar ore.

The ore of the Alabama consists of galena, chalcopyrite, and pyrite, in a gangue consisting of quartz and barite.

Poughkeepsie mine.—This property is situated half a mile up the gulch from the Alabama and on the same side. It was located in 1874 and has produced in all about $12,000. The strike of the vein is N. 62° E., and it is practically vertical. It is a strong and apparently fairly simple vein, frozen to the walls of rhyolitic flow-breccia. It produced a silver ore carrying bismuth. The pyrite in the vein is also said to have carried good values. The tetrahedrite, on the contrary, was poor ore. Argentite is reported [1] as occurring in the ore as mined in 1875. The mine was worked through two tunnels. It has been idle since 1891 and inspection was confined to the surface.

The ore in the bins showed chalcopyrite, pyrite, and tetrahedrite in a quartz gangue.

Amador mine.—This mine, formerly worked through two tunnels, lies on the west side of the gulch. It was deserted in 1899 and little could be ascertained of its history or product. The lower tunnel was originally run in on a small vein striking N. 30° E., but it soon turns and cuts a vein of which the course is N. 80° W. This vein dips north about 60° and is a solid and fairly regular but not heavily mineralized mass of quartz. The lode has been stoped and a shaft sunk from the drift, evidently some years ago. The upper tunnel, 100 feet above, is run directly on the main lode, which is here nearly vertical, and shows from 3 to 5 feet of solid white quartz in the croppings.

[1] Raymond: Mineral Resources west of the Rocky Mountains, 1875, pp. 383-386.

The country rock, which is rhyolitic flow-breccia, of rather andesitic aspect, is sheeted and traversed by veinlets parallel to and for a distance of several feet from the large lode. A small mill, equipped with crusher and rolls and run by water power, was formerly operated at the mouth of the upper tunnel.

Saxon mine.—This lode is apparently the same as that first followed in the lower tunnel of the Amador. Its strike is N. 30° E., with a steep southeast dip. The mine was working in 1879, and in 1890 is credited with a production of over $30,000. The workings were not accessible in 1899. The ore on the dump showed galena, sphalerite, tetrahedrite, chalcopyrite, and pyrite, in a quartz gangue.

Alaska mine.—Situated about a third of a mile a little north of west from Lake Como, this mine is best known from the occurrence in it of an argentiferous galenobismuthite named alaskaite by Koenig.[1] The mine was working in 1879, and in 1881, when Professor Koenig collected his specimens, had two adits. The alaskaite occurs with tetrahedrite and chalcopyrite in a gangue of quartz and barite, the latter being abundant in the richer ore. Zinc blende and galena were not seen by Koenig, but they are abundant on the dump and in portions of the lode now accessible. Kaolin also was noted, intimately associated with the ore minerals. In 1887 the Mint reports credit the mine with a production of $35,658, but in 1888 it was not producing. I was informed on reliable authority that about $90,000 had been produced in all, chiefly from one large pocket. The Alaska lode strikes about N. 75° E., with a dip of about 75° to the southeastward. At least three veins intersect at or near the old shaft through which the mine was worked; and another lode, the May Belle, which has been superficially prospected, lies just west of and nearly parallel with the Alaska lode. All these lodes carry some rhodonite and rhodochrosite crystallized with the quartz. The country rock is andesite.

In 1900 the mine was being reopened and it was possible to see something of the old workings, although still partly filled with ice. Like many other lodes in this region, the Alaska is less prominent under ground than the croppings would indicate. It contains much altered country rock traversed by stringers of quartz. It is in the soft altered country rock, near the quartz, that the richest ore, alaskaite, occurs as bunches of varying size. Pieces of this ore may carry as much as 3,000 ounces of silver per ton. The solid vein quartz, on the other hand, is low grade, carrying pyrite, galena, sphalerite, and some tetrahedrite, associated with rhodonite, barite, and rhodochrosite.

No ore was extracted from the main workings in 1900, but good bismuthiferous ore, in small bunches, was being taken out from a lower tunnel on the Acapulco claim, which joins the Alaska claim on the east and is under the same ownership.

---

[1] Proc. Am. Philos. Soc., Vol. XIX, pp. 472–477; also Vol. XXII, p. 211.

Bonanza mine.—In 1879 this mine was opened on the north side of Lake Como, through a tunnel 250 feet in length, and is said to have produced an ore carrying gray copper (tetrahedrite) and barite. In 1881 it was reported that 700 tons of tetrahedrite and galena ore had been extracted and piled on the dump, and in 1883 some 700 feet of development work had been accomplished. The lode was also exploited by a small shaft near the edge of Lake Como. No great amount was produced and the property has been idle since 1890. The ore seen on the dump in 1899 showed galena, sphalerite, pyrite, chalcopyrite, and tetrahedrite, in a gangue of quartz, barite, and a little rhodonite. The tetrahedrite appeared to be closely associated with the barite. The ore shipped is reported to have carried from 40 to 45 ounces of silver. There are several other idle prospects about the lake, but they present no features meriting special description. A project was at one time set on foot to drain the lake by a tunnel, in order to secure the rich "float" which it was supposed would be found in its bottom. Fortunately for the projectors and for the lovers of the picturesque, this scheme fell through.

Other mines and prospects.—The Red Rogers claim is on a very prominent lode just southwest of Lake Como and crossed by the trail from the lake to Cement Creek. This property, which was well known in the early eighties, has been idle for some years. It was developed by surface cuts and one or more small tunnels, and probably never produced much paying ore. The course of the lode is N. 43° E., with nearly vertical dip. It is the most westerly of the numerous large lodes that converge from the south toward the lake. It does not appear to reach the latter, however, as it curves strongly eastward just north of the Red Rogers cabin (which is on the croppings), and joins the Seven Thirty lode. The Seven Thirty lode is one of the strongest and most persistent lodes of this region. Its course is about N. 85° E. The dip is approximately vertical. Toward the west the lode passes over the ridge into Grey Copper Gulch.

LODES OF THE UNCOMPAHGRE CANYON.

General.—The Uncomphagre River, below the mouth of **Poughkeepsie** Gulch, has cut the greater part of its picturesque gorge deep into the Algonkian schists. The San Juan andesitic breccias rest upon the uneven eroded surface of these schists, slates, and quartzites, and tower in lofty cliffs above the river. These cliffs are traversed by fractures and veins, which are particularly conspicuous when they cut the San Juan breccias. The cliff faces of the latter are sometimes directly due to the falling away of great masses of rock along one of these fracture planes. Some of the fractures carry ore, and have been worked with varying success.

Michael Breen mine.—This mine, more familiarly known as the Micky Breen, is on a small, nearly east-and-west lode (N. 85° W.)

dipping about 85° to the north.   The mine has been quite extensively worked through five tunnels.   Most of the pay ore has been extracted from the four upper tunnels.   The lower one was a more ambitious venture, opened in connection with a mill in an attempt to work the property on a large scale.   Although the lode has produced some good ore, it is doubtful whether it is sufficiently abundant, or accompanied by enough lower-grade ore, to justify an extensive plant. The Mint reports credit the mine with a total product of $105,751 during the years 1890–1892.   In 1899 the second and nearly third levels above the mill adit were being worked by Italian leasers at a very small profit.   The country rock is the andesitic breccia of the San Juan formation.

.The lode filling is usually less than a foot in width and frozen to the walls.   It is accompanied, however, by considerable parallel sheeting and veining of the adjacent country rock, and the lode itself contains thin sheet-like horses.   The fissures are clean cut, regular, and quartz filled, carrying greater or less amounts of ore.   Where stoped in the upper of the two tunnels worked in 1899, the lode consists of numer. ous small stringers, frozen to the country rock and separated by thin sheets of the latter.   It is an excellent example of a deposit filling empty spaces in a sheeted zone of much regularity.   The pay streak is usually only a few inches wide and lies on the south wall of the vein.   It carries tetrahedrite, galena, sphalerite, pyrite, and chalcopyrite.   The tetrahedrite is not rich in silver.   Small vugs and comb structure are common, often with crystals of tetrahedrite and rhodochrosite projecting with the quartz into the open spaces.   The country rock is impregnated with pyrite and chalcopyrite, but is sharply distinguishable from the vein filling.   Fluorite was noted on the dump of the uppermost tunnel.

The Micky Breen lode is crossed by a conspicuous system of approximately parallel fissures striking about N. 20° W. and dipping east about 80° to 85°.   These fissures usually carry small quartz veins, often showing comb structure.   Two narrow and conspicuous zones of sheeting, accompanied by considerable veining, one striking N. 20° W. and the other N. 7° W., are indicated on the map.   These converge toward the north and form the lode of the Silver Link mine, and possibly also the Silver Queen lode, in Bear Canyon, on the northern border of the quadrangle.   The relation of these fissures to the east-and-west Micky Breen lode could not be determined.

The Micky Breen mill, on the Uncompahgre River, is equipped with a Blake crusher, 2 sets of rolls, 1 5-foot Huntington mill, trommels, and 7 4-foot Frue vanners.   It was run by water power and steam. The mill had been idle some time when visited in 1899.

Happy Jack mine.—This property is situated just west of the Micky Breen, and apparently on the same vein.   It at one time shipped a little ore and then erected a mill, but the ore gave out.   Mine and mill were abandoned, and in 1899 the workings were not accessible. The mill is equipped with roasting furnace, crusher, two sets of rolls, trommels, three sets of mixing troughs, and vats. The ore was apparently treated by the Augustin or a similar process.

Silver Link mine.—This mine is situated on the east side of the canyon, about a mile north of the Micky Breen lode. Its principal adit is a crosscut tunnel 1,700 feet in length. The course of the lode is about N. 12° W. It is nearly vertical. This tunnel was run in the early eighties. According to the Mint reports, the total product of the mine for the years 1888, 1890, and 1891 was nearly $50,000. The mine had been idle since 1873, but was reopened in 1899.

The tunnel mouth is in Algonkian quartzite and schist, but the crosscut soon passes into the overlying San Juan breccia. The latter lies upon a clean and very uneven erosion surface, showing local variations in relief to the extent of 100 feet or more. Sometimes a breccia of schist and quartzite fragments mingled with andesitic material rests directly upon the Algonkian and grades upward, without break, into the typical andesitic San Juan breccia. There is no trace of movement or mineralization along this contact plane.

The lode is a sheeted zone, 20 feet or more in width. The ore is chiefly bornite and tetrahedrite, carrying silver and chalcopyrite, and occurs as bunches along some one of the fissures. It is found in quartz, which is largely silicified San Juan breccia. It contains from 20 to 30 per cent of copper and sometimes as much as 300 ounces of silver per ton. The best ore is said to occur in solid quartz. There has been considerable post-mineral movement along the lode, resulting in soft gouge, and where the disturbance is pronounced it is regarded as unfavorable for ore. The ore usually occurs near the west (called the foot) wall, but it is far from continuous. The country rock on both sides of the lode is traversed by numerous parallel fissures, along which there has frequently been some late movement, as attested by the presence of wet clay gouges.

An old shaft was sunk on the lode for 60 feet below the tunnel level, but apparently did not reach the bottom of the andesitic breccia. Thus it is impossible to gain from this occurrence any light on the important question of what changes, if any, take place in an ore-bearing lode as it passes from the Tertiary breccia to the Algonkian schist. In 1900 this old shaft was being reopened, and it is possible it may be carried down into the schists.

The country rock in the vicinity of the Silver Link mine is intersected by numerous fissures, with a general trend of a few degrees west of north, and therefore approximately parallel with the Silver Link vein. Many of these fissures can be seen extending southward across the Uncompahgre to Mount Abrams. Their average strike, as seen on the surface, appeared to be about N. 5° W. Their dip is 80° to 85° to the east. At the mine these fissures, which do not, as a rule, carry conspicuous lodes, have cut the San Juan breccia into huge, nearly vertical slabs (Pl. VII), which, becoming undermined at their bases, occasionally fall in chaotic ruins at the base of the cliff. A second and less prominent set of fissures, usually carrying quartz stringers, has a general strike of about N. 60° W. The relative ages of the two sets of fissures could not be determined.

Silver Queen mine.—This is a small mine nearly a mile north of the Silver Link, in the canyon of Bear Creek, and on the northern boundary of the quadrangle. The lode, which is possibly a continuation of the Silver Link, here strikes N. 3° E., and is practically vertical. It is a breccia zone, 10 or 12 feet wide, in San Juan breccia, with a clay gouge on the east wall. The pay streak, from 3 to 6 feet wide, lies next the clay seam on the east wall. It is worked through a tunnel 300 feet in length and a 40-foot shaft with short drifts below the tunnel level.

As far as known, the ore occurs in three distinct bodies or shoots, pitching south. In such shoots there is practically no quartz, the ore, sometimes with a little barite, filling the spaces between the fragments of country rock, which is altered to a soft, putty-like material, locally called "talc," but in reality impure kaolin. The northern ore shoot, nearest the tunnel mouth, consists of low-grade copper ore, chiefly chalcopyrite, with subordinate galena and tetrahedrite. It is separated from the next ore shoot on the south by 12 or 15 feet of solid quartz, carrying too little ore to work. This second ore body was rich in a bismuthiferous silver ore (argentiferous sulphobismuthite of lead), associated with galena, at the tunnel level, but changed into chalcopyrite in stoping upward. The third ore body, also separated from the second by a mass of relatively barren quartz, is almost wholly galena ore. South of this ore body the main lode is cut, but not noticeably faulted, by a younger, nearly east-and-west vein, carrying galena, chalcopyrite, and zinc blende and assaying fairly well in gold. South of this intersecting vein the main lode is poor and grows smaller. It is to be noted that not sufficient work has been done to establish the fact that the three small ore bodies cut in the tunnel are really distinct and characterized throughout by different kinds of ore. The fact that the second one changed into chalcopyrite when followed upward would rather tend to negative this idea of the characteristic individuality of these ore bodies.

Post-mineral movement is indicated by the clay gouge along the east wall of the vein and by similar sheets of gouge within the vein, generally nearly parallel to the walls. Contrary to the experience on the Silver Link, the best ore is said to occur in soft clayey ground.

The mine was working a small force in 1899, the ore being packed to the toll road on burros and thence hauled to Ouray in wagons. In 1900 it was idle. The total output has been between $15,000 and $20,000.

LODES OF CANYON CREEK.

General.—Under the above head it is proposed to describe the lodes of Imogene, Richmond, and Silver basins, Potosi Peak, and other portions of the Canyon Creek drainage included within the bounds of the Silverton quadrangle. The only really important mine at present being worked in this area is the Camp Bird. The Revenue tunnel, it is true, has its entrance at Porter, in the Silverton quadrangle, but the lodes worked through it are within the Telluride quadrangle, and have been described by Purington.[1]

Canyon Creek, from near Ouray to the mouth of Richmond Gulch, is crossed by a system of 'fractures conspicuously visible in the lofty cliffs of the San Juan breccia, and resembling those already noted on the Uncompahgre River. These fractures appear to have an average strike of about N. 35° W., and dip steeply northeastward. They divide the San Juan formation and the underlying sediments into huge, nearly vertical slabs. As far as observed there is no visible vertical displacement along these fissures. They sometimes carry quartz veins, which have been prospected in a small way.

Camp Bird mine.—The Camp Bird lode crosses the head of Imogene Basin, with a strike of N. 80° W. and a dip of 65° to 85° to the south; the average is probably about 70°. Toward the west the lode has been traced on the surface into Marshall Basin, and identified with great probability with the Pandora lode, in the Telluride quadrangle. Toward the east the lode is less easily followed on the surface.

Considerable work had already been done in Imogene Basin prior to 1880. But no high-grade ore had been found, and the locality had fallen into disrepute. The Una and Gertrude claims, on the Camp Bird lode, had been worked for silver-lead ore, but in 1896 had been abandoned, like the other mines in the basin. The presence of high-grade gold ore was first discovered by Thomas F. Walsh, who for many years had industriously leased and prospected a great number of claims scattered over the entire quadrangle. The rich ore was first struck in the Camp Bird, then an undeveloped claim. It was found by Walsh on the dump of the Una and Gertrude claims, which were purchased by him in 1896 for $10,000. At the same time he secured control of practically all the lodes in the basin, much of the necessary surveying being accomplished in the deep snow of winter. The following year development was actively pushed. The substantial tramway connecting the mine with the mill at the mouth of the gulch, 2 miles in length, was completed in forty-nine days. In 1899 the vein had been opened by a crosscut tunnel and about 4,000 feet of drifting, with stopes extending up to the old workings. In 1900 this tunnel had already been superseded as the main adit

---

[1] Preliminary report on the mining industries of the Telluride quadrangle, Colorado: Eighteenth Ann. Rept. U. S. Geol. Survey, Pt. III, 1898, pp. 836-838.

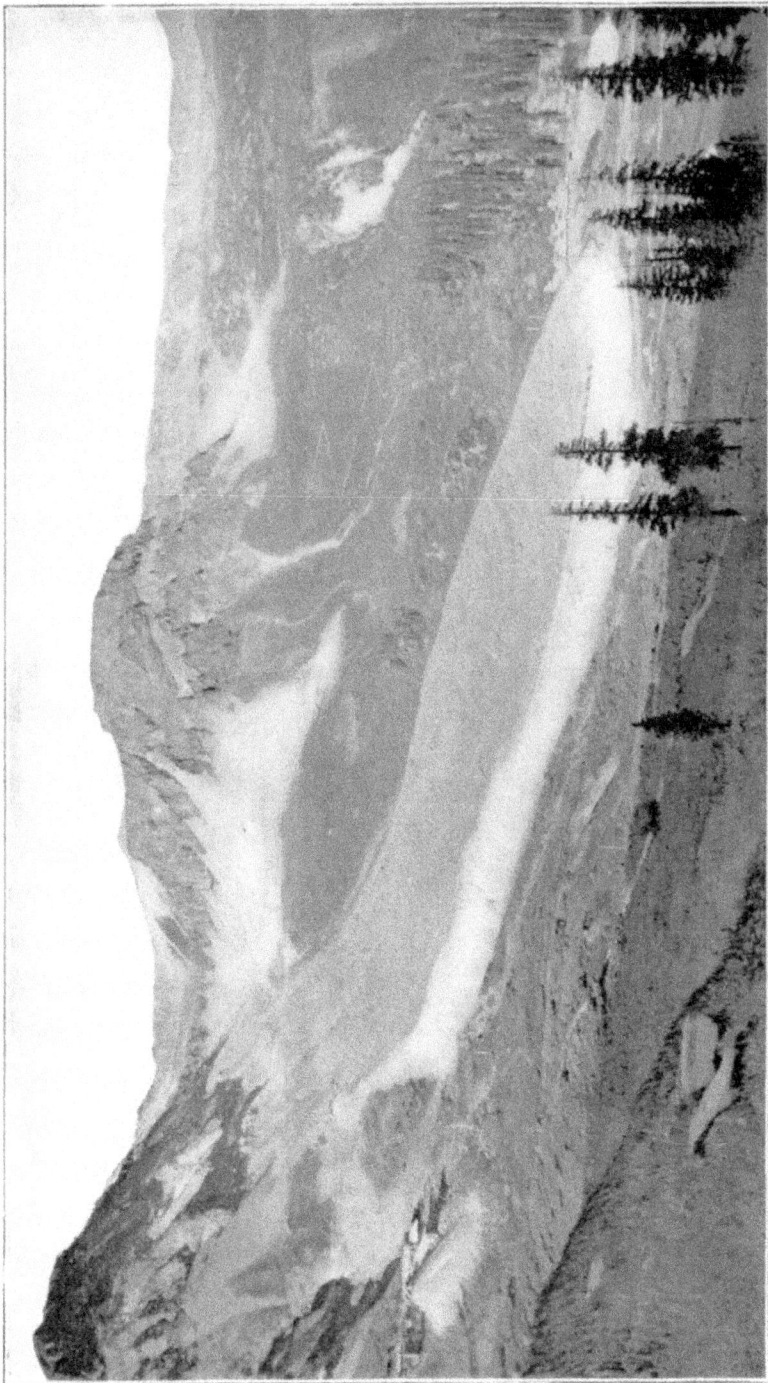

VIEW OF THE SURFACE WORKINGS OF THE CAMP BIRD MINE.

The older workings are shown on the left; the present main adit is on the right and runs under the large lobe of talus in the middle of the picture.   Photograph by Arthur C. Spencer.

by a much longer tunnel 300 feet below it. This is a crosscut about 2,200 feet in length, driven S. 15° W. It is furnished with a single well-laid track, and in October already connected with over 1,200 feet of drift, which length was being rapidly augmented. A view of the surface features of the Camp Bird mine and of the large lobe of "slide-rock" beneath which the main tunnel is run is shown in Pl. XIV.

The country rock of the Camp Bird is the andesitic breccia of the San Juan formation. As this rock extends to the mouth of the gulch and some way down Canyon Creek, it is probably considerably over 2,000 feet in thickness below the croppings of the lode. It thus insures practical uniformity of volcanic country rock to a rather unusual depth, and is so far favorable to regular and persistent ore bodies.

Nearly all the ore produced under the present ownership has come from the level 300 feet above the new adit. The lode where intersected by this crosscut adit is a mere fracture, the thickness of a knife blade, accompanied by a few small quartz stringers on the footwall side. The fissure soon widens to the east and west, and the lode has an average width of 4 or 5 feet. When typically developed it is a sheeted zone in the rather fine San Juan breccia, made up of narrow stringers of gangue and ore alternating with sheets of country rock, and having fairly well-defined walls. The hanging (south) wall is the more regular, and a very thin seam of clay gouge shows that there has been slight movement along it since the lode was formed. Near the foot wall there is uniformly a streak of lead-silver ore, carrying galena and sphalerite. This is sometimes contiguous with and apparently an integral part of the main lode. At other times the two are separated by a thin sheet of country rock. The vein is unusually adherent to a somewhat irregular foot wall, which is often traversed by irregular quartz stringers, or may be brecciated and cemented by quartz. Although the lode is a sheeted zone, the country rock, outside the recognized limits of the deposit, is not notably fissured. The foot-wall streak of galena ore has been partly brecciated and recemented by stringers of white quartz. This later quartz is not to be distinguished from the quartz of the auriferous portion of the lode and may be of the same period of deposition. The silver-lead foot-wall streak was the ore mined twenty years ago, before the high-grade ore lying alongside it was discovered. The richest ore occurs near the hanging wall, often directly against it, and the richest portion of the pay streak is indicated by dark, narrow, undulating lines of ore minerals. These lines define the course of small stringers, usually an inch or less in diameter, which are of later origin than the mass of the vein. The latter is a pale greenish-white mottled aggregate of quartz, fluorite, sericite, and calcite, thickly sprinkled with small crystals of pyrite, with some sphalerite, galena, and possibly other ore minerals in minute specks. It is probably in part altered country rock. The rich stringers, although very irregular in detail, are remarkably persistent along the hanging-wall side of the main lode. They possess, moreover, a regular internal structure. Next their walls there is usually a thin layer of fluorite. Then follows a narrow band, from one-sixteenth to one-eighth of an inch in width, of rosin-yellow sphalerite and galena, with a few minute specks of chalcopyrite. The medial portion of the stringer is occupied by white quartz carrying a little sericite and calcite. The gold occurs chiefly free in this white quartz, especially close to the dark ore lines near the walls of the stringer. It is firmly embedded in the quartz and is often distinctly visible, although not occurring in large masses.

When the ore is crushed and panned, the concentrates contain abundant free gold in irregular hackly particles, averaging about 0.25 mm. in diameter. Microscopic examination shows that the vast majority of these gold particles have the characteristics of original gold that has been embedded in vein quartz. A very few small, rod-like particles suggest a possible derivation from a telluride of gold. Chemical tests reveal the presence of a very small amount of some telluride in the ore, but it has not been detected by the eye and is not present in sufficient amount to determine its mineral species. Microscopic and chemical investigation shows that the value of the ore is chiefly in free gold and not in telluride of gold.

The quartz in portions of the Camp Bird lode is often very glassy in appearance and shows little trace of mineralization, even when carrying some $20 per ton. Ore below $8 or $10 is not removed. The average value of all the material removed from drifts and stopes on the lode is probably not far from $40 per ton. The average on good stopes, however, may be three or four times this. According to a statement by Mr. J. W. Benson, manager of the Camp Bird mines, the material removed in merely running the main drift, waste included, ran in places about $1,600 per linear foot. Much of the vein quartz on the upper level (300 feet above main adit) is shattered and stained with black oxide of manganese. The shattering of the brittle quartz does not appear, however, to be associated with any very pronounced post-mineral movement, nor has it any known effect on the tenor of the ore. A little rhodochrosite and considerable calcite occur in the lode, the latter often filling vugs or comb structure in the quartz. A soft, white, putty-like substance, which proves to be chiefly kaolin, is also common in the quartzose portions of the lode.

Not enough development has yet been reached to determine the shape of the pay shoots. The ore body on the west drift of the main level is about 1,200 feet in length. Other pay shoots occur along the vein, separated by pinches. They are said to be increasing in length as they are followed down. The ore bodies east of the crosscut are generally wider, but of lower grade than the large one west of it.

About 1,700 feet west of the crosscut tunnel on the main level the lode suddenly contracts and splits into a few small stringers. Beyond this point the country rock on the south side of the drift is lighter colored than elsewhere and is full of irregular quartz stringers, some of which appear to strike southerly into the hanging wall. I was unable in the field to discover any contact or fault in the exposures in the drift, but microscopic study shows that the country rock at this point is a devitrified and altered rhyolite. It is probably an intrusive mass or dike cutting the San Juan breccia. It is exposed in the south wall of the drift for 15 or 20 feet, but has never been found on the north side. West of this contraction no ore was found on this level for nearly 900 feet. A barren stringer was followed for a portion of this distance, which finally led into the main lode, which was found to be offset into the hanging wall. The rhyolitic dike is older than the Camp Bird lode, but its presence has probably caused a local deflection of the generally straight simple fissure. The newly found portion of the lode has been drifted on for 600 feet, and is being followed back toward the rhyolite and point of original deflection. This irregularity is apparently not due to faulting, but is an example of linked-vein structure. The western deflected portion of the lode has an average width of 9 or 10 feet of good ore. The dip is slightly flatter than in the eastern portion of the workings. In the present bottom level of the Camp Bird the galena ore, commonly found on the foot wall above, is less abundant and more bunchy. The dark, curving lines described in connection with the rich ore are less regular and continuous.

Power drills are used almost exclusively for drifting, crosscutting, and stoping. The air compressors are run by electricity from Ames, in the Telluride quadrangle. The ore from the present main level is carried on a temporary Huson wire-rope tram to the tram house near the mouth of the new adit. Thence it goes over the permanent Bleichert tramway to the mill, situated at the mouth of the gulch, on Canyon Creek. This tramway is about 2 miles in length and has a transfer station at the turn in the gulch, in Richmond Basin. In 1899 about 45 buckets were running over it, carrying about 700 pounds of ore each. One man can dispatch 350 bucket loads in a day, of which the mill, then running 30 stamps, required from 325 to 330. All ore goes direct to the mill without sorting. The mill, as equipped in 1899, contained 2 Blake crushers, 40 800-pound stamps, dropping 90 to the minute, and 20 Frue vanners. In 1900 20 more stamps were being added. The pulp from the batteries passes through a 40-mesh screen over silvered copper plates 17 feet long. Thence it passes through classifiers onto the vanners. The latter are so arranged that all the slimes must pass over two vanners before escaping as tailings. The ore passing through the mill is often worth $200 per ton. The average, when visited in 1899, was said to be nearly $150. About 75 to 80 per cent of the total gold is amalgamated on the plates.

The concentrates rarely exceed $4 or $5 per ton of ore. The tailings from the main mill average from $3 to $5 per ton. They were formerly treated in a small auxiliary mill fitted with Wilfley tables. A cyanide plant was, however, nearly completed in 1900, and will treat the slimes directly from the mill. Power for the main mill is supplied by a Pelton water wheel or a 75-horsepower engine. The full capacity with 40 stamps is about 130 tons.

Hidden Treasure mine.—This mine, just north of the Camp Bird, was opened in 1875, but, like other properties in Imogene Basin, has lain idle for many years. It is now owned by Mr. Walsh, and will probably be again worked. It has at present about 2,000 feet of drift, and has produced some ore from its stopes. It is opened by a crosscut tunnel about 150 feet in length. The lode strikes about N. 50° W. and dips northeast at approximately 60°. It is a regular lode, in part occupying a sheeted zone and in part a solid vein of quartz and ore. The country rock is San Juan breccia. There has been some later movement along the plane of the vein, as shown by thin clay gouges. The ore formerly extracted is said to have consisted chiefly of galena and tetrahedrite, and to have contained 100 ounces of silver and 42 per cent of lead. The present workings show a low-grade milling ore consisting of galena, abundant sphalerite, and some chalcopyrite and pyrite.

Hancock mine.—This property lies just east of the present Camp Bird workings, and is also owned by Walsh. The lode strikes N. 55° W., and is very similar in general character to the Hidden Treasure. It joins the Camp Bird lode on the southeast. It has been idle many years, and is not now worked. In 1884 it was said to have furnished in all about 80 tons of ore, but has never been a large producer.

Yellow Rose mine.—This is another old prospect, situated in Richmond Basin, upon which considerable work has been done. In 1881 a "strike" of ore, consisting chiefly of galena, with chalcopyrite, pyrite, tetrahedrite and sphalerite, argentite, and some native silver, was reported on level 2, 470 feet from the entrance to the tunnel. But the mine has never produced much. The vein strikes N. 47° W. and dips NE. 65°. The ore, as seen in 1899, is low grade and very similar in general mineralogical character to that of the Hidden Treasure.

U. S. Depository mine.—This mine, with the Caribou, Grand Trunk, and other claims, is regarded as being probably on the same vein as the Yellow Rose. Considerable work was done on this mine years ago, and in 1887 it was credited with an output of nearly $18,000, but it never was thoroughly successful and has been idle several years. A tunnel some 700 feet in length was run in from the small mill, now dismantled, on the road up Richmond Basin. The vein, as opened at this level, is in part a fairly regular sheeted zone in San Juan breccia and in part a less regular stringer lead. Such ore as the mine produced came from upper workings, 400 or 500 feet higher up the slope, and was carried down to the mill on a light and primitive wire-rope tramway. The ore is a low-grade galena-sphalerite ore, resembling in general character that of the Yellow Rose. Some freibergite has also been reported.[1]

Wheel of Fortune mine.—This lode occupies a strong and persistent fissure, with nearly northeast-and-southwest strike, crossing Canyon Creek just east of the Revenue tunnel. It thus crosses such lodes as the Yellow Rose and Hidden Treasure nearly at right angles, and is the most prominent member of several approximately parallel lodes which may be seen traversing the southern slope of Potosi Peak.

The Wheel of Fortune was located about 1877, and has produced some very rich ore. According to Mr. Krisher, now foreman of the Revenue Tunnel, the mine, prior to 1880, shipped some ore to Black Hawk containing 20 ounces of gold and 800 ounces of silver. In 1879 the mine had about 500 feet of drift and shaft, and was said to have produced $20,000. Most of the work was done in the early eighties. During the latter six months of 1882 the mine shipped about 62 tons of ore, averaging 176 ounces of silver and $8 in gold per ton.[2] The ore appears to have been irregular or pockety in its occurrence, and is said to have contained freibergite and stephanite,[1] with galena and ruby silver. In 1899 the mine was being reopened after several years of idleness.

The local strike of the Wheel of Fortune lode is N. 30° E., and the dip 65° to the northwest. The country rock is San Juan breccia. The lode is chiefly a stringer lead, but it grades on the one hand into a breccia zone and on the other into a fairly regular sheeted zone. It has been faulted along a nearly vertical transverse fissure, the southern portion of the lode being thrown about 15 feet to the west. There has also been movement along the foot wall, as shown by clay seams or gouges. The croppings of the lode extend strongly up into Silver Basin, and the country rock for a distance of 100 feet or more southeast of the lode shows pronounced sheeting unaccompanied by perceptible faulting. There was no opportunity in 1899 for studying the ore in place. In 1900 the mine was reported to be shipping ore, but was not visited.

Bimetallist mine.—This property, situated on Potosi Peak, and probably on the same lode as the Wheel of Fortune, has produced from $40,000 to $50,000 from a single rich pocket near the surface. The ore of this pocket ran usually about 10 ounces of silver and 1 ounce of gold. Chemical examination of a specimen of this ore shows it to be a silver-copper sulphantimonic arsenite (tetrahedrite?), in which the gold is probably combined with the silver. Small portions, however, were very much richer. The workings were idle in 1899 and were not visited.

---

[1] Tenth Census, Vol. XIII, 1885, p. 84.   [2] Report of the Director of the Mint, 1882.

LODES OF SAVAGE BASIN.

General.—The upper portion of Savage Basin, although within the Silverton quadrangle, was included by Purington[1] in his study of the lodes of the Telluride area. In 1899 two mines, the Tomboy and Japan, were working in the upper basin, while in 1900 some work was also in progress on the Argentina and the Columbia. The basin contains several large lodes and many smaller ones, all possessing a common northwesterly strike.

Tomboy mine.—The character of the Tomboy lode was so well described by Purington[2] in his report, written in 1897, that his account will be quoted almost entire, and supplemented by the observations made in 1899 and 1900. He says:

It is only within the past four years that the property known as the Tomboy has made any considerable product; especially in the last two years the output of the mine has surprisingly increased. The workings now on the vein comprise the length of two claims, very nearly, namely, the Belmont and the Tomboy, and several more claims have been located on the vein in its southern extension. The history of this mine is somewhat remarkable, considering its present large output.

The original location on the vein was the Belmont claim, made in 1880; and in 1886 the Tomboy claim, to the south of the Belmont, was located. Since the original discovery of the lode, and up to the year 1892, no considerable product was obtained; in fact, the vein was generally regarded as of little value and interests in the claim were sold and resold at very low figures. The Tomboy Gold Mining Company, which bought the mine in 1894, had it profitably working by April, 1895. In that year the product was nearly $600,000, and in 1896 the product was nearly $800,000. Recently the mine has again changed hands, and it is stated that the present monthly output averages about $90,000.

The main adit is at 12,130 feet elevation, very near the head of Savage Basin. The greater portion of the work has been done on one level, consisting of a drift over 3,000 feet in length, with stopes extending up to a height of 200 feet above the drift. As indicated on the map, a second adit, 381 feet vertically below the upper one, is being run to cut the vein. At the time of my visit this was 1,128 feet in length. Its starting point is directly above the site of the Tomboy mill. All the portions of the vein worked during 1896 were in the upper augite-andesite, although the contact of this with the top of the San Juan breccias lies not far below. The lower adit is certainly in the breccias, but the exact level at which the contact occurs it is impossible at present to state. A specimen collected in a winze sunk 100 feet from the main level, although it contains fragments, is of the normal color characteristic of the augite-andesite, and is altogether so decomposed that its microscopic determination is impossible. It was stated that a dike accompanies the Tomboy vein for a part of its length, as now developed. Although I can not say that the observations in the mine confirmed this statement, nor that the microscopic examination revealed any rock differing from that of the country in general, yet in view of the limited examination which was made, it can not be denied that a dike may exist.[3]

The peculiarities of the Tomboy vein, considered with reference to the fissure systems which it follows, have been described and illustrated.[4] So constantly does the vein shift from one to another set of fissures that it is impossible to get,

[1] Preliminary report on the mining industries of the Telluride quadrangle, Colorado: Eighteenth Ann. Rept. U. S. Geol. Survey, Pt. III. 1898.
[2] Op. cit., pp. 838-841.
[3] No dike was seen in my own examination. F. L. R.
[4] See Purington, op. cit., p. 778.

at any one point in the workings, a compass reading which represents its average course. This is, however, as may be measured from the surveys plotted on the map, N. 41° W., while the dip averages 77° to the southwest. The limits of the narrowly cleaved zones, which have been filled by ore, are fairly well defined, and it is only occasionally that stringers accompany the main lode, separated by any considerable amount of wall rock. Such stringers have been found, however, at as great a distance as 10 feet from the main vein. The width of the main lode is variable, but is generally from 4 to 7 feet. It sometimes attains the width of 12 feet, and in one place pinches to less than 1 foot. This vein presents more the character of a solid filling between two walls than most of those seen in the district. The linked-vein type is very well shown in some places, but is subordinate. Angular fragments of the country rock are frequent, and are especially well shown in the faces of the stopes and drifts on account of the whiteness of the surrounding quartz. Figure 73 shows such a fragment included in the quartz. As before stated, the angular character of these fragments is one of the strongest pieces of evidence that the ore is the filling of open spaces, unaccompanied by molecular replacement of the country rock.

The vein minerals are remarkably few in number, and, with the exception of the quartz, do not appear to bear much relation to the values, which are almost entirely in gold. The quartz is by far the largest constituent, is white, and may be described as coarsely saccharoidal, with the spheroidal arrangement of crystals noted above. It is remarkable that much of the quartz which is commonly referred to as "bony," and is ordinarily barren of values, is, in the Tomboy vein, rich in gold, and often contains free gold visible to the unaided eye. Accompanying the quartz there is much blackish, soft material, which has been determined as peroxide of manganese. Although no manganese carbonate has been found in this vein, its general occurrence in neighboring veins makes it probable that it was formerly present, and the oxide is in part the result of its decomposition. The vein, on the whole, presents an unusually decomposed appearance, as does the wall rock in contact with it. Much white, clay-like material, usually of a sticky consistency, from its saturation with water, is of universal occurrence in the vein. This is, indeed, present in such quantities as to give considerable difficulty in the milling of the ore. It is usually known as "talc." Doubtless much of it is merely material residual from the trituration of the country rock along the fissured zones. Some is, however, of more definite composition, and chemical tests prove it to consist largely of finely divided sericite—one of the potash micas. Kaolinite has been shown to occur with it, its presence being probably due to surface decomposition of portions of the wall rock included in the lode. The greater or less amount of this material in the vein does not appear to bear any relation to the value of the ore.

Calcite, and probably siderite, occur. The occurrence of fluorite has already been described at some length. Iron pyrite is present in small amount, and the gold values do not appear to be associated with it to any extent. More often than otherwise it occurs associated with galena and zinc blende in a narrow streak, which crosses from one wall to the other in the vein, and is fairly continuous through the workings. Small quantities of sulphurets are, however, disseminated throughout the vein. The gold is more than three-fourths free, its total value being about $20 to the ton. The concentrates run $40 to $45. The gold averages 0.735 fine. The gold has not been found in the iron pyrite except in small amount. Much of it is extremely finely divided, and hitherto difficulty has been experienced in saving it at all.

Since Purington's observations were made, the lower tunnel has been extended through the main lode to what is known as the "North vein," or Iron vein, some 300 feet beyond. The main lode was cut through and considerable drifting done on the low-grade North lode before the fact was suspected that the main lode had been overshot. These two veins show a very interesting relationship, which will be further described later on.   This main adit connects, through stopes and raises, with the drifts of the "300 level," 140 feet below the "200 level." A shaft has also been sunk for 300 feet below the 600 level (main adit level), and drifts run in both directions on the main lode at the 700 and 800 levels, while drifting was about to begin also on the 900 level in 1900.

Purington states that the "vein presents more the character of a solid filling between two walls than most of those seen in the district." To one coming to the study of this lode from the Silverton region, to the east and southeast, its striking feature is the regular and narrowly spaced zone of sheeting and the fact that the lode is composed of a series of more or less regular plates of quartz separated by sheets of country rock.   In fact, this structure, combined with that characteristic of the linked-vein type, are the two most prominent structural features of the Tomboy lode.   The peculiar intersections, at small angles, of various fissured zones referred to by Purington[1] appears to be a relatively subordinate feature, and was recognized by me at only one place.   There is one curve in the main vein which is found on all the levels.   At this curve on the 600 level a few small stringers were noted running out into the walls.   However, in any but ideal exposures it is manifestly a most difficult matter to distinguish the intersections described by Purington from what I have preferred to regard as an example of linked-vein structure.

The dip of the lode varies from 50° to 80° to the southwest—usually about 70°. It is commonly accompanied by a wet seam of clay gouge along the hanging wall.   Along the foot wall is a stringer, or vein, somewhat more solid than the auriferous portion of the lode, carrying galena and zinc blende and "frozen" to the wall.   This foot-wall streak is sometimes in contact with the higher-grade gold-bearing quartz, sometimes separated by a plate of country rock.   Its solidity and fresh appearance is often in great contrast to the rusty auriferous ore alongside.   It has been brecciated in places, and the ore solidly healed with white quartz.   The foot wall alongside it has been rather irregularly fractured and veined.   I was unable to ascertain whether this foot-wall streak was distinctly different in age from the auriferous portion of the lode.

The crushed and decomposed aspect of the gold-bearing part of the lode, often accompanied by black oxide of manganese, kaolinite, sericite, and copper stains, has been referred to by Purington.   That these portions of it have been subjected to some movement subsequent to the original deposition of the quartz is shown by the presence of clay seams traversing the ore.   Subsequent fractures filled with wet

---

[1] Op. cit., p. 778, fig. 66.

clay gouge and crushed quartz frequently enter the lode obliquely from the southwest. Some of the richest bunches of ore said to have been found were where they intersect the main lode. The best ore usually occurs in the steeper portions of the vein. The pay shoots are stated to lie flat, one above another, and to extend diagonally across the lode from northwest to southeast.

The richest ore is invariably found in the shattered and stained parts of the lode, generally alongside a soft gouge of crushed and decomposed andesite. The gold usually occurs free and very finely disseminated. When visible, it is embedded solidly in the unpromising-looking vitreous quartz, although Mr. Roscoe Wheeler, the assayer at the mine, informed me that he had found one piece of gold in a fracture plane in the quartz. In the more recent workings, below the 300 level, the vein is, as a whole, less shattered and stained. Even at the 800 level, however, portions of the lode exhibit the character above described, accompanied by clay seams, and such portions are always rich. On this level the vein is sometimes 10 to 12 feet wide, with 2 to 3 feet of galena ore on the foot wall, the latter ore being more abundant here than in the upper levels.

The description thus far has been of the main Tomboy lode. There remain to be described the Iron lode and its relation to the Tomboy. The two lodes are nearly parallel in strike, the Iron vein having a slightly more westerly course than the Tomboy, which is about N. 40° W. But the Tomboy lode dips to the southwest, whereas the Iron vein dips northeast. As a consequence of this diversity of dip the two veins intersect, so that the Iron vein, which above the 300 level was encountered on the northwest side of the Tomboy, on the lower levels is found on the northeast side. Near the main adit this intersection takes place at about the 300-foot level, but the line is not horizontal. The slight difference in strike of the two lodes gives their intersection a pitch to the northwest of about 6° from the horizontal. The intersection in vertical plane is shown in the accompanying cross sections (figs. 14 and 15), drawn from notes of surveys made by Mr. John Herron. The relation of these lodes is such that at each

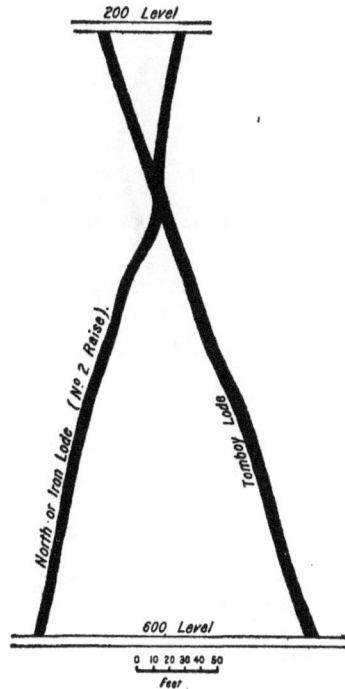

FIG. 14.—Cross section of the Tomboy and Iron lodes through No. 2 raise, between the 600-foot and 200-foot levels. Platted from notes of surveys by Mr. John Herron.

level, if carried far enough, there should be found a place where the two lodes cross in a horizontal plane. The confluent portions, however, must lie much farther northwest, theoretically about 950 feet for each additional 100 feet in depth. The workings as yet are not sufficiently extensive to determine whether this feature of their intersection actually holds true on all levels. The intersection of the lodes led to a curious error when the present main adit tunnel (600 level) was run. In all the old adits above the 300 level the relatively barren Iron vein had been cut before reaching the Tomboy lode, the two being 220 feet apart on the 200 level. In the 600 adit, however, the

Tomboy was first cut, and as it was not productive at this point, it was supposed by those then in charge of the exploitation to be the Iron vein reversed in dip. Consequently the adit tunnel was continued 185 feet farther to the Iron vein, which was then worked on the supposition that it was the Tomboy. It is generally barren, but on this level shows an ore body about 15 feet wide and over 100 feet in length. It is, however, a galena ore, carrying about 40 ounces of silver, and totally different from the Tomboy ore.

The exact nature of the intersection of the two lodes is not determinable from present exposures. In raising from the 600 level to the 200 level on the Iron vein for the purpose of connecting with the upper drifts on the Tomboy lode, it was found that these raises came

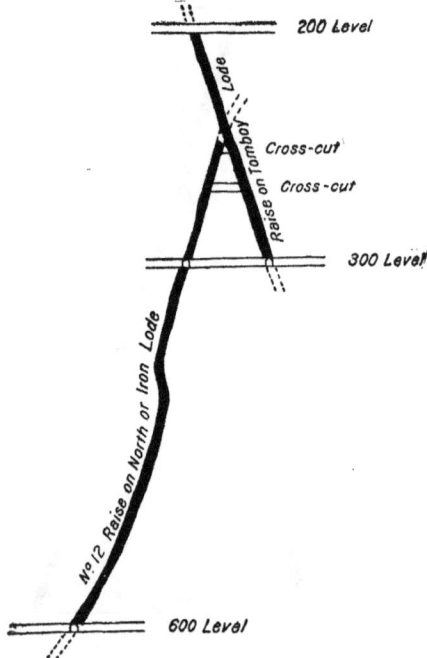

FIG. 15.—Cross section of the Tomboy and Iron lodes through No. 12 raise on the Iron lode and a neighboring raise (projected) on the Tomboy lode. Platted from notes of surveys by Mr. John Herron.

out as much as 50 feet within the hanging wall of the latter lode and connection had to be made by crosscutting (figs. 14 and 15). It was noted that each of these raises passed at some point through a great mass of somewhat shattered quartz, usually about 12 feet wide, and with no regular walls. This was probably the intersection of the lodes, as indicated by subsequent surveys. One of the best pay shoots in the mine occurred at a point which was afterwards discovered to be on or near this line of the intersection. On the 300 level, northwest drift, the gradual convergence of the Tomboy and Iron lodes has been traced by crosscutting at intervals from the former to the latter. At the point where the intersection probably takes place is a confused mass of quartz, altered country rock, and ore over 12 feet wide, with irregular walls. From this ore body two veins again diverge, which may represent the northwest continuations of the Tomboy and Iron lodes. They are here small and have not yet been much drifted upon. It is at present impossible to determine the relative ages of the Tomboy and Iron lodes. It is not unlikely that their ages are substantially the same.

The contact between the San Juan breccia and the overlying massive andesite belonging to the Silverton series occurs in the Tomboy at about the 300 level, but the two rocks are not very readily distinguished under ground. It does not appear that the change in country rock has effected any noticeable change in the character or quantity of the ore.

In the general sheeted character of the fissuring, in the vitreous nature of most of the auriferous quartz and in its shattered and stained appearance, in the presence of a low-grade galena streak near the foot wall, and in the almost exclusively gold output, the Tomboy lode shows much similarity to the Camp Bird. About nineteen-twentieths of the Tomboy product is gold.

The Tomboy mill is equipped with 2 Union and 1 Blake crusher, 2 sets of rolls, 8 Huntington mills, and 24 Frue vanners. The pulp from the Huntingtons is passed over silvered amalgamating plates, which retain over 75 per cent of the gold. The Huntingtons are the equivalent of about 60 stamps, and the capacity of the mill is about 200 tons daily. The power is electricity transmitted from Ames, 12 miles distant, to a 150-horsepower motor. A Corliss engine of 120 horsepower is held in reserve.

The water supply of the Tomboy is drawn from a small lake on the eastern side of the divide, in Ouray County. It is pumped to the mine by electric power from Ames.

The Tomboy has produced from $4,000,000 to $5,000,000.

Japan mine.[1]—The Japan vein is nearly parallel with the Tomboy and lies to the southwest. It is recognized as crossing the main Tomboy adit tunnel at about 900 feet from its mouth, and therefore about 835 feet southwest of the Tomboy lode. It has been drifted on for a short distance, but in the Tomboy ground is too small to work. The workings of the Japan mine lie about one-fourth mile northwest of the Tomboy adit, and the two mines are not connected. Not much work had been done on the Japan prior to 1894.

The course of the Japan lode is about N. 47° W. The general dip is to the southwest at from 70° to 80°, although on level No. 1 the exceptionally low dip of 50° was noted.

---

[1] Since this description was written, readjustment of the primary triangulation of the United States Coast and Geodetic Survey has resulted in placing the Japan and Columbia mines outside of the Silverton quadrangle. The descriptions, however, are retained as originally written.

A crosscut tunnel of about 250 feet, running a little west of north, gives access to the vein on the No. 1 level. Below this level, at intervals of 125 feet, are levels 2 and 3, worked through a shaft from No. 1 level. The lode as a rule is an unusually regular plate of ore, having an average width of about 18 inches and generally frozen to the walls. Less frequently it is a regular sheeted zone, 3 or 4 feet wide, the stringers of ore and quartz showing comb and vug structure. Even where the lode is a simple solid vein a foot in width, the walls, especially the foot wall, may show conspicuous sheeting for some distance from the ore. The hanging wall is usually smooth and regular. The country rock of the portions of the vein now worked is andesitic breccia of the San Juan series.

The ore consists of galena and sphalerite, with a little free gold, argentite, and wire silver, in a gangue of quartz, rhodochrosite, and a little fluorite. A banded structure is often very noticeable. A typical section of the vein as seen in one of the stopes is shown in fig. 5 (p. 69). In this case the stringers of barren-looking quartz next the walls appeared to be of later age than the banded ore between them.

The average value of the Japan ore is about $25 per ton. Nearly one-half of this is in gold, the other being equally divided between silver and lead. All the ore, except a little streak of rich lead ore, which often occurs in veins, is milled.

About 150 feet northeast of the Japan vein is a second parallel lode—the Morning vein—with the same general dip as the Japan. Where cut by the Mikado crosscut from the No. 1 level of the Japan, this vein is only 5 inches wide and carries galena ore, frozen to the walls. Elsewhere it is wider and often carries much coarsely crystalline pale-green and white fluorite. This lode is not worked.

About 750 feet northeast of the Morning vein the Mikado crosscut passes through a third parallel vein, here not much more than a fissure carrying some gouge. In all, the Mikado crosscut has been driven nearly 1,900 feet to the northeast in the hope of cutting the continuation of the Tomboy lode. Several small fissures, generally parallel with the Japan lode, have been cut, but none of them carry ore in commercial quantities. Apparently the Tomboy lode does not continue so far to the northwest, or else is diminished in size and changed in character so as to be unrecognizable.

The Japan and Morning veins are both faulted by the Flora vein, with a course of N. 65° E. and a dip of 55° or 60° to the southeast. This is a barren quartz vein about 2 feet wide filling a fault fissure which has thrown the southeastern portion of the Japan and Morning lodes about 16 feet to the southwest. This same vein probably faults the Tomboy lode also, in ground not yet explored, and may be connected with the failure of the Tomboy to appear in the Mikado crosscut. The position of the Flora vein is well marked on the surface, as its outcrop determines the presence of a little ravine.

About 150 feet southwest of the Japan lode lies the Whale vein, striking N. 40° W., and therefore a member of the same generally parallel system of veins to which belong the Tomboy, Iron, Japan, and Morning lodes, with others yet to be described. The dip is southwest at 75°. The Whale is a small vein, seldom over 6 inches wide, and consists of solid white quartz carrying some free gold. It has been worked near the surface and has produced some rich gold ore. It is cut in the Climax tunnel, an unused upper level lying east of the Flora vein. The Climax crosscut also continues into the Japan lode (here called the Climax) and on to the Morning vein, 170 feet beyond. The andesitic country rock between these veins shows pronounced parallel sheeting.

The Japan mill is equipped with 1 Gates crusher, 2 sets 12 by 20 Davis rolls, 6 3-compartment Harz jigs, 1 5-foot Huntington mill, 2 6-foot Wilfley concentrators, and 2 4-foot Triumph concentrators.

Power is furnished by a 50-horsepower Westinghouse motor supplied with electricity from Ames. The capacity of the mill is about 60 tons. The total product of the Japan has been about $600,000.

Argentina mine.—This is located on a very prominent lode cropping near the bed of Savage Creek, just southwest of the Tomboy and Japan mines. The strike of the lode is about N. 35° W., and the dip 75° to 80° to the southwest. The croppings, often stained black by oxide of manganese, may be traced over the ridge to the southeast into Ingram Basin and beyond the edge of the quadrangle to the northwest into Marshall Basin. It is one of the prominent group of strong, continuous, generally parallel lodes of this region, to which belong the Smuggler, Cimarron, Tomboy, Japan, and countless smaller veins. As early as 1882 and 1883 several shafts were sunk on the Argentina claim and some rich free-gold ore was taken out near the surface, but for many years the mine has been idle. In 1900, however, the Tomboy Company took hold of the property and drove a tunnel on the vein. In August a length of 800 feet had been attained, and the whole width of the lode was being milled.

The lode nowhere shows less than 5 feet of ore, and in places is 12 and 15 feet wide. The country rock is the San Juan andesitic breccia, with some massive andesite. There is no gouge and no regular walls. Stringers from the main vein-filling traverse the hanging wall irregularly, and the best ore is usually found in this mixture of stringers and country rock.

The ore shows galena, sphalerite, chalcopyrite, pyrite, and sometimes free gold, in a gangue of quartz and rhodonite. It is variable in tenor, but as a whole carries about $16, chiefly in gold, with less important values in silver and lead.

The present work is considered merely as prospecting, and the ore removed in drifting has been treated in the Tomboy mill.[1]

A tramway has since been erected, connecting the Argentina with the Tomboy mill.

Columbia mine.—This property is also on a prominent lode which lies southwest of the Argentina, and which is the southeastern continuation of the Cimarron vein in the Telluride quadrangle. The Columbia mine is at present not extensively or continuously worked. The vein as seen underground is much like the Japan, but less regular. Its strike is about N. 30° W., with a variable dip of 60° to 75° to the southwest. The ore is sometimes 2 feet in width, but more often about 10 inches. In places it is solid, consisting of quartz, galena, sphalerite, and pyrite, with little or no gouge; but often it is crushed and pinched by post-mineral movement along the fissure. As in the Japan, parallel sheeting of the walls of San Juan breccia is a noticeable accompaniment of the vein. The stringers of quartz which often fill the fissures in the sheeted country rock are practically barren.

All the ore is concentrated. The best of it carries about 1 ounce of gold, 70 ounces of silver, and 30 per cent of lead.

North Chicago mine.—This property, which has been idle for some time, lies in Savage Basin near its head, and is perhaps on the Tomboy lode. It was worked through a shaft.

ORE DEPOSITS OF THE RED MOUNTAIN REGION.

General.—Under this head will be described the various ore deposits occurring about Ironton Park and Red Mountain. They include ordinary fissure lodes and replacement deposits of more than one kind. But the bodies of ore which gave this particular region its fame are of a type less commonly met with in other districts. The ore occurs in so-called "chimneys"—i. e., nearly vertical bodies of ore roughly circular or elliptical in plan. Unfortunately, owing to the idleness of these mines, satisfactory study of the ore bodies is not at this time possible. In a few cases access was had to the old upper workings, but for information in regard to the deeper portions of the deposits recourse must be had to a few published descriptions, the annual reports made by superintendents to the directors of companies, and to information gathered from those who were familiar with the deeper mines before they shut down. The latter is sometimes conflicting, but by judiciously weighing its various sources and by making allowances for different points of view an endeavor has been made to supply as nearly as possible the unavoidable lack of observation at first hand.

The country rock in which the Guston, Yankee Girl, and neighboring mines lie is a part of the Silverton formation. It is an andesitic breccia, usually somewhat decomposed near the surface, and apparently originally of rather homogeneous character. With it are associated some flows of massive andesite. Locally the rock has been so altered by thermal action as to obliterate its original features. It is changed to a white or yellowish siliceous mass. in which the breccia structure is usually obscured or lost. Quartz is abundantly visible in it as fine veinlets and irregular bunches, and the mass generally contains much finely disseminated pyrite. The nature of this altera- tion has already been fully described on pages 124–131.

The Silverton breccia is cut by numerous dikes, sills, and irregular intrusive masses of andesite. Intrusive plug-like masses of mon- zonite-porphyry, carrying conspicuous phenocrysts of orthoclase and quartz in a fine gray groundmass, are also common. Such a mass is exposed at the roadside between the Yankee Girl and Guston mines.

The general character and distribution of the typical Red Mountain ore deposits have already been discussed (pp. 103–104), and it is only necessary in this place to proceed directly to the detailed descriptions of the various mines.

Yankee Girl mine.—The Yankee Girl ore body was discovered in the autumn of 1881 by John Robinson.[1] In 1882 it was being opened by two shafts, each about 50 feet deep. At that depth the ore is said to have been about 9 feet wide, consisting chiefly of galena with bunches of chalcopyrite, and carrying as much as 80 ounces of silver and 65 per cent of lead. The ore body was rapidly opened up and proved large and rich. In 1883, with a thousand feet of drifts and shafts, about 3,000 tons of ore were extracted, with an average value of nearly $150 per ton. The product for this year is given in the Mint report as $400,000, and the ore is said to have carried a high percentage of lead. In 1884, according to the same authority, the mine was pro- ducing about 40 tons a day, which, at $150 per ton, would be something over $2,000,000 for the year. This, however, is obviously an excessive estimate. In 1887 the output is not known, but was probably much less than $200,000. In 1890 it is credited with $1,352,994, the silver, as usual, being given at its coinage value and no return being made for copper or lead. In 1891 the product is given as $601,465 in gold and silver, and in 1892 it had fallen to $95,445. Of this amount $5,200 was in gold, $48,333 in silver (coinage value), $3,632 in lead, and $38,280 in copper. Thus these fragmentary records show that in the course of ten years' working the ore changed, within a vertical dis- tance of 1,000 feet, from one carrying chiefly galena to one rich in copper.

This has probably been the most widely known and most produc- tive mine in the Red Mountain district, although closely rivaled by the Guston. But it was an expensive mine to operate on account of the irregular form of its large ore bodies, the abundance and corrosive activity of its waters, and the necessity of hoisting and pumping through deep shafts. These adverse conditions, in conjunction with a falling off in the value of the ore and the decline in silver, finally caused the mine to shut down about 1896.

The Yankee Girl ore body lies in Silverton breccia (possibly extend-

---

[1] T. A. Rickard (Trans. Am. Inst. Min. Eng., Vol. XXVI, 1896, p. 842) states that the discovery was made in August, 1882, by Andrew Meldrum. This is apparently erroneous.

ing below into the San Juan breccia), which, in the vicinity of the mine, is decomposed and weathered near the surface. Such croppings as may formerly have been visible have been covered by the shaft house and other mine buildings. The mine was formerly worked by a vertical shaft over 1,000 feet in depth, which in 1899 was nearly full of water, rendering the workings wholly inaccessible.

T. E. Schwarz [1] described the Yankee Girl ore in 1883 as occurring "in four chimneys, three near the south side of a dike and the fourth near the north side of another dike some distance away. The ore in each chimney is similar and often occurs in bowlder form." In 1886 Mr. S. F. Emmons made a brief visit to the mine and recorded his observations in the following notes, not hitherto published:

The main shaft is sunk about on a level with the road, but the old tunnel runs in from the bottom of the valley below, which slopes here very steeply; the tunnel cuts the shaft about 75 feet from the surface. From the top of the shaft down to the tunnel level the ore was lead ore, mainly galena and pyrite. From the tunnel level down the ore began to change, copper coming in and lead disappearing; below the third level is practically no lead. The rich ore is now a purple copper ore with stromeyerite, copper glance, copper pyrite, and some gray copper and barite. Outside the ore body the country rock is impregnated with fine-grained pyrite. In the Orphan Boy, one of the Yankee Girl claims, is some manganese spar, but it is not at all common.

The mine has six levels, the adit making the first. The sixth level is at a depth of 500 feet. The shaft has been sunk 113 feet deeper, and a station made for a seventh level, which has not yet been opened.

The ore is found in several cylindrical or elliptical chimneys, called, respectively, the Yankee Girl, Orphan Boy, North, West, and South chimneys. The main shaft was sunk about in the middle of one of these chimneys—the Yankee Girl. The discovery shaft, about 50 feet north of the main shaft, was 10 feet down in this body when the mine was finally purchased for $125,000. In the first and second levels the main shaft is still in this chimney, but below it is cut at an ever-increasing distance south. The chimney is nearly cylindrical in shape, 20 to 30 feet in diameter, and consists of white quartz rock, more or less impregnated with mineral. Around this is country rock, somewhat decomposed, and impregnated with pyrite—a sort of transition material; beyond this is unaltered breccia. Dike rock, according to Schwarz, is always within 15 feet or less of the ore shoot. The west drift on each level, running off as a rule a little south of the shaft, generally cuts through the "dike rock" and finds another chimney beyond it. But the shape of the dike is very irregular; in one drift it is over 100 feet wide, in the next below almost wanting, and comes in again in the next below. According to Schwarz it sometimes runs quite flat for a considerable distance. He recognizes two dikes running in his ground, and another running from the Guston into the Robinson. I observed many joint planes or cross fractures, some showing slickensided surfaces, and these generally form walls to the ore bodies. Although these have a prevailing direction northwest, there are many other directions, and their surfaces are sometimes curving. It is my impression that such planes have admitted or given form to the ore bodies, and Schwarz admits that there is always a prominent one running with the longer axis of the ore body, which often tapers up into it when running out.

Kaolin occurs, presumably pure, but resembling Chinese talc, although not so translucent. Bournonite and enargite occur in the Yankee Girl; also polybasite in the upper part.

[1] Proc. Colo. Sci. Soc., Vol. I, 1883-84, p. 132.

In a paper published in 1888 Mr. Emmons[1] says:

In the Yankee Girl there are several chimneys in which the ore occurs. They are of elliptical outline, the longer axis corresponding in direction with a main system of fractures running through the region. Although the striated surfaces of these planes show that there has been movement along them, there is but little evidence left of actual brecciation of the country rock, the ore solutions having completely replaced the andesitic country rock between the fracture planes which gave access to them; in places a siliceous skeleton is left, the basic constituents being replaced by vein material; in other places a solid body of metallic minerals is found, while the country rock adjoining the body of pay ore is impregnated to a considerable distance with low-grade sulphurets.

This same year (1888) Mr. Emmons paid a second visit to the Yankee Girl. He found[2] that on the sixth level, 432 feet deep according to Schwarz, the Yankee Girl ore body was connected by ore with the Orphan Boy, a small ore "chimney," lying in the upper levels about 150 feet south of the Yankee Girl. On this level the Yankee Girl ore body was elliptical in plan, 10 feet wide, and 32 feet long. At the fifth level it was nearly cylindrical, 8 to 10 feet in diameter, and nearly solid galena, passing below into "gray copper ore." Just below the sixth level occurred a "floor" or horizontal fault plane. This varied from 3 inches to a foot or more in thickness, of clayey attrition material, and frequently contained rounded masses of ore with rock centers. It dipped to the westward, sometimes as much as 25°. There seemed to him to have been some movement along this floor to the westward or downhill toward the valley, but that the ore bodies had not been much displaced. A third or "west ore body" lay to the westward of the Yankee Girl shoot, lenticular in shape, with its longer diameter nearly north and south. On the sixth level this was 2 or 3 feet wide and 8 or 10 feet long, but with rather poorly defined boundaries, the ore fading off into mineralized country rock.

Mr. Emmons records that—

Around each chimney of ore is an envelope of "quartz." This is a blue siliceous material, sometimes granular looking, sometimes compact and almost jaspery, and yet in the mass full of cracks, and sometimes of vesicles which are filled with white kaolin. It is generally impregnated with fine-grained pyrite. The ore, moreover, fills little fine cracks or seams down to an eighth of an inch in width. These are sometimes large enough to be worth taking out in the vicinity of the ore body. Outside the "quartz" envelope comes the "white rock," which is evidently altered porphyry, which passes into a hard, compact blue rock, which is probably the less altered variety.

What was called quartz at that time in the mine, Mr. Emmons considers at the present time to have been a jasperoid or siliceous replacement of the country rock. He also states that replacement is very abundant in the volcanic breccia, the matrix being first replaced by ore, which finally coats the pebbles or fragments of the breccia.

---

[1] Structural relations of ore deposits: Trans. Am. Inst. Min. Eng., Vol. XVI, 1888, p. 804.
[2] Unpublished notes.

Mr. T. E. Schwarz, for several years superintendent of this mine, has kindly furnished some important data in regard to the ore bodies. The ore in the Yankee Girl shoot, down to No. 1 level, a distance of 75 feet, was chiefly galena, and averaged 77 ounces of silver and 36 per cent of lead. From No. 1 level to No. 2 level, 80 feet, it averaged 242 ounces of silver and 29 per cent of copper, lead not being given. The galena, however, was disappearing, and the characteristic mineral from the second to the sixth levels was stromeyerite, associated with chalcopyrite, pyrite, and occasional occurrences of argentite and other rich silver ores, some of which carried bismuth. Similar changes took place in the Orphan Boy shoot, although the galena continued to greater depth. On the sixth level (432 feet deep) these two ore shoots came together, a straight drift in continuous ore connecting them. A few feet below this occurred a heavy, strong, nearly horizontal seam or slip carrying much "talcy" material.

The rich ores of the Yankee Girl disappeared at this seam, no stromeyerite or other distinctive rich silver mineral being found below it. The main Yankee Girl ore shoot continued down through it without perceptible change of position, and was large and strong from No. 6 to No. 8 levels. The character of the ore changed, however, and "peacock copper," with pyrite and chalcopyrite became its constituents. Bornite I took to be the principal ore for large masses. To show grade of selected ore produced from richest stopes, I may note a shipment of 10 tons which carried 3,270 ounces of silver and 29 per cent of copper. A smaller lot of 7,360 pounds returned 5,301 ounces of silver and 28.75 per cent of copper.[1]

Speaking from memory, Mr. Schwarz believes that the ore in the bottom levels of the Yankee Girl was low grade and carried more pyrite than above.

These various descriptions of the ore bodies are given at some length, since they are not in perfect agreement on all points, and yet they furnish cumulative evidence of the truth of certain important features.

Mr. Schwarz[2] has also published a description of the ore bodies of which the Yankee Girl is the type. He says, in part:

The ore occurs in "chimneys," so called, having in some cases an elliptical or a circular cross section, but more generally long in proportion to their width. The greatest length of ore body so far observed has been about 60 feet.

The immediate envelope of the ore chimneys is "quartz," which is sometimes of considerable extent, while the whole is inclosed in an area of greater or less extent of andesite.

The andesite is metamorphosed in proportion to its nearness to the ore chimney, showing more kaolin and quartz and less of its original structure. It thus merges into a fine-grained brown or grayish quartz, sometimes very hard and flinty, and again quite porous. In depth the quartz matrix becomes less characteristic.

[1] Personal letter from Mr. Schwarz. It is not clear, in view of the figures just given by Mr. Schwarz, just what he means by saying that no rich silver ore occurred below No. 6 level. He probably refers merely to the mineralogical change in the ore from stromeyerite to bornite, or he may mean that these high-grade lots came from above level 6. The high copper percentage, however, would not indicate this.
[2] Notes on the ore occurrences of the Red Mountain district: Proc. Colo. Sci. Soc., Vol. III, 1888, pp. 77-85.

In the case of six chimneys occurring on the Yankee Girl, Guston, and Silver
Bell properties, being the only ones on which depth has been obtained, a marked
increase in the silver content of the ores occurred from the surface down to 300 to
400 feet of depth.

Changes of character of ore with depth are noticeable in several chimneys, but
notably so in the Yankee Girl chimney. In this case the distinctive minerals, in
order of depth, have been galena, gray copper, stromeyerite, and bornite. In those
chimneys in which enargite has been the surface ore no depth has yet been
obtained.

The chimneys frequently change their pitch, sometimes quite suddenly. * * *
A jump of 15 to 25 feet along some horizontal plane is not infrequent. * * *

Every chimney is connected with one or two cleavage or fracture planes, which
bound a portion of the ore body, while on the other sides the ore merges into the
quartz or andesite. These planes usually show striations, slickensides, and more
or less gouge material, but are not of great persistence laterally.

Horizontal planes occur of great strength, which have indirectly had marked
influence on the ore bodies, changes of character occurring at such horizons.

In the Guston and Yankee Girl the increase in the amount of the copper ore
with depth is notable, as is also the fact that the silver is confined mainly to such
ores.

Mr. Schwarz supposed the ore to have been brought in by solutions
moving chiefly along horizontal planes, perhaps bedding planes, and
to have been derived from the andesite of the region. His explanation is essentially that of a rather narrowly confined lateral secretion.
The maximum length of 60 feet was probably exceeded by the ore
bodies worked in the Yankee Girl and Guston subsequent to 1888.
One of them is reported by one who saw it to have been 140 feet in
length and 20 feet wide. According to Schwarz, several carloads of
ore from the Yankee Girl carried from 1,500 to 3,000 ounces of silver
per ton, and one lot of about 6 tons contained 5,300 ounces of silver
per ton. Such ores carried 30 to 33 per cent of copper.

According to Mr. J. Owen, who formerly worked in the Yankee
Girl and is at present superintendent of the Silver Lake mines, the
rich bodies of stromeyerite, which had been so productive prior to
1888, began to change, at about 500 feet in depth, to bornite, which,
although lower grade, was yet good ore. At this level there was a
fault encountered which threw the lower portion of the ore body to
the west. Below this the deposit, according to Owen, has more the
character of a vein 10 or 12 feet wide and running "north and
south." It consists, near the bottom of the shaft, of large bodies of
bornite, with some chalcocite, assaying as much as 57 per cent of copper. (As pure, nonargentiferous bornite contains only 55.5 per cent
of copper, this assay, if correct, would indicate the presence of considerable chalcocite.) Two carloads of this ore were shipped by Owen
from a winze 40 feet below the bottom of the present shaft, and it is
said to be abundant at that depth. From the eleventh level ore was
shipped worth from $500 to $700 per ton in silver and copper. The
latter averaged about 20 per cent. No large bodies of iron pyrite
were found in the Yankee Girl, according to Owen.

At present the Yankee Girl shaft is about 1,050 feet in depth. A plan of the extensive levels shows an intricate maze of workings in which no linear system is discernible. The mass of the workings lie just west of the shaft, and in plan may be roughly inclosed in an irregular triangle. A smaller extent of workings lies just east of the shaft.

Inspection of the dump, as well as inquiry, shows that there was never much vein quartz associated with the ore. The "quartz" of the miners is very largely the bleached and silicified country rock adjacent to the nearly solid bodies of ore. Where vein quartz occurs it usually carries iron pyrite. Barite, in small masses and crystals, occurs embedded in the bornite. The chalcocite is generally intimately associated with small amounts of chalcopyrite. Some specimens show that the ore has been fractured and recemented by veinlets of calcite. The ore minerals observed on the dump were galena, sphalerite, chalcocite (stromeyerite), bornite, chalcopyrite, and pyrite. Cosalite was recognized in 1884, therefore, in the upper part of the deposit, and was analyzed by A. H. Low,[1] as follows:

Analysis of cosalite from Yankee Girl mine.

Constituent.	Per cent.	Constituent.	Per cent.
Bismuth	36.22	Iron	4.48
Silver	8.70	Sulphur (by difference)	18.64
Lead	28.22		100.00
Copper	3.74		

Proustite and polybasite also occurred occasionally in the Yankee Girl ore bodies.

All accounts of the Yankee Girl mine unite in emphasizing the chemical activity of the underground waters encountered in the workings. According to Schwarz[2] they contained "24 grains per gallon of sulphuric acid (SO_3)." Candlesticks, picks, or other iron or steel tools left in this water become quickly coated with copper. Iron pipes and rails were rapidly destroyed, and the constant replacement of the piping and pumps necessary to handle the abundant water was a large item in the working expenses. All agree in stating that the water entered far less abundantly below the sixth level. Some who worked in the mine express it as being "not so bad" below that level. But closer inquiry usually elicits the information that it was less abundant but as much or even more corrosive than at the upper levels.

The product of the Yankee Girl is roughly estimated at about $3,000,000.

Robinson mine.—This mine is really a part of the Yankee Girl, both being worked together. It has produced some good ore, but

---

[1] Proc. Colo. Sci. Soc., Vol. I, 1884, p. 111.
[2] Trans. Am. Inst. Min. Eng., Vol. XVIII, 1889–90, p. 139,

little could be ascertained concerning the details of its occurrence. According to Mr. J. Owen, the ore changed to iron pyrite, but the depth at which the change took place was not stated by him. The shaft of the Robinson is sunk alongside one of the mounds of siliceous altered breccia characteristic of this region.

It is stated by Mr. Owen that a distinct vein or fissure was followed from the Guston into the Robinson, and was demonstrated by drifting to extend into the Genesee-Vanderbilt. The same fissure was cut by an east crosscut in the Yankee Girl.

Guston mine.—This ore deposit, just north of the Yankee Girl, was opened shortly after the latter, in 1882. The ore body as exposed at that time is said to have been 5 feet wide and carried 58 ounces of silver per ton. It was chiefly galena, with some copper mineral, probably tetrahedrite. In 1883 the mine produced, according to the Mint reports, $57,500. In September, 1887, steps were taken to form the present New Guston Company, Limited, of London, England, and the mine was closed until June, 1888, when work was resumed with Mr. T. E. Schwarz as superintendent, and extensive bodies of ore were found on the third, fourth, and fifth levels, yielding in carload lots up to 450 ounces of silver per ton. The depth of the mine at this time was only 288 feet, while the Yankee Girl was down 1,050 feet. The product for 1888 was about $130,000. In 1889 ore was produced to the value of over $390,000, and work was carried to the sixth level, 378 feet from the surface. Mr. James K. Harvey this year succeeded Mr. Schwarz as superintendent. At the close of 1889 the mine had already paid dividends amounting to £56,374 (about $274,000). In 1890 the shaft was sunk to the seventh level and extensive stoping carried on in the upper levels. Ore to the amount of 4,469 tons was sold, which realized $491,336 (£101,098), or an average of nearly $110 per ton after deducting treatment charges. Several carloads of this ore carried from 600 to 1,400 ounces of silver per ton. The product this year was principally from the stopes of the fifth and sixth levels. The ore carrying most silver was encountered on the fifth level north of the shaft, and contained 12 per cent of copper. At this time there were from 200 to 300 men employed in the mine, and the small settlement of Guston had grown up in the vicinity of the shaft. The water pumped from the mine this year was about 35 gallons per minute, and is said to have been charged with sulphuric acid from oxidation of pyrite. The Mint reports credit the mine with a total output of 864,768 ounces of silver and $57,500 in gold for 1890. The product in copper is not given. In 1891 the main shaft had reached the ninth level, about 678 feet below the surface, and the water had increased to about 50 gallons per minute, being of the same corrosive character as that found in the upper levels. The ore raised this year amounted to 11,723 tons, which realized $824,467, or an average of over $70 per ton, net value. The cost of mining was nearly $19 per ton, while during the preceding year it had averaged about $32 per ton. The decrease in the price of silver was already affecting the profits of the company. In 1892 the ore raised amounted to 14,291 tons, which realized approximately $247,232, or an average of only about $17.40 per ton. The cost of mining was about $13 per ton. This great decrease in value was due not only to general fall in the price of silver, but to a falling off in the grade of the ore. The shaft had been sunk to the tenth level, but no important bodies of ore were discovered in the 200 feet between the eighth and tenth levels, and the product was chiefly from the nearly worked-out stopes of the upper levels, from No. 3 to No. 7. In 1893 the shaft was carried down 136 feet below the tenth level and a crosscut started on the eleventh level. The ore raised amounted to 7,280 tons, which sold for $18.40 per ton. The water now pumped from the mine amounted to 60 gallons a minute. In 1894 a large quantity of ore was extracted, chiefly from the ninth to the twelfth levels, of which the greater part, being low grade, went to the Silverton matte smelter, then beginning operations. In 1895 the maximum tonnage for any year—14,833 tons—was extracted from the mine, but the average value was lower than ever before. The annual tonnage, average sales value of the ore per ton (gross value less cost of treatment), average cost of mining per ton, and amount of dividends paid, covering the eight years in which the mine was in most active operation, are shown in the following table, compiled from the annual reports of the directors of the company:

Tonnage, ore values, cost of mining, and dividends of Guston mine.

Year.	Ore raised.	Average sale value per ton.	Average cost of mining per ton.	Amount of dividends paid.
	Tons.			
1888	315	$363.25	$53.50	$77,941.90
1889	2,882	131.50	32.60	181,875.00
1890	4,469	109.80	32.25	315,250.00
1891	11,723	70.65	18.90	426,800.00
1892	14,291	17.40	13.10	146,712.50
1893	7,280	18.40	17.30	
1894	*13,334	12.80	9.60	
1895	14,833	10.70	10.20	
Total	69,127	a 91.81	a 23.43	1,148,579.40

a Average.

In 1897 the mine was compelled to cease operating, the low grade of the ore and the decline in the value of silver having acted together to render further work unprofitable. The total output since 1888 has considerably exceeded $2,500,000, of which nearly half was distributed in dividends. In all, 14 levels had been worked and a depth of nearly 1,300 feet attained. To free the mine from water at this depth the pumps were required to raise over 100 gallons a minute and great care and constant attention was required to prevent the destruction of the pumps and column by corrosion.

When visited in 1899 the mine was filled with water to the second level, 114 feet below the surface. The level of this water fluctuates with the seasons, and in spring sometimes rises to the first level, 50 feet from the surface. As the deeper workings are wholly inaccessible, the following description of the ore bodies is drawn from the annual reports of the superintendents, Messrs. Schwarz and Harvey, and from personal conversation with men who worked in the mine.

In their general shape the ore bodies resembled those of the Yankee Girl. They were roughly ellipsoidal masses standing nearly vertical, and consisting of nearly solid ore with very little gangue, surrounded by altered country rock impregnated with pyrite. The actual sizes of these masses of ore are difficult to ascertain at the present time. Some of the stopes were nearly 200 feet in length and several hundred feet in height. The width of the ore varied greatly in different bodies and in different portions of the same ore mass. Widths of from 20 to 30 feet appear to have been not uncommon.

The ore-body stope on the seventh level was, in 1892, 45 feet in length and 94 feet average height. The width of the ore for a distance of 20 feet was 14 feet; elsewhere less, and irregular. The ore from this stope carried from 20 to 80 ounces of silver and 0.3 ounce of gold per ton, with from 5 to 8 per cent of copper. The middle stope on the same level had a length of 116 feet, an average height of 87 feet, and an average width of ore of 22 feet. Its ore carried from 15 to 35 ounces of silver and 0.2 ounce of gold per ton. These are given simply as examples of typical stopes which produced large amounts of rather low-grade ore.

These ore bodies usually occurred along what was termed the "ore break," a zone of altered, disturbed, and more or less broken country rock, intersected by slip planes and containing much kaolin, as well as soft attrition material. This ore break is said to traverse the Guston claim from end to end, which would give it a course of from 20° to 30° east of north and to be, in places at least, nearly 60 feet in width. From levels 2 to 7, a depth of 364 feet, this ore break dips steeply to the east; but just below the seventh level a slip or fault was encountered dipping 45° to the west, which cut off the ore entirely. By following the soft clay gouge which occupied this fracture plane the ore break was found again below, lying from 30 to 50 feet to the westward. It here dipped steeply westward and did not carry much ore. It was not until the tenth level was reached that its dip again became easterly and it resumed the character which it had had above the seventh level. The ore in this level also was better, or at least in larger bodies, than anything hitherto found below the seventh level.

The ellipsoidal or irregular ore bodies were nearly always found in connection with this ore break, but they were most irregularly placed and were found only by persistent and thorough prospecting, partly

with the diamond drill.   In general the longer axes of the ellipsoids
appear to have been nearly parallel with the trend of the ore break.
Individually, however, the ore bodies show much variation.   Thus, on
the thirteenth level a streak of ore is recorded as bearing N. 30° W.
They were commonly found by drifting along the "soft ground,"
presumably fracture planes associated with attrition material and
kaolin.   A plan of the underground workings of the Guston mine
(Pl. XV) shows a highly intricate and irregular maze of drifts and
crosscuts running out in all directions from the shaft.   The map,

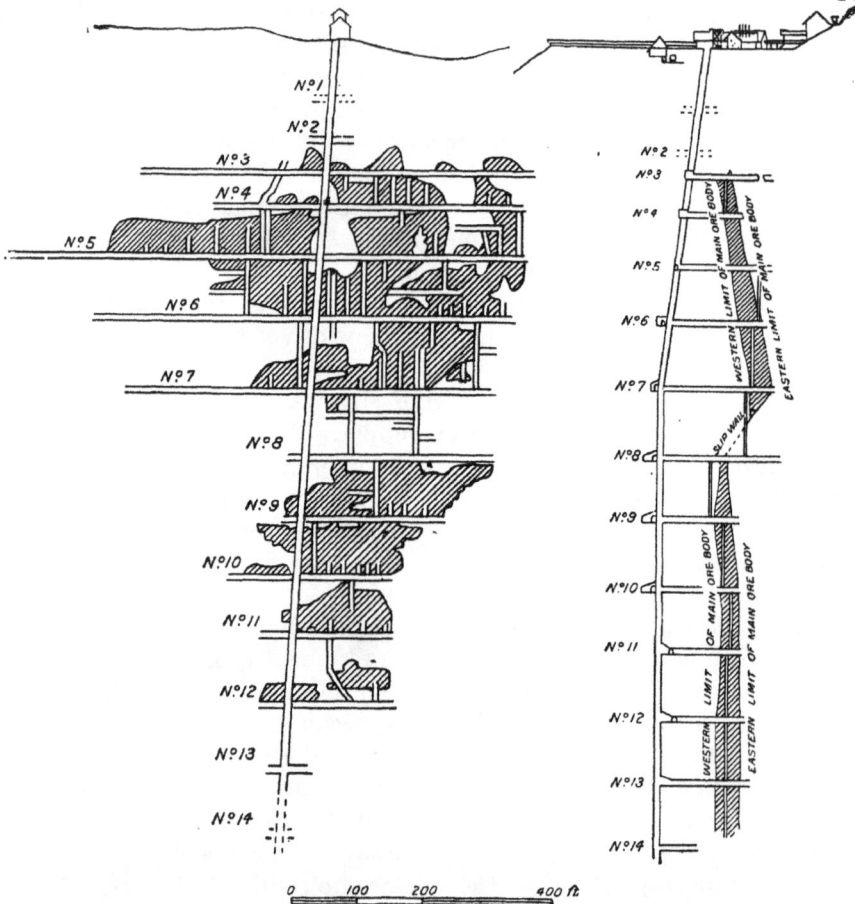

FIG. 16.—Longitudinal and transverse sections through the Guston mine.

however, shows distinct evidence of a general elongation of the ore bodies, or of the ore-bearing ground, along an axis bearing about N. 20° E.
As shown by the longitudinal section of the mine, the stopes attained
their greatest aggregate length, about 600 feet, on the fifth level, and
grew shorter below.   Longitudinal and transverse sections of the
"main ore body," compiled from a mine map made in 1890 and from
the annual reports of the company, are shown in fig. 16.   The transverse section was evidently intended to show the general position of
the so-called ore break, and not of the actual ore bodies.

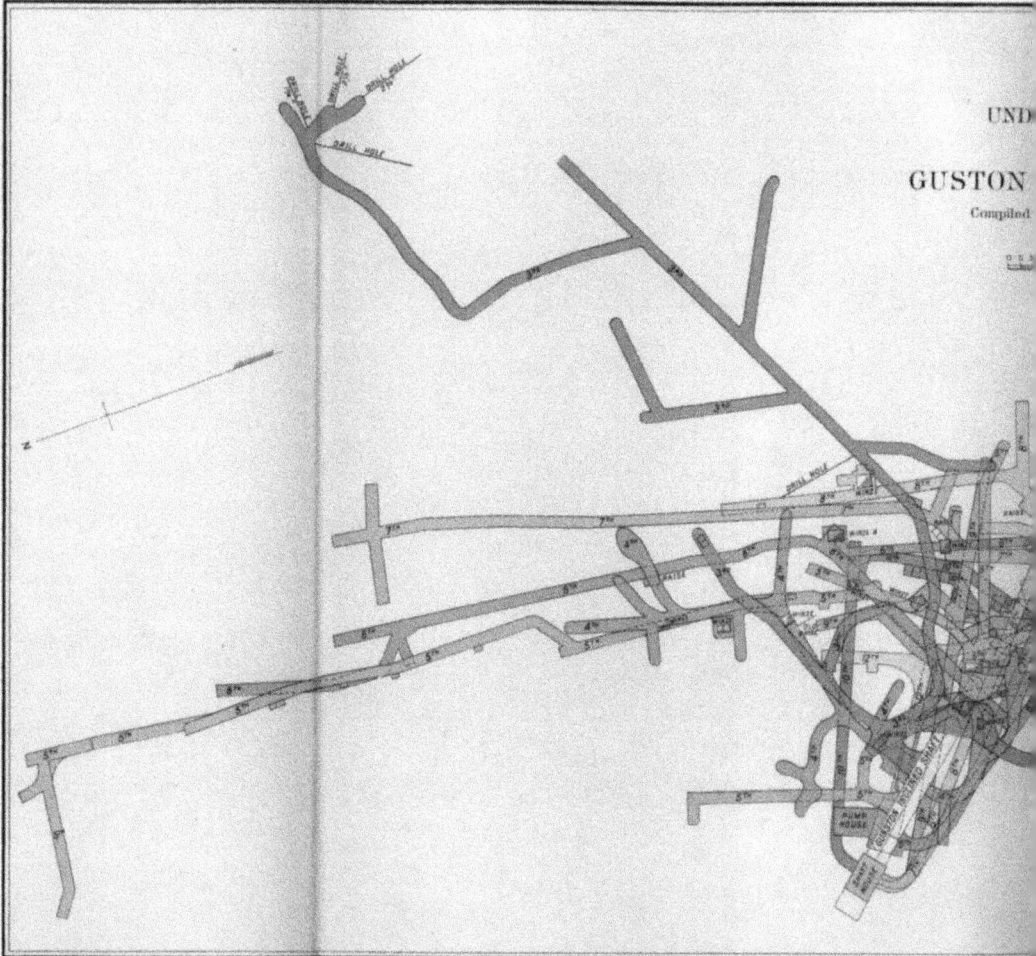

MAP OF A PORTION
OF THE
ERGROUND WORKINGS
OF THE
AND ROBINSON MINES
from mine maps made in 1890 and 1895
Scale of feet

One of the most interesting and important facts in connection with the Guston, as well as with other mines of this district, is the change in character of their ores with depth.   The galena ore of the main ore body, carrying from 50 to 60 per cent of lead and from 30 to 40 ounces of silver, continued down to the second level, or a little beyond, where it is said to have changed rather abruptly to high-grade stromeyerite. Galena, however, occurred in the mine in considerable quantities down to the fifth level, 288 feet in depth, and in decreasing amounts to greater depths.   Some heavy galena, carrying from 100 to 175 ounces of silver, is known to have occurred at a depth of 212 feet.   With increasing depth stromeyerite, or argentiferous chalcocite, became the characteristically rich ore of the mine, associated with pyrite and chalcopyrite.   Some of this was very rich, and in 1891 some ore from the sixth level (378 feet) carried as much as 15,000 ounces of silver and 3 ounces of gold per ton.   The lowest-grade ore in the deeper workings was pyrite, which usually carried less than 20 ounces of silver and a fraction of an ounce of gold per ton, and from 3 to 5 per cent of copper, probably from admixture of chalcopyrite.   The chalcopyrite was higher grade, usually carrying from 12 to 45 ounces of silver and a fraction of an ounce of gold per ton, and from 8 to 12 per cent of copper.   Even in the best stopes of the fifth and sixth levels there was always more or less chalcopyrite and pyrite associated with the rich stromeyerite, the lower-grade ores apparently often prevailing on the peripheries of the larger ore masses.

Below the slip plane, which occurred just below the seventh level, at about 500 feet in depth, the ore was harder and more compact and stromeyerite was not found in any quantity, the ore down to the tenth level consisting chiefly of large masses of low-grade pyrite, with some bornite and chalcopyrite.   The pyrite rarely carried as much as 20 ounces of silver, with a little gold and copper.   Some bornite found on the ninth level (678 feet) carried from 125 to 450 ounces of silver and one-fourth to one-half ounce of gold per ton, with 25 to 50 per cent of copper.   This was associated with chalcopyrite carrying 16 to 75 ounces of silver and gold up to 1 ounce per ton, with from 8 to 15 per cent of copper, and with pyrite carrying up to 15 ounces of silver per ton, with a little gold and copper.   In both the ninth and tenth levels small quantities of ore were frequently met with carrying free gold, assays from level 9 giving as much as 29 ounces of gold per ton.

The ore of the eleventh level was similar to that of the tenth, but, as in all these lower levels, the rich bornite bore a smaller proportion to the low-grade pyrite and chalcopyrite than did the stromeyerite and galena of the upper workings.   The ore in the stopes connected with these levels varied in width from 18 inches up to 25 feet.   Some free gold was seen in barite, associated with the bornite.   On level 12 ore occurred in large solid masses, one being known to be 72 feet in length,

Bull. 182—01——15

with a maximum width of 37 feet. But it was chiefly low-grade pyrite containing occasional nodules of bornite and chalcopyrite, usually with some barite. Substantially identical ore was found on the thirteenth level.

Through the kindness of Mr. Bedford McNeill, of London, liquidator of the New Guston Company, the following notes, dated March 13, 1900, were obtained from Capt. James K. Harvey, formerly superintendent of the mine:

Galena ore was most abundant in the superficial workings, it having been met with almost at surface, and continued down to the fourth level. The ore contained from 8 to 50 per cent lead; silver, 9 to 30 ounces; gold, trace per ton.

At the third level chalcopyrite and chalcocite were discovered, and continued down to where the fault was encountered, a few feet underneath the floor of the seventh level. Contents of ore: Copper, 5 to 15 per cent; silver, 25 to 700 ounces; gold, 0.1 to 3 ounces per ton. A few cars were shipped (about 10 tons to the car) that ran over 1,200 ounces silver per ton, with a corresponding gold increase.

From the third to the sixth level the ore occurred in the form of a chimney, its dimensions being from 12 to 15 feet in length and width at the third, fourth, and fifth levels, to 20 to 30 in length and width at and over the sixth level. Underneath the sixth level the ore lengthened south approximately 200 feet, with a width of from 3 to 30 feet.

Solid bodies of iron pyrites were discovered at the sixth level, and continued to the deeper workings. Contents: Silver, 4 to 20 ounces; copper, 1 to 3 per cent; gold, trace to 0.20 ounce per ton.

Bornite was met with in quantity between the ninth and twelfth levels. Contents: Copper, 18 to 25 per cent; silver, 60 to 175 ounces: gold, one-fourth to 1½ ounces per ton. Free gold was found occasionally with the bornite; that is to say, on the fluorspar [1] associated with the bornite. No free gold was seen above the ninth level, or in any other instance.

There was but little change in the volume or character of the water. It was destructive to pumps and column if they were not properly protected.

Generally speaking, the inclosing porphyry was more decomposed where the ore was most compact and richest in its metallic contents.

From the foregoing it appears that there was first an increase in the value of the ores down to the sixth or seventh level, and then a sudden falling off in value at the slip-plane below the seventh level. Below this occurred a diminution in the richness of the ore down to the bottom of the mine. The characteristic ore minerals, in the order of the depth at which they prevailed, were galena, stromeyerite (pyrite?), bornite, chalcopyrite, and pyrite. But this succession is true only in general terms. Sudden and perplexing changes in the value and mineralogical composition of the ore occurred within the individual ore bodies at all levels, and the various minerals named overlapped in most complex fashion, although the bornite apparently had a definite upper limit and the stromeyerite a definite lower limit. Of all the ore minerals pyrite appears to have had the greatest vertical range, and it undoubtedly formed an increasing proportion of the ore with depth. The highest values in silver and gold were found in connection with high percentages of copper.

---

[1] Barite ? (F. L. R.)

According to Mr. Terrill, at present in charge of the mine, good ore, consisting of argentiferous bornite, was struck in the fourteenth or bottom level, but was never extracted. Had there been any considerable body of high-grade ore revealed, however, it is hardly probable that the mine would have shut down, no considerable loss in its operation having been incurred and no assessments having been levied on the shareholders when work ceased.

Mr. S. F. Emmons made a brief visit to the mine in 1888 and has recorded his observations in unpublished notes. He visited a new find of ore on the third level, which had been overlooked by the former owners. This was a solid mass of chalcopyrite mixed with tetrahedrite and galena and carrying up to 750 ounces of silver per ton. He found that this ore body, with a strike of N. 30° E., and dipping from 30° to 45° to the northwest, was limited on its foot-wall side by a distinct plane showing fault movement. According to Mr. Emmons, the sheeted structure of the original country rock, which had been replaced for a width of about 10 feet by ore, was still recognizable in the latter. The ore body, which was not more than 10 or 15 feet long, graded into quartz, clouded with snow-white kaolin, and this in turn into altered country rock or "white rock."

Genesee-Vanderbilt mine.—The original Vanderbilt shaft was sunk just southwest of the Yankee Girl, in one of the characteristic siliceous knolls of the district. This consists of a hard, quartz-like rock, in which can still be detected outlines of porphyritic crystals of feldspar, now wholly decomposed. In some places the original rock is practically wholly replaced by quartz. In the Mint report for 1882 it is noted that the Genesee, then opened by two shafts 15 feet and 20 feet deep, showed a body of galena 3 feet thick, associated with "copper," the whole carrying from 15 to 40 ounces of silver per ton. The "copper" was probably chalcopyrite or tetrahedrite. In 1888 the product of the Vanderbilt as derived from the same source was $72,672. It was subsequently found that the Vanderbilt and Genesee ore bodies were identical, or at least so closely associated that it was impracticable to work them separately, and the mines were therefore consolidated. The present workings (fig. 17), begun in 1891, consist of a tunnel 820 feet in length, connecting with the old shaft through stopes, and a second shaft, 700 feet in depth, sunk near the end of the adit tunnel. Levels were run at 300, 400, 500, 600, and 700 feet below the tunnel level. In 1899 this shaft was filled with water to within a few feet of the collar. Although the mine has produced some good ore, it is said to have been on the whole unprofitable. The ore was of a lower grade than that in the Guston and Yankee Girl, and the profits accruing from the exploitation of one ore body were usually more than expended in searching for the next one. Work was finally suspended in 1896.

In 1899 a little ore, consisting chiefly of galena and sphalerite, was being taken out by leasers above the adit level. Access was gained

to one large stope extending nearly to the surface where the original
Vanderbilt shaft was sunk.   The ore body that once filled this stope
was evidently of lenticular form, the maximum dimension nearly vertical, and the mean diameter nearly north and south.   Around the
periphery of the lens the ore body wedged out completely.   I was unable
in the accessible portions of the stope to recognize any dominant fissure of which the lens of ore might be considered as filling an enlarged
portion.   Nevertheless it was by following the so-called slips or gouges

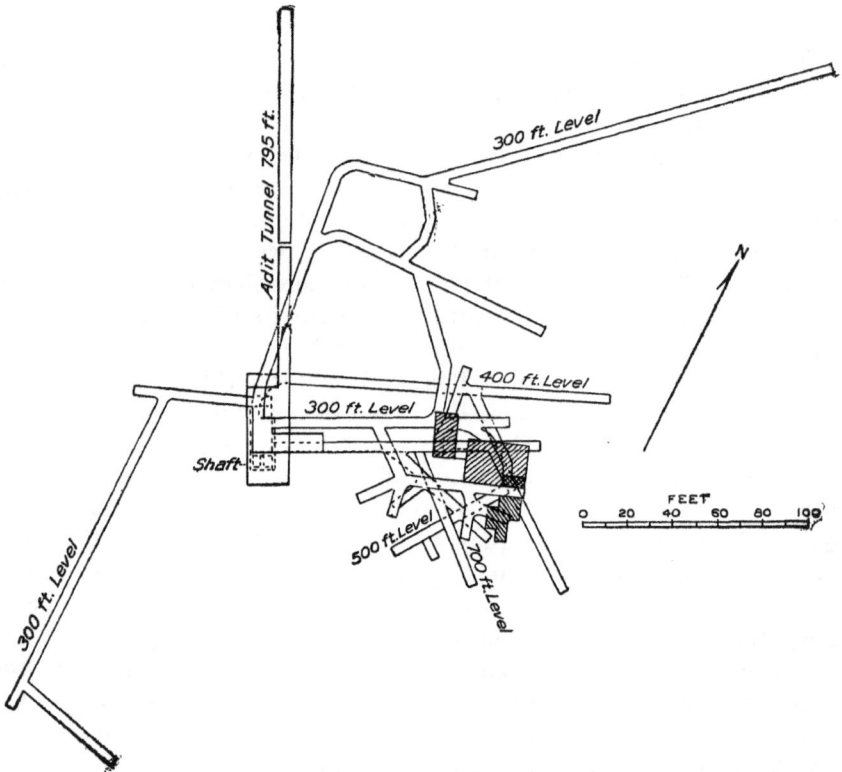

FIG. 17.—Plan of a portion of the underground workings of the Genesee-Vanderbilt mine.

in the country rock that the ore bodies were always discovered when
not found by drilling (fig. 18).

Mr. T. E. Schwarz has pointed out that the principal "ore break"
of the Genesee-Vanderbilt turned through an angle of about 140° in
passing from the surface down to the 600 level.   This is indicated in
fig. 19, taken from Mr. Schwarz's paper.[1]

According to Mr. Otto J. Schulz, of St. Louis, at present trustee of
the Genesee-Vanderbilt mine, there were four separate pay shoots on
the surface, only one of which was explored to the 700 level.   The best
ore occurred at a depth of about 250 feet below the collar of the main
shaft, and extended down to within about 20 feet of the 300 level.

---

[1] Trans. Am. Inst. Min. Eng., Vol. 26, 1896, p. 1058.

This ore was solid and contained little waste. Its value is shown in the following table, compiled from data furnished by Mr. Schulz:

Assay contents per ton of ore from Genesee-Vanderbilt mine.

Year.	Ore.	Gold.	Silver.	Lead.
	Tons.	Ounce.	Ounces.	Per cent.
1893 a	1,525	0.2	30.4	1.6
1894 a	2,153	.4	45.9	3.3
1895 b	5,939	.2	39.8	3.9
1896 c	3,005	.07	18.1	.06

a From 300 level.
b About two-fifths from 500 level and rest from upper workings.
c From 500, 600, and 700 levels. Output for 2 months only.

It is rather difficult to account for the low percentage of lead and the very small amount of copper in these assay returns, as the silver was probably associated with one or both of these metals.

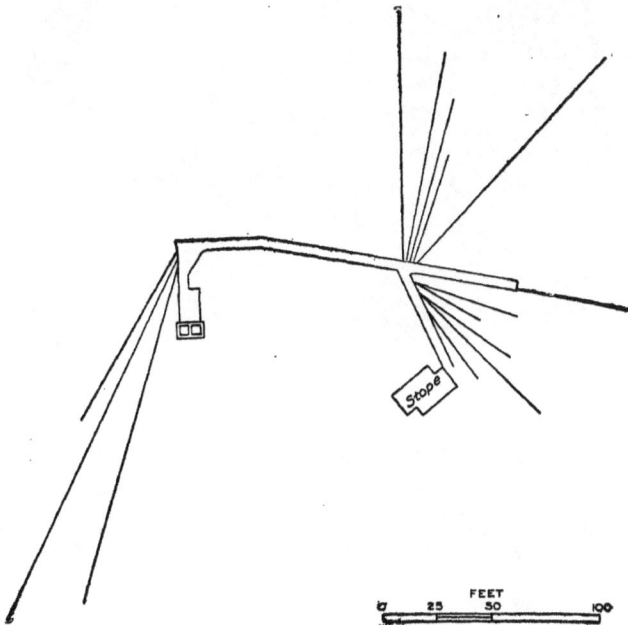

FIG. 18.—Plan of fourth level of the Genesee-Vanderbilt mine, showing method of prospecting for ore bodies with the diamond drill.

The value of the ore, never high grade as a whole, fell off with increase of depth below the 300 level, depreciating rapidly below the 500 level, and changing near the 700 level to large bodies of low-grade pyrite.

Near the surface, in the old workings, the ore is said to have been oxidized and soft, changing to galena at about 100 feet in depth.

Particulars concerning the mineralogy of the ore of the Genesee-Vanderbilt are now difficult to obtain.   It is said to have contained considerable lead from top to bottom, which statement, however, is not fully borne out by the records furnished by Mr. Schulz.   Cosalite is recorded as having been an important constituent of the ore.   Some small caves containing stalactites of pyrite occurred in the ore bodies, but it is not known at what depth.   Specimens of ore said to have come from this mine and preserved in the mine office showed some very argentiferous bornite, a slightly argentiferous sulpho-bismuthite of lead containing a little copper and corresponding in

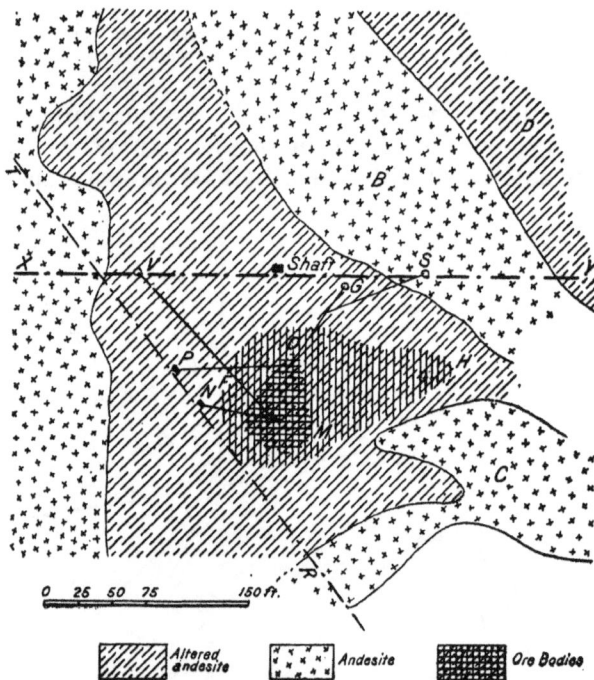

FIG. 19.—Sketch plan showing changes in position of the Genesee-Vanderbilt ore bodies, with depth. X-Y, surface outcrop of ore break; V, G, S, surface outcrop of ore bodies; M, O, position of same ore bodies on 300-foot level; P, N, position of same ore bodies on 600-foot level; T, R, ore break on 600-foot level (after T. E. Schwarz).

physical properties and appearance to cosalite, and some small crystals of ruby silver (proustite).

The total output under the old Vanderbilt Company of Red Mountain is not known, although the Mint reports credit the mine with $72,672 in 1888.   The output of the Genesee-Vanderbilt Company from 1891 to 1896 amounted to about 12,622 tons of ore, containing 2,884 ounces of gold, 436,675 ounces of silver, 702,183 pounds of lead, and 30,520 pounds of copper.

The water encountered in the Genesee-Vanderbilt was of the same corrosive character as that met with in the Yankee Girl and Guston.

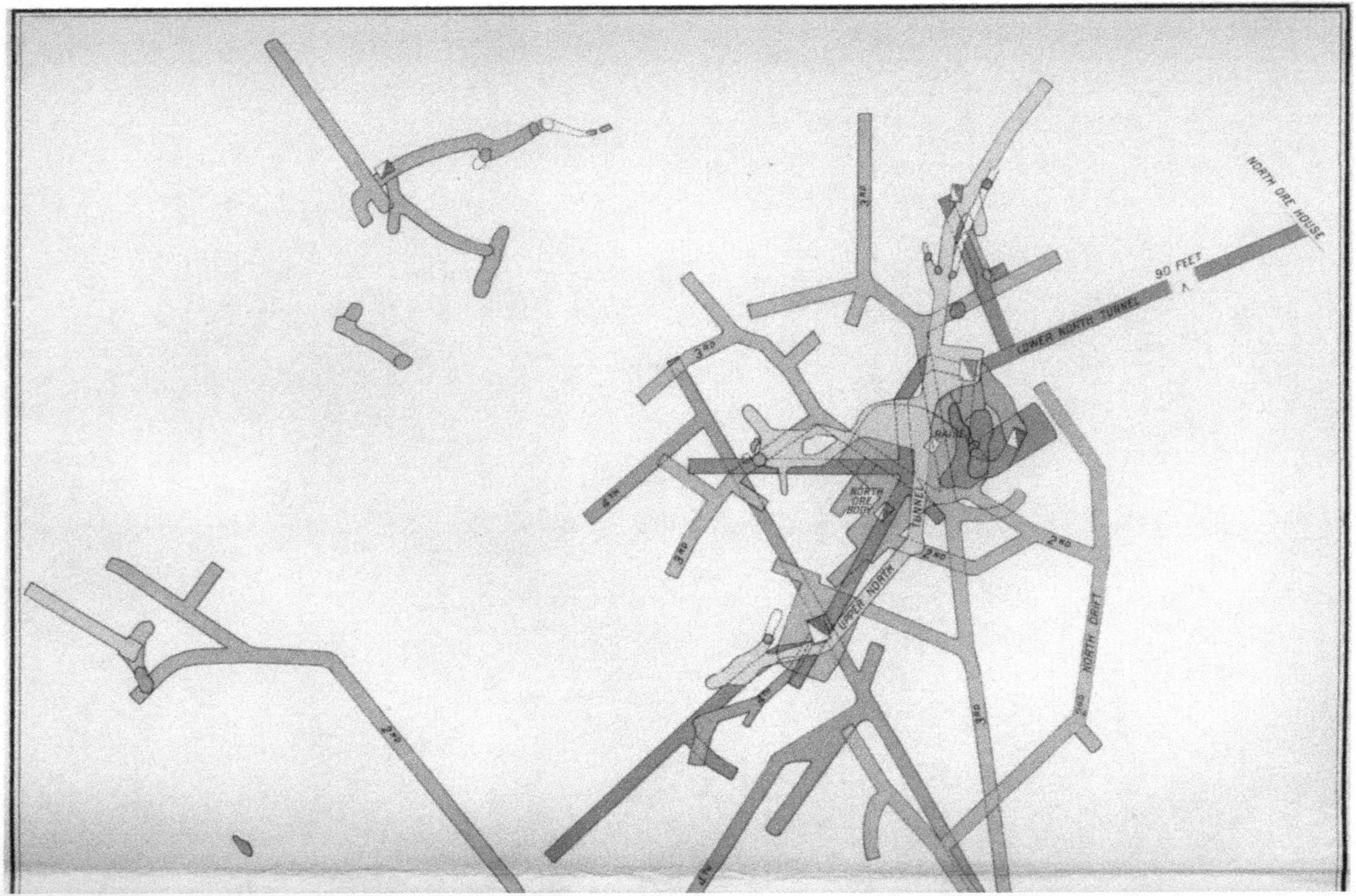
NORTH ORE HOUSE
90 FEET
LOWER NORTH TUNNEL
UPPER NORTH TUNNEL
NORTH ORE BODY
RAISE
2ND NORTH DRIFT
2ND
3RD
4TH

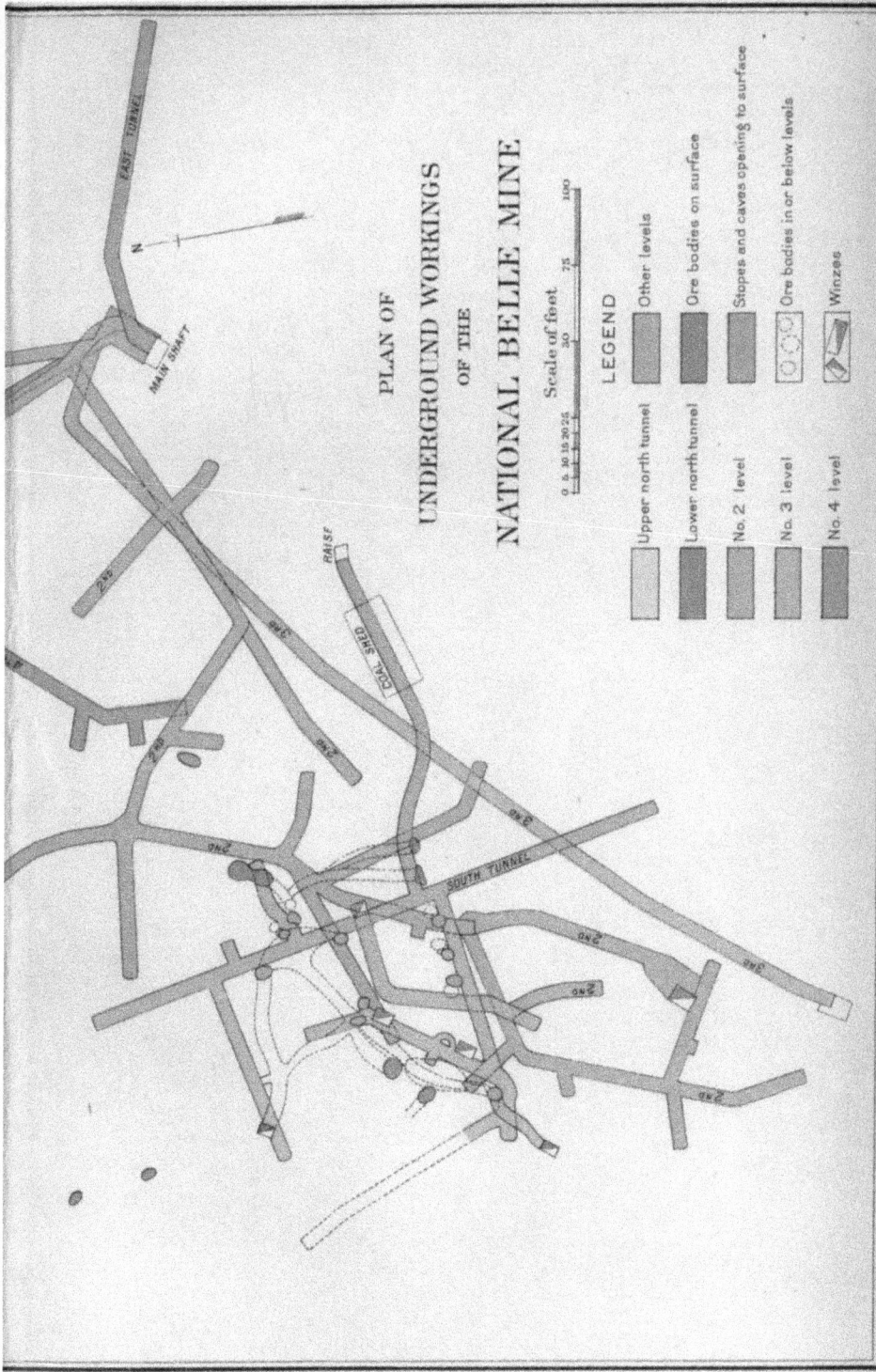

PLAN OF
UNDERGROUND WORKINGS
OF THE
NATIONAL BELLE MINE.

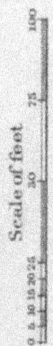

Scale of feet

0 5 10 15 20 25   50   75   100

LEGEND

Upper north tunnel	Other levels
Lower north tunnel	Ore bodies on surface
No. 2 level	Stopes and caves opening to surface
No. 3 level	Ore bodies in or below levels
No. 4 level	Winzes

EAST TUNNEL

MAIN SHAFT

RAISE

COAL SHED

SOUTH TUNNEL

Alexandra mine.—This is situated on the slope of Red Mountain, a few hundred feet above the Guston. The earlier workings are in one of the siliceous mounds common in this region, this one forming a small, elongated ridge, striking S. 85° W. The upper workings are reached through a tunnel. These seem to have two or more of the usual lenticular, nearly vertical ore bodies. One small one, stoped at the tunnel level, was elliptical in plan, the longer diameter striking N. 35° W. The ore pinched at the ends, but a small fissure, filled with kaolinite, can be seen continuing on into the country rock. The ore left in this stope is a fine-grained aggregate of pyrite, galena, and sphalerite, with specks of kaolinite. The country rock near the ore bodies is much altered, contains much kaolinite, and is thickly impregnated with fine pyrite. The ore is, in part at least, a replacement of the andesitic country rock.

About 400 feet below this workings is a second tunnel, from which issues a strong stream of highly ferruginous water. This tunnel is now only partly accessible. It runs through a mass of bleached and altered andesitic breccia, now changed in places to an aggregate of barite, kaolinite, and pyrite. As the mine was abandoned in 1899 or 1900, no information could be gained concerning its output or the portions of the workings not now open.

Grand Prize mine.—This property, a few hundred yards north of and about 150 lower than the upper Alexandra tunnel, is abandoned and at present wholly inaccessible. Its dump shows the usual materials common to the mines of the vicinity.

National Belle mine.[1]—The croppings of the bleached and silicified country rock associated with this ore body form a prominent knoll, over 200 feet in height, on the northwestern outskirts of the settlement of Red Mountain. The principal mine buildings, adits, and main shaft (490 feet deep), are on the eastern side of this knob, close to the railway. The mine was closed and deserted in 1897, and the superficial workings only were accessible. A plan of the workings, from a mine map made in 1890, is shown in Pl. XVI.

The first extensive development was in 1883, when about 980 tons of ore were produced, valued at nearly $70,000. This occurred chiefly in an oxidized form in caves, as carbonate of lead, one of which, occurring about 22 feet below No. 2 level (fig. 20), is known to have been 40 feet long, 20 feet wide, and 10 feet high. This cave ore carried from 5 to 7 ounces of silver and from 15 to 60 per cent of lead. For 1890 the product is given as $87,591, of which $4,500 was in gold, $29,091 in silver at coinage value, and $54,000 in copper. It is evident that the output for this year was chiefly from the lower workings,

---

[1] This mine is now often called the "American Belle mine," from the name of the company which last operated it, but the company's reports always refer to it under its original name, National Belle. The attempts sometimes made to change the names of mines with change of ownership are usually to be deplored, especially when, as in this case, the old name is well known in mineralogical literature.

in unoxidized, relatively low-grade enargite and pyrite.   In 1891 the property came into the possession of the American Belle Mines, Limited, a London company, which operated the mines until 1897.   The report of the Director of the Mint for 1891 credits the mine with a product of $145,721 in gold and silver only, with no returns for copper.   Since the sale of ore for this year yielded the company in the neighborhood of $110,000, the foregoing output probably includes both copper and lead.   At this time, however, the lower-grade copper ores could not be handled, as the nearest point at which they could be smelted was Denver, 500 miles distant.   It was not until 1892 that these ores could be treated in the Standard Smelter, at Durango.

Mr. S. F. Emmons visited the mine in 1886, and the following facts in this paragraph are taken from his unpublished notes.   He found

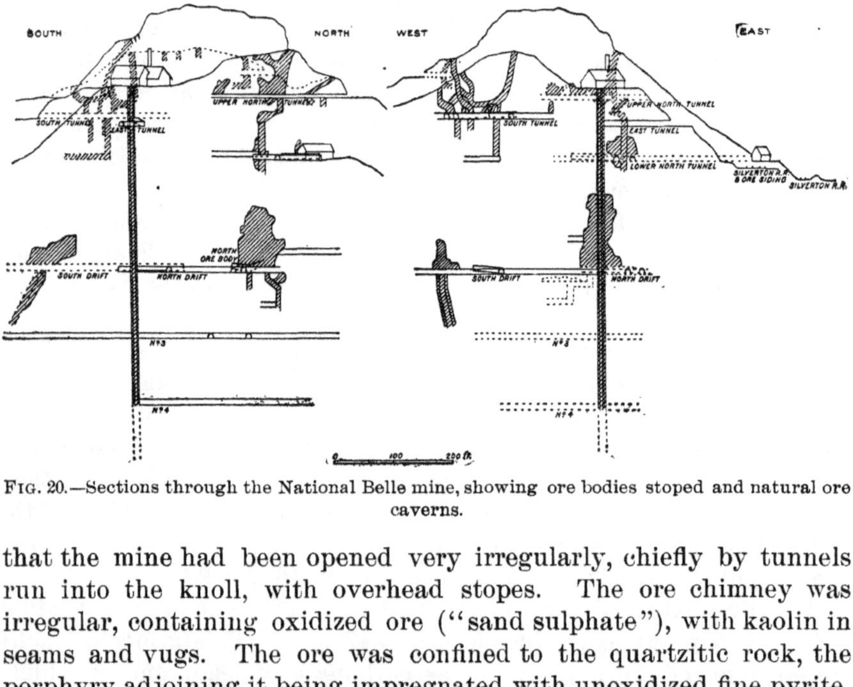

FIG. 20.—Sections through the National Belle mine, showing ore bodies stoped and natural ore caverns.

that the mine had been opened very irregularly, chiefly by tunnels run into the knoll, with overhead stopes.   The ore chimney was irregular, containing oxidized ore ("sand sulphate"), with kaolin in seams and vugs.   The ore was confined to the quartzitic rock, the porphyry adjoining it being impregnated with unoxidized fine pyrite, as in the Yankee Girl.   It seemed to follow two general fracture planes running at right angles to the main direction of fracturing, viz, N. 10° W.   He suggested that the copper sulphates had been leached out first, leaving the relatively insoluble lead salts and open spaces.   Mr. Emmons did not visit the lower part of the mine, which he was informed produced a low-grade enargite carrying chalcopyrite. The caves characteristic of the upper workings were said not to occur in the lower workings.   Schwarz,[1] in the paper already referred to, gives

---

[1] The ore deposits of Red Mountain, Ouray County, Colo.: Trans. Am. Inst. Min, Eng., Vol. XVIII, 1889-90, pp. 139-145.

a description of the secondary or oxidized ores of the American Belle and neighboring mines. They "occur above a former water line, either attached to walls of caves, as broken detached masses, or as a bed of clayey mud or sand, more or less completely filling the cave. The cave formation is identified with the massive outcroppings or knolls of silicified andesite ordinarily termed 'quartz.' These knolls * * * present a rough mass of quartz, cut up by cross fractures, and showing small vugs and cavities on the exposed cliff faces. The ore-bearing caves, which ramify throughout the mass, generally come to the surface along the cliff base." The caves are stated to come together in depth. The ores were mainly carbonates of lead and iron, with iron oxide, lead sulphate, and arsenates. Kaolinite and zinc blends were common. Galena occurred, but usually as a residue core. Such were the ores of the National (American) Belle, Grand Prize, and Vanderbilt mines, according to Schwarz. He briefly summarizes the characteristics of these secondary ores as follows:

1. The secondary ores are richer than the sulphide ores occurring below them.
2. The ores of adjoining or connecting caves are sometimes greatly different in grade.
3. In some cases the formation of the caves along fracture or cleavage planes is evident, but in others all traces of such planes are quite obliterated.
4. The cave walls are a porous sandy quartz, the sand from the disintegration of which forms part of the cave filling.
5. The line of change from oxidized to unoxidized ores, or the former water level, is very marked. It varies as much as 100 feet in elevation in properties within 1,000 feet of each other, rising to the south and west. The quartz outcrop rarely rises more than 200 feet above it.
6. In isolated cases may be found masses of the unoxidized ore, the enargite (i. e., in National Belle), above the line of change, in the vicinity of the secondary ores.

Mr. Schwarz ascribes the formation of the caves and the secondary ores within them to "surface waters" which "dissolved" sulphide ores and country rock as they moved along fracture planes.

In 1899 no thorough examination of the mine was possible, the croppings and an adit tunnel constituting the only accessible places for observation. The knob, or "blow-out," as the miners commonly term it, is in many places plainly an altered igneous rock, probably andesite or andesitic breccia. But the material in which the original porphyritic structure is still recognizable passes with no discoverable line of demarcation into a rock consisting chiefly of quartz, in the condition of a fine-granular crystalline aggregate. On the northwestern side of the knoll the country rock was originally a monzonite-porphyry mass of the same kind as that exposed just south of the Yankee Girl, or forming the cliffs just east of Summit Station. This intrusive rock has been altered in a similar manner to the andesitic breccia. The comparatively fresh facies in which the orthoclase and quartz phenocrysts lie in a fine greenish groundmass may be traced into a medium stage of alteration in which the orthoclase phenocrysts are still recognizable from their outlines, although changed to kaolin and diaspore, and finally into a white siliceous rock in which nearly all, or quite all, traces of original structure are lost.   The exact character of this alteration has already been discussed on page 126.   The silicified material of the knob is most irregularly and thoroughly fissured. The larger fissures often expand into irregular caves, which, as far as could be seen with the limited opportunities for observation in 1899, did not run in any single dominant direction.   A few of the smaller caverns only, near the surface, could be seen.   These were lined with crumbling, rust-colored, oxidized ore, beneath which there was often unaltered galena.   Sometimes these caves are lined and partly filled with a light, porous, pumiceous sponge of quartz.   This is not the cellular, honeycombed quartz resulting from the removal of ore by oxidizing waters, but is an original spongy crystallization of minute quartz crystals, with some interstitial kaolin.   It forms a light friable mass that is strongly suggestive of pumice.   It is analogous to the "sugar quartz" occurring in vugs, with sericite, in some of the gold-quartz veins of California, although not to the "sugar quartz" which results from a crushing of the vein.

The National Belle mine has long been noted for the abundance and purity of the kaolin found in its workings.[1]   It occurs as a soft, white filling of fractures in the altered country rock, or in brecciated zones as a matrix inclosing fragments of the wall rock.   It appeared to be, in part at least, of more recent origin than the sulphide ores, but no opportunity was available to study its occurrence in the deeper workings or in connection with the unoxidized ores.   When dry this kaolin forms a pure white powder which the microscope shows to consist of crystal scales, usually hexagonal in outline and about 0.1 mm. in average diameter.   The optical properties of these crystals have been described by Reusch.[2]   Two chemical analyses of this material are here reproduced, (I) by Hillebrand[3] and (II) by Hiortdahl:[4]

Analyses of kaolinite from National Belle mine.

Constituent.	I.	II.
SiO_2	46.35	45.57
Al_2O_3	39.59	41.52
Fe_2O_3	.11	
H_2O	13.93	13.58
F	.15	
	100.13	100.67
Less O for F	.06	
	100.07	

[1] R. C. Hills: Am. Jour. Sci., 3d series, Vol. XXVII. 1884, p. 472.
[2] Neues Jahrb. für Min., etc., 1887, Vol. II, pp. 70-72.
[3] Bull. U. S. Geol. Survey No. 20, 1885, p. 98.
[4] Neues Jahrb. für Min., 1887, Vol. II, p. 70.

As shown by E. B. Hurlbut,[1] however, not all of the white crystalline powder found in the National Belle is kaolin, but alunite also occurs in nearly identical form, clearly associated with enargite. Hurlbut's analysis is here quoted:

Analysis of alunite from National Belle mine.

Constituent.	Per cent.	Constituent.	Per cent.
SiO_3	38.93	H_2O	13.35
Al_2O_3	39.03	Insol	.50
K_2O	4.26		100.48
Na_2O	4.41		

No study could be made in 1899 of the occurrence of the unoxidized ore, chiefly enargite, of the deeper workings of the mine. Some of the enargite occurred lining caves below the zone of oxidation, forming beautiful radiating clusters of orthorhombic prisms several centimeters in length. These prisms are usually covered with a thin moss-like film of malachite and often partly incrusted with minute crystals of quartz.

From reports made to the company by the superintendent, Mr. James K. Harvey, for the years 1891–1896, the ore appears to have occurred in large masses irregularly distributed in the altered country rock and irregular in shape. The greatest recorded length of one of these bodies is about 75 feet, and they commonly varied in width from 1 to 4 feet, although occasionally as much as 60 feet wide. The greatest dimension was commonly nearly vertical, and in some cases the chimney-like form was pronounced (see fig. 20). The unoxidized ore was in part enargite, carrying at its best as much as 30 ounces of silver and 40 per cent of copper. It was associated with both copper and iron pyrites, some tetrahedrite, and galena in diminishing amount down nearly to the No. 4 level. The latter, at the second level, occurred as streaks in the copper and iron pyrites, and carried up to 44 ounces of silver and 60 per cent of lead in 10-ton carload lots. In a winze 50 feet below the third level galena occurred in small nodules only. In the National Belle ore the gold rarely amounted to more than a tenth of an ounce per ton, and was usually much less.

Practically no paying ore was found below the third level, explorations on the fourth level resulting only in finding small bunches of ore and masses of crumbling iron pyrite carrying less than one-tenth ounce of gold and about 5 ounces of silver per ton, with from 1 to 3 per cent of copper. The large north ore body, which on level 3 was a nearly solid mass of ore 31 feet long and 24 feet wide, carrying as a whole 4 to 7 ounces of silver, 4 to 6 per cent of copper, and .04 ounce of gold, and which had been stoped up over 100 feet above level 2, was not encountered in the bottom level. On level 2 this ore body was 75 feet long by 60 feet wide, and carried from 7 to 25 ounces of silver per ton and from 7 to 20 per cent of copper. It is said to have been chiefly iron, pyrite, and chalcopyrite, with some galena and enargite. This ore was firmer and of higher grade than that at the level below. From the annual reports of the company it appears that masses of iron pyrite were not limited to the deeper workings, but occurred above the second level, sometimes immediately beneath the carbonate ore. Much of the higher-grade ore, such as enargite, galena, and tetrahedrite, occurred as bunches of various sizes in the pyrite.

Owing to the decline of silver to almost half its former value, the general low grade of the National Belle ore, and the lack of facilities for its economic treatment, it was found impossible to operate at a profit, and the mine was closed in 1897. The ground below the fourth level is as yet unexplored, but the diminished value and amount of ore at that level deterred the company from further sinking.

The water in the National Belle mine was apparently not abundant or troublesome, as no mention is made of it in the superintendent's reports. From the location of the mine and its moderate depth (490 feet) no great amount would be expected.

White Cloud mine.—This mine, now abandoned, lies less than a quarter of a mile north of the Guston. It appears never to have struck a paying ore body. The shaft is sunk on the northern end of an elliptical knob of bleached and silicified San Juan breccia. The length of this knoll is about 50 feet and its trend about N. 10° E. It terminates rather abruptly, with no sign on the surface of any vein in the line of its major axis. The silicified material of the knob is impregnated with small crystals of iron pyrite.

Paymaster mine.—This also is an abandoned property lying a quarter of a mile northeast of the White Cloud. About 16 carloads of ore are said to have been shipped from one ore body, but the mine was never a large producer. It is credited with an output of $14,434 in 1890. In 1899 the shaft was filled with water to within about 200 feet from the surface, and the first level only was accessible. The country rock at this depth is all much altered and irregularly fissured and the fissures are frequently filled with seams of kaolin or talc, which is evidently a product of the alteration of the adjacent rock. Drifts and crosscuts run in all directions from the shaft, but apparently no ore was found on this level. The ore seen on the dump and in the ore bins was essentially a galena ore, associated with much pyrite.

Silver Bell mine.—Lying about a quarter of a mile north of the Paymaster, this, at one time one of the better-known and productive mines of the district, is also dismantled and inaccessible. The shaft, about 700 feet in depth, is now filled with water to within 100 or 200 feet from the surface. Sections through the mine are shown in fig. 21. According to Schwarz[1] this mine had, in 1888, produced large amounts of ore carrying from 500 to 1,000 ounces of silver per ton. This rich

---

[1] Proc. Colo. Sci. Soc., Vol. III, 1888. p. 84.

ore is said to have been bismuthiferous. The product for the three years that have been recorded in the Mint reports is as follows, the silver being quoted at coinage value and lead or copper not given:

Product of Silver Bell mine.

Year.	Gold.	Silver.	Total.
1887	$2,400	$77,600	$80,000
1888		72,404	72,404
1890	1,512	24,211	25,723

A specimen of the rich ore, obtained from a miner who had worked in the mine, proved on chemical examination to be an argentiferous lead sulphobismuthite. Kobellite, a sulphobismuthite of lead, in which

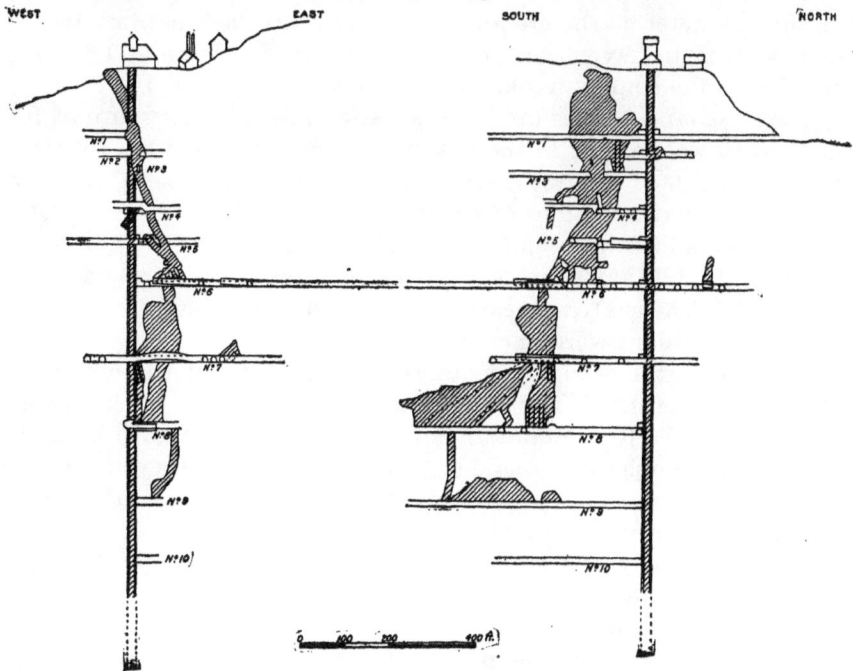

FIG. 21.—Sections through the Silver Bell mine, showing ore bodies stoped.

part of the bismuth is replaced by antimony, has been described by H. F. Keller [1] from this mine, associated with chalcopyrite and barite. He gives four analyses, of which the mean is here quoted:

Mean analysis of kobellite from Silver Bell mine.

Constituent.	Per cent.	Constituent.	Per cent.
S	18.39	Fe	1.50
Bi	28.40	Zn	.39
Sb	7.55	Gangue	.45
Pb	36.16		98.74
Ag	3.31		
Cu	2.59		

[1] Zeitschrift für Kryt., Vol. XVII, 1890, pp. 67-72.

In 1890 the Silver Bell, with other mines, came into the possession of the American Belle Mines, Limited, of London, and in 1891 the mine was worked through a vertical shaft 706 feet in depth, with ten levels. At this depth about 65 gallons per minute of strongly corrosive water had to be removed by the pumps. According to Superintendent Harvey, this water was "strongly impregnated with sulphuric acid generated from the pyrites in the ore bodies."

The rich ore found in the upper levels changed in depth to large bodies of low-grade ore consisting chiefly of pyrite and carrying less than 20 ounces of silver. Most of the high-grade ore seems to have been extracted when the mine came into the possession of its new owners, and after some attempt to work the low-grade ores in the face of the falling price of silver the mine was abandoned in 1894. Details of the character of the ore and its changes are lacking, but the ore bodies seem to have been generally similar, in shape and in their relation to the country rock, to those of the Yankee Girl.

Congress mine.—This ore body, situated about a mile south of Red Mountain, was located in the autumn of 1881, and sold the following year for $21,000. In 1883, according to the Mint reports, the mine produced about 2,500 tons of ore, valued at $220,000. It was worked through an adit tunnel and a shaft 350 feet in depth sunk from the tunnel. In 1899, after some years of inactivity, the mine was being reopened. The shaft was carried to the surface and new buildings and hoisting plant were in course of erection.

The old workings were not accessible in 1899, but the ore bodies appeared to constitute a chimney-like stock deposit of the kind characteristic of the Red Mountain mines. The ore occurs in irregular bunches, although there is said to be some indication of a vein running nearly north and south. The richest ore was found in a soft, decomposed streak near the surface and carried from $12 to $20 in gold. The ore is chiefly enargite, but there are also bunches of galena, which are said to occur on the peripheries of the ore bodies. The enargite as seen near the surface had been partly brecciated and recemented by a very compact, soft aggregate apparently largely composed of kaolin. In 1900 the mine was again idle.

St. Paul mine.—This property, lying a short distance south of the Congress, shows certain points of resemblance to the latter. Some ore was formerly extracted from a shaft 75 feet deep. The present workings comprise a vertical shaft 200 feet deep, with some drifts at the 200-foot level. The ore is low grade and consists chiefly of enargite with some galena, forming a fine-grained aggregate. It occurs in irregular bunches in much altered andesitic breccia of the Silverton series, which, near the ore, is heavily impregnated with fine iron pyrite, associated with small knots and veinlets of kaolin. No definite line can be drawn between ore and country rock, the former being largely a replacement of the latter. The ore bodies are found by following small seams of gouge or kaolin in the country rock. There is no regular direction to these seams, and they are themselves curved and irregular. The mine has not produced much, but was shipping ore in 1900.

Carbonate King mine.—This property, in Corkscrew or Humboldt Gulch, is opened by a shaft 395 feet deep. After lying idle for some years, work was resumed in 1900. In some ways this deposit is a connecting link between the stock or chimney deposits of Red Mountain and the lodes of the rest of the quadrangle. It resembles in this regard the Henrietta mine, which is described on another page. The ore of the Carbonate King occurs in a distinct fissure, striking about N. 5° W. and dipping west about 50°. The lode is a zone, 4 to 5 feet wide, of crushed country rock, apparently altered andesite, in which the ore occurs in bunches. There is usually a soft gouge on the foot wall of this crushed zone. The ore is sometimes frozen to the hanging wall, sometimes separated by gouge. Between the walls is much kaolin, occurring in the interstices between fragments of altered unmineralized rock. It also occurs as seams in the wall rock.

There is a second fissure of similar character which branches off from the main lode. It strikes N. 20° W. and dips southwest at 60° to 65°. It consists of a zone of crushed rock 4 to 5 feet wide, containing abundant kaolin, and separated by gouge from uneven walls. This branch fissure contains bunches of galena ore.

The ore of the main lode is an intimate mixture of pyrite, one of more sulphobismuthites, and probably tetrahedrite, with barite as the common gangue. No vein quartz was seen. The ore streak may vary from a few inches to 3 feet in width. It is often absent altogether.

The bismuthiferous varieties carry as much as 500 ounces of silver, with lead and copper. Ore worth less than $20 per ton can not be shipped.

Midnight mine.—This mine, situated in Corkscrew Gulch, a short distance below the Carbonate King, has not been worked for some years. It contained some good ore, but its extraction was rendered expensive by the abundant water met with. It is wholly inaccessible at present.

Other mines.—Many other mines, such as the Senate, Hudson, Charter Oak, and Humboldt, were all formerly active in this region, and have in some cases produced considerable ore. In those named the ore, when it occurred, appears to have had the form and characteristics already described as common to the typical Red Mountain deposits. All these mines had been abandoned in 1899 and consequently were not accessible for study, and very little information regarding them could be obtained. The Hudson, owned by the American Belle Company, was worked through a vertical shaft 253 feet in depth. The ore does not appear to have occurred in large bodies. On the first level ore consisting of pyrite and chalcopyrite (and probably enargite) carried from 20 to 26 ounces of silver per ton and 6 to 40 per cent of copper. The mine was closed in 1891. In 1895, 20 tons of enargite ore, carrying about 20 per cent of copper, were collected from the dump and realized $18.77 per ton. In its general character the ore of this mine seems to have been very similar to the low-grade ore of the National Belle, carrying enargite and copper and iron pyrite.

Other mines and prospects, showing more or less ore occurring in typical Red Mountain form, occur at the heads of Prospect and Dry gulches on the eastern slope of Red Mountain. The Mineral King, in Prospect Gulch, has a shaft 150 feet deep, sunk to develop a "chimney" of ore which outcropped in the bed of the creek. This ore shoot was cylindrical in form and nearly vertical, with a diameter of about 6 feet. The ore was a highly argentiferous copper ore, which is said to have contained as much as 1,300 ounces of silver per ton. It was surrounded by the usual envelope of silicified andesitic breccia, which graded outward into less altered forms of the same rock. This ore body pitched slightly to the south, and was followed down for 110 feet, when it pinched out. Forty feet deeper, drifts have been run in all directions in an endeavor to find a continuation of the pay shoot, but without avail. The rock at this level is traversed by numerous fissures, usually carrying some kaolin and accompanied by kaolinization of the wall rock. They sometimes widen into "pockets" filled with soft, unctuous yellow clay. One prominent fissure has a course of N. 80° W., another of S. 5° W., and still others run in various directions. It is probable that the original ore channel was formed by the intersection of two or more of these fissures. All of the andesitic breccia exposed in the artificial openings of this vicinity show some mineralization, with frequent small bunches of galena and other ore, but it was apparently only under exceptionally favorable conditions of fissuring that this diffuse mineralization was concentrated into ore bodies of workable size.

Near the head of Prospect Gulch the Galena Queen, with a shaft about 300 feet deep, shows a dump of similar character to that of the St. Paul. It is no longer worked.

Active prospecting was in progress in 1900 in the neighborhood of the Galena Queen, many of the shallow openings showing some ore.

The Henrietta mine, in the lower part of the same gulch, is described with the Cement Creek mines, on page 257.

The Webster, at the head of Dry Gulch, is worked through an inclined shaft 320 feet deep. The shaft is sunk on a fissure carrying a little gouge, in Silverton breccia. This fracture strikes about S. 10° W. The dip at first is about 75° to the south, but at 200 feet it flattens to about 45°. A little argentiferous copper ore has been found in bunches.

Saratoga mine.—On the eastern side of Ironton Park are several ore deposits which are of different types from those just described. The conspicuous alteration which has given the rocks of Red Mountain their brilliant tints is limited on the northeast by Grey Copper Gulch, and it is to this area of altered rock that the peculiar chimney-like deposits of the Yankee Girl type are confined.

The Saratoga was located in 1883, and has produced in all about $125,000. A far greater sum than this has been expended in development and in the erection of an expensive lixiviation plant. Considerable work was done in 1886, and in 1889 and 1890 about 100 men were employed. Since 1890 work has been confined to extracting pockets of ore with two or three men.

The lowest formation exposed at the Saratoga mine is a massive quartzite, white or light buff in color, which outcrops near the mill, with no recognizable bedding. Without much doubt this quartzite represents a portion of the Algonkian series, here rising a little above the level of Ironton Park. Lying unconformably upon this quartzite is a locally nearly horizontal formation of limestone, sometimes thinly bedded, which is probably of Devonian age and to be correlated with the limestone occurring south of Silverton. The thickness of this limestone is not known, but is apparently somewhat variable, and in places it certainly reaches a thickness of 75 feet.[1] Some thinly bedded portions at its top appear to have been silicified, forming a material resembling quartzite at first glance, but composed of vein quartz containing numerous little vugs. Above the limestone there occurs a thin stratum of telluride conglomerate, originally composed largely of limestone pebbles, but now silicified and with its conglomeratic character obscured. This appears to represent the rapidly thinning eastern margin of this conglomerate, for it is not recognizable a short distance within the hill, in the mine workings, where the volcanic breccias of the Silverton series, containing some massive andesite, lie upon the limestone.

The ore body of the Saratoga occurs in the limestone, and probably in part in the telluride conglomerate, and is very irregular. Near the surface of the hill it is a mass of soft, rust-colored, oxidized material, easily worked with pick and shovel, and traversed by numerous irregular seams of soft clay gouge. The pay ore occurs along these seams or gouges and is chiefly a carbonate of lead carrying silver, probably as chloride. Residual kernels and masses of white crystalline limestone are of frequent occurrence, surrounded by the rusty decomposed material of the ore body. The best ore is found resting on the oxidized upper surface of a mass of iron pyrite. This pyrite varies greatly in thickness, and in what is known as the lower tunnel it is nearly 100 feet thick—a mass of nearly pure, crumbling pyrite, resting on the quartzite, or occasionally on limestone, when the latter has not been wholly replaced. This pyrite carries, according

---

[1] According to Kedzie, loc. cit., its mean thickness is 140 feet.

to Mr. W. R. Fry, part owner and former superintendent of the
mine, from 3 to 4 ounces of silver and 0.4 to 0.5 of an ounce of
gold, being thus too poor to work. Several tunnels have been run
into the hill, but owing to the easterly dip of the ore they all pass
through the ore body and, if continued far enough, come into the
volcanic rocks of the Silverton series. This oxidized ore is not capped
by the overlying andesite, but passes upward into the loose material
and soil of the present surface of the hillside. When, however, the
ore is followed far enough to the eastward it passes beneath andesitic
breccia and changes to a sulphide ore, consisting of pyrite and a little
chalcopyrite, with occasional bunches of partly oxidized pay ore, carrying galena and argentite. This ore is the result of a replacement of
the limestone to varying depth along the contact between it and the

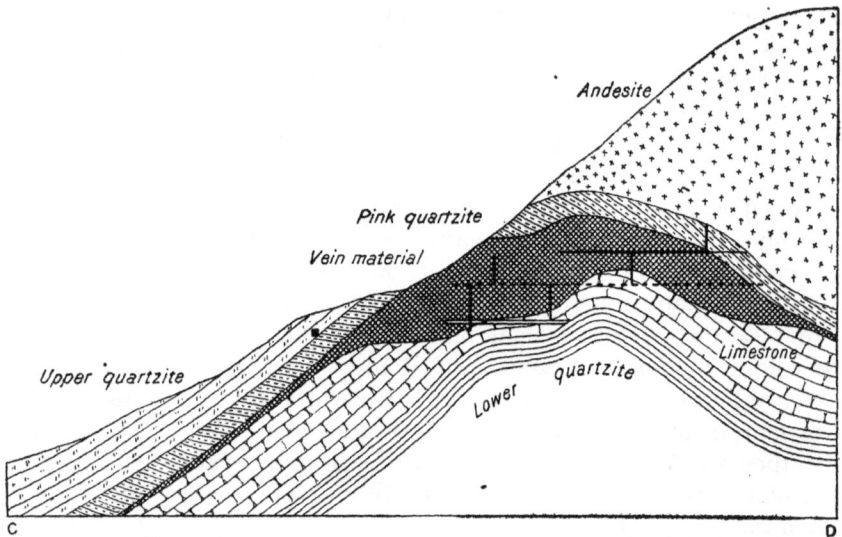

FIG. 22.—East-and-west section through the Saratoga mine. (After G. E. Kedzie.)

overlying andesitic breccia. It varies in thickness from a few inches
up to 3 feet. The overlying volcanic breccia is usually much altered,
thickly impregnated with pyrite, and traversed by veinlets of the
same mineral. The ore is not distributed at random through the contact plane between limestone and breccia, but the bunches of ore thus
far found lie along a line bearing about N. 30° E., as if the mineralization had been controlled by some fissure or fissures having this
direction. As this is the dominant direction of fissuring in the vicinity, this seems not unlikely. The actual fissure which determined
the mineralization is, however, not readily identified. It may lie to
the westward, in which case its intersection with the ore horizon has
been removed by erosion. It may be the large lode, presently to be
described, which lies to the eastward, or it may be some less conspicuous fissure. A small fissure having this strike of N. 30° E. was

noted, which faulted the ore body, throwing down the southeast portion a few inches. This, however, appeared to be a post-mineral fault and is probably a late manifestation of the same faces which produced the earlier dominant fissuring of identical trend.

The upper tunnel of the Saratoga has been run 1,000 feet into the hill. It passes through the ore and limestone into a fine-grained, massive andesite, and about 600 feet from its mouth cuts a strong lode carrying pyrite, in a gangue of vuggy, white quartz. This vein is from 2 to 3 feet wide, striking N. 30° E., and dipping southeast at about 85°. It is too low grade to work. Beyond the lode the tunnel passes through about 50 feet of iron pyrite, limestone, and quartz—all much shattered. This appears to be a mineralized fault breccia, disturbed by later movement. On the southeast of this shattered zone is andesitic

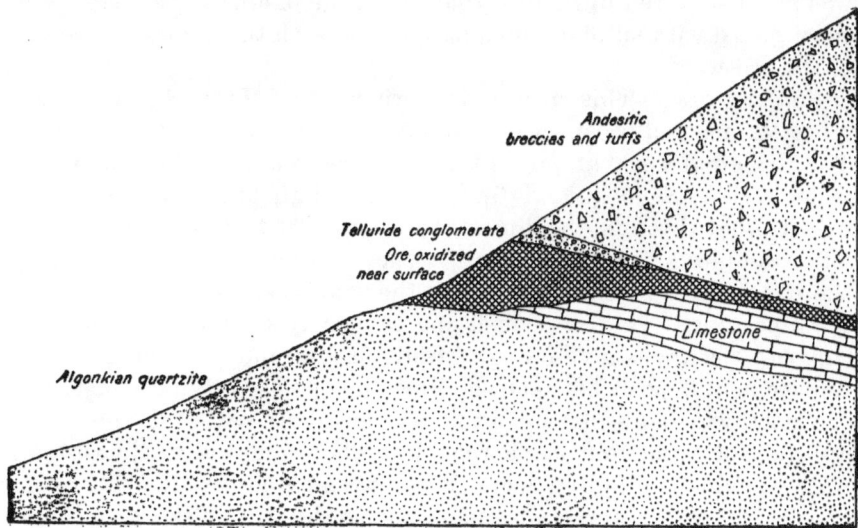

FIG. 23.—Diagrammatic east-and-west section, illustrating in a general way the occurrence of ore in the Saratoga mine.

breccia, but owing to the gas in the end of the tunnel it was found impossible in 1899 or 1900 to make any further observations.

It is to be noted that G. E. Kedzie[1] has interpreted the form of the Saratoga ore body in a different manner from that set forth in the foregoing pages. His section is herewith reproduced (fig. 22). This same section is also given in Phillips and Louis's Treatise on Ore Deposits, although these authors remark that there seems to have been hardly enough work done to pronounce very positively as to the character of the ore body.[2] At the time of Kedzie's visit the mine was more accessible throughout, but not enough ground had been exploited upon which to base all the details of his section. Thus his "upper

[1] The bedded ore deposits of Red Mountain mining district, Ouray County, Colo.: Trans. Am. Inst. Min. Eng., Vol. XVI, 1887-88, pp. 570-581.
[2] A Treatise on Ore Deposits, J. Arthur Phillips, second ed, rewritten and enlarged by Henry Louis, London, 1896, p. 163.

quartzite" appears to be the same as the massive quartzite here included in the Algonkian. His "pink quartzite" is evidently the silicified telluride conglomerate, while his "lower quartzite" could not be identified in the Saratoga workings. He states that it is penetrated by the Jackson tunnel, about 800 feet southwest of the Saratoga, but this tunnel is no longer accessible. A truer representation of the ore body and its relations to the inclosing rocks is believed to be afforded by the rough diagram, fig. 23.

In 1900 a drift was being run along the ore-bearing horizon toward the northeast with the object of ascertaining whether any ore has been deposited at the intersection of this horizon by the large vein cut in the tunnel above. This seems an advisable piece of prospecting. Hitherto the development has been rather unsystematic, and it was not discovered until 1900 that the large oxidized ore bodies were continuous, with sulphide ores passing beneath the andesitic breccias to the eastward.

Baltic group.—This group, comprising the Maud S., Mono, and other claims, is a prospect in Brooklyn Gulch, south of the Saratoga. The principal work at present is in connection with the Mono lode, which strikes N. 43° E. and dips northeast at about 75°, and is reached by a tunnel through the limestone about 400 feet in length. This lode occupies a fault fissure of at least 100 feet throw, which drops the overlying andesitic breccia on the southeast until it lies against the limestone. The ore, consisting of nearly solid pyrite and chalcopyrite, occurs as a vein in the fault fissure and as a relatively flat replacement deposit between the limestone and overlying andesite on the northwest side of the fault. Where both walls of the fissure are andesitic the vein is narrow; but where one wall is limestone it widens, evidently by replacement of the calcareous wall. There has been some post-mineral movement along the fault fissure, as shown by the presence of gouge and slickensides.

The ore is low grade, carrying at its best about 30 ounces of silver and 12 per cent of copper. Ore of this general character is very abundant both here and in the Saratoga mine, but can not at present be economically worked.

It is evident that the ore bodies have been only partly prospected. No attempt has been made to find the ore which very likely exists at the top of the limestone on the downthrown side of the fault. This would seem to be an excellent opportunity to employ a diamond drill, provided the known ore bodies indicate that the ore on the other side of the fissure is worth searching for.

The Maud S. is also on a fault fissure, which crosses the gulch near the mouth of the Mono tunnel. Its general strike is N. 30°–40° E., and its dip southeast at about 85°. This also is a normal fault, letting down the andesite against the limestone on the southeast for at least 25 feet. Some argentiferous copper ore was formerly taken from this claim, the ore occurring in replacement bunches in the limestone alongside the fault. The ore is associated with a soft, black alteration product or residue of the limestone. Between the Maud S. and the Mono fissures is a third fault exposed on the Mono tunnel, of variable northeasterly strike, dipping southeast at about 70°. Here also the andesitic tuff or breccia is brought down on the southeast so as to be in juxtaposition with the limestone. The fault is well shown in a long drift run on it to the southwest. The fissure contains soft gouge and shows slickensides, but is about destitute of ore.

The Brooklyn tunnel, 200 feet below the Mono tunnel, enters in limestone, which is probably over 200 feet thick in this gulch. The tunnel has several branches, the longest about 800 feet in length. It cuts through at least two large masses of crumbling iron pyrite, formed by replacement of the limestone. These masses are bounded on the southeast by a northeast fault plane, which brings them against a downthrown mass of andesitic breccia. This fault is supposed to be the one already described between the Maud S. and Mono fissures. The only one of these pyrite masses which could be reached in 1900 is at least 6 feet in thickness (the height of the drift), and probably considerably more. Northwest of the fault it extends irregularly into the limestone for over 100 feet. Northwest of these masses of pyrite is a long northeast drift, which appears to follow a fault for most of its course. On the southeast side of the drift is limestone with bunches of iron pyrite. On the northwest side is an irregularly laminated and much disturbed quartzite, which may probably be the "lower quartzite" of Kedzie. The general strike of this laminated quartzite, i. e., of the laminæ, is about northeast. The general dip is northwest, but exceedingly variable. This quartzite belongs presumably below the limestone, but whether it is the same as the massive quartzite at the Saratoga Mill or is a younger quartzite not there seen, and perhaps of Devonian or Silurian age, is not at present known.

Jay Eye See mine.—This also is a prospect situated in Parole Gulch, south of the Maud S. A shaft was formerly sunk on a heavy lode or bunch of pyrite, associated in this case with a little quartz gangue. No pay ore was ever found in this shaft. Subsequently a tunnel run in just west of the shaft cut a lode which in places carried tetrahedrite, pyrite, and chalcopyrite up to 4 feet in width. The ore occurred in lenses connected by a barren seam of clay gouge in soft decomposed andesitic tuff-breccia, and the mine never paid steadily. The tetrahedrite sometimes carried 400 ounces of silver per ton. The bodies of pyrite cut in the original shaft are quite probably part of the Mono lode. But parallel fault fissures of southeasterly dip and normal heave are here so common, however, that identification of any single one is almost impossible.

Barstow mine.—This mine, formerly known as the Bobtail, was the only mine in the Red Mountain district working and producing ore in 1899 and 1900 on any considerable scale. It is situated about a mile west of the Guston, on a prominent lode running up Commodore Gulch, on which the Iranstan, Bobtail, Hector, Saratoga, and Moscow claims are located. The Iranstan and Bobtail were worked early in the eighties, but have been idle for the last few years, until work was resumed by the Barstow Mining and Milling Company. The mine is opened by an oblique crosscut tunnel and all the work is on one level.

The lode, which has a general northwest-and-southeast strike, and dips southwest at about 50°, is a large one, 30 to 40 feet wide, but the ore is confined to one or two rich streaks in its middle portion. The lode contains much soft, altered country rock and is traversed by clay gouges, making it exceedingly treacherous and difficult to work by overhand stoping. The mass of the lode contains pyrite, chalcopyrite, sphalerite, and some bunches of low-grade galena. The gangue is chiefly quartz, often cavernous and drusy, with a little fluorite. The best ore is a light-gray metallic mineral called "gray copper" at the mine, which, when pure, assays as much as 2,000 ounces of silver and 25 ounces of gold per ton. The gold always accompanies and is proportional in amount to the silver. Chemical examination shows that the so-called "gray copper" is in this case an argentiferous lead sulphobismuthite containing almost no copper, probably the mineral cosalite. But the present material obtainable (sp. gr. 7.04 at 30° C.) proves on chemical investigation to be a mixture of the lead sulphobismuthite with pyrite and some telluride. Under these circumstances it was inadvisable to spend time in making a complete quantitative analysis. From his preliminary tests, however, Dr. W. F. Hillebrand has kindly supplied the following rough approximation to the composition of the mixture:

Approximate composition of mixture from Barstow mine.

Constituent.	Per cent.	Constituent.	Per cent.
Pb	45.5	Cu	.3
Bi	22.5	Fe	3.5
Ag	6.0	S	14.1
Au	.06	Insol	1.5–2.0

The lead of cosalite is known to be sometimes partly replaced by silver and copper. In this particular case the silver is evidently accompanied by some gold, which is probably combined with tellurium and not directly with the sulphur and bismuth. The sulphobismuthite is often closely associated with chalcopyrite, which carries little value. The rich ore streaks are more or less lenticular in form and are usually separated from the mass of the lode by wet clay gouges. The best ore occurs nearer the footwall. As the workings are confined within the mass of the lode, the relation of the latter to the wall rock could not be studied.

Silver Ledge mine.—This property, situated about half a mile north of Chattanooga, and operated by the Ledge Mining and Milling Company, is worked through a shaft about 400 feet in depth, sunk in the bed of Mineral Creek. Most of the work at present is on the fourth level, at a depth of 360 feet, and in the stopes immediately above it. The workings are all in rhyolite. This rhyolite forms a considerable area extending to the westward of the mine with a rather irregular boundary. Its eastern limit is very close to the mine workings. For more detailed accounts of this rhyolite mass the reader is referred to the forthcoming work of Mr. Cross. The conclusion reached by him at the present writing is that it represents an area of the Potosi rhyolite[1] which has been faulted down into the surrounding rocks. If this view is correct the rhyolite is limited in depth. There are no data now available for even a rough estimate of the thickness of this rhyolitic mass, but if mining be pushed deep enough it will be passed through. If, on the other hand, the rock is intrusive, it may extend to practically unlimited depth.

There are certain features about the rock itself which sustain Mr. Cross's hypothesis. It has many of the characteristics of the surficial rhyolite flows of the Potosi series, and, as far as could be seen in its altered condition near the mine, has not the crystalline structure that might naturally be looked for in an intrusive mass of such size. Moreover, on the 400 level of the Silver Ledge occur masses of a compact, slightly pinkish rock, which are sometimes fairly distinct from the white altered rhyolite and sometimes appear to grade into it. This pinkish rock occurs as isolated patches or nodules in the rhyolite, and at one point forms a considerable mass, of which the bottom is not yet exposed. Under the microscope this rock shows numerous small grains of quartz in a very minutely crystalline sericitic groundmass. It is, without very much doubt, a devitrified glassy rhyolite or rhyolitic tuff. Such material could hardly occur in such a place if the rhyolite were intruded. It strongly suggests volcanic or effusive conditions. It is even possible that the appearance of this material indicates an approach to the bottom of the rhyolitic mass.

The work of Mr. Cross has established the existence along the eastern border of the rhyolite mass of a fault which, just southeast of the mill, brings the rhyolite against andesite on the east. The line of this fault would carry it not far east of the main shaft, and it was thought that a crosscut which had been driven 100 feet to the eastward from the 400 (360 feet) level might be found passing through this fault fissure into andesite. About 70 feet east of the shaft the crosscut does pass through a strong gouge-filled fissure, but the

---

[1] See the Telluride folio, No. 57.

rock, although all much altered, shows no change either in hand specimens or under the microscope. It all appears to have been originally rhyolite.

The ore, consisting of galena and sphalerite, occurs replacing and impregnating the rhyolite in the vicinity of two or more fractures, which have a general strike of about N. 27° E. and dip southeast at a high angle. These fissures contain crushed country rock and ore, and there has evidently been late movement along them since the ore was deposited. They are probably merely a portion of a pronounced fault zone, which here follows the general trend of the creek. The ore bodies are found by following these fractures, but are not confined within their walls. They extend irregularly into the rhyolite for 30 or 40 feet to the west, the low-grade galena, associated with considerable sphalerite, sometimes forming large masses wholly replacing the rhyolite, but more often impregnating it thickly in small nests and isolated particles. There are no walls to the ore bodies, and they are followed into the rhyolite until the proportion of ore to gangue becomes too small for profitable working. Very little ore has thus far been found to the east of the fissures, although the workings have as yet revealed no change of country rock in that direction.

The rhyolite has been altered in connection with the ore deposition to a finely crystalline mass of sericite, kaolin, and quartz in varying proportions, serving as gangue for the ore.

The principal fractures of this mine appear to be comparable to the so-called ore break in the Guston mine. It is very probable that the zone of fractures extending up Mineral Creek and passing just east of Summit forms part of the fissuring found in the Lake and National Belle mines near Red Mountain.

The ore of the Silver Ledge is concentrated in a small mill run by steam. It is equipped with 1 Gates crusher, 2 sets of wells, 4 three-compartment jigs, 10 stamps, and 4 double-deck and 2 single-deck Wilfley tables. The capacity is about 60 tons a day. The principal output of the mine is a low-grade galena, which is shipped in the form of concentrates. A little free gold, associated with a telluride supposed to be calaverite, was found on the 300 level as a thin skin on a slickensided fracture surface in the usual impregnated rhyolitic country rock. This was the only occurrence of free gold known in the mine.

The future development of this mine is likely to be of great interest. It is evident that if the interpretation of the rhyolite as a down-faulted sheet or flow be correct, then an entirely different set of conditions will be encountered when the bottom of the rhyolite is reached. In all probability the ore deposit will become a simple fissure vein or a series of parallel veins. Moreover, further exploration of the country east of the shaft may reveal ore bodies yet unknown. The crosscut already begun in this direction might perhaps be extended with advantage until the andesitic rock of the Silverton series is reached.

Magnet mine.—This, an old location, is situated on the east side of Mineral Creek, near Burro Bridge. It produced a little rich silver ore in the early eighties, but owes its interest at present to the occurrence of a telluride of silver, probably hessite.

The workings comprise a tunnel about 540 feet in length and a few superficial openings. The lode strikes N. 85° W. and dips from 60° to 80° to the south. It is a compact stringer lode, in places 2 feet wide, which contains considerable mineralized and altered andesitic breccia, which forms the country rock. The stringers are filled with quartz, "frozen" to the walls and containing bunches of ore. A radial structure of the quartz is common and vugs are numerous. The ore, consisting of argentiferous galena, a silver telluride (hessite?), a little free gold, sphalerite, chalcopyrite, and pyrite, occurs partly in the quartz and partly replacing the included fragments of breccia. The richest ore, containing the hessite, is usually associated with the radial quartz and often occurs in the interstices between the outer ends of the crystals. The ore is unevenly distributed through the lode and has been found only in small bunches.

Brobdignag claim.—This abandoned prospect is of interest as the locality whence some zinkenite was formerly obtained, which was described and analyzed by Dr. Hillebrand.[1] No ore was visible in 1900. The country rock was apparently originally an andesite, but is altered to a white aggregate consisting chiefly of quartz and barite.

Zuñi mine.—This mine, well known in mineralogical literature as the source of guitermanite and zunyite, is situated near the summit of Anvil Mountain, at an elevation of about 12,000 feet. The ore body was worked from the surface by an open cut and a shaft and by two tunnels. The upper tunnel branches, the two breasts being 156 and 180 feet in from the surface. A second tunnel, 200 feet below the upper one, is 540 feet in length. There is no connection between these two tunnels.

The surface ore of the Zuñi, discovered in 1881, was chiefly massive sulphate of lead (anglesite) containing abundant small tetrahedral crystals of unaltered zunyite. This soon changed to unoxidized guitermanite in characteristic association with zunyite. This was the chief ore of the upper tunnel, and occurred in a mass about 60 feet long and 15 feet wide. Below the tunnel level this bismuthiferous lead ore decreased in width to about 6 feet and changed gradually to a copper ore, consisting chiefly of enargite, with pyrite, kaolin, and a little barite. The enargite carried about 216 ounces of silver and 40 per cent of copper. In the lower tunnel the ore body, about 40 feet long and 12 feet wide, is a soft mass of kaolin and embedded pyrite, with nests of enargite. The pyrite occurs chiefly in octahedra and is said to carry a very little gold. The ore at this depth contains many fragments of the immediate country rock, a much altered

[1] Bull. U. S. Geol. Survey No. 20, p. 93.

porphyry, apparently originally an andesite or latite. The contact between the ore and these fragments is perfectly sharp, and the latter often have sharp, angular corners.

Small masses of similar ore, showing no apparent connection with the main body, were encountered at several points in the tunnel. They often contain small caves or vugs lined with crystals of quartz, barite, enargite, and pyrite. The country rock around such caves is brecciated and the crevices are filled with ore. The ore in the lower tunnel, consisting chiefly of pyrite, is too low to work. The longer diameter of the Zuñi ore body apparently strikes about N. 15° W., and the dip is from 75° to 80° to the west.

There are several smaller prospects of the general type of the Zuñi on Anvil Mountain. These are usually small irregular bunches or "chimneys" of ore of small horizontal and vertical extent. They sometimes occur immediately at the surface, as in a claim owned by John Roland, where masses of galena, showing only superficial oxidation, occur at the grass roots. In some cases the surface ore is argentiferous enargite (Dewey claim). There can be little doubt that many undiscovered small irregular masses of ore occur within the altered andesites and other igneous rocks of Anvil Mountain. But such is their irregularity in occurrence, and so limited are they in extent, that small inducements are held out for extensive deep prospecting.

Meldrum and Hammond tunnels.—The success of the Revenue tunnel has encouraged the prosecution of other extensive tunneling ventures in the vicinity. The Meldrum tunnel is being driven on a coarse about 4° S. of W. from a point on Red Creek opposite the Silver Bell mine through to Bridal Veil Basin in the Telluride quadrangle, a distance of over 4 miles. The tunnel is 12 by 12 feet and had run about 900 feet in 1899. Work was also in progress at the western end of the tunnel line. In 1900 all work was suspended. The Hammond tunnel also enters from Red Creek, nearly west of the Yankee Girl mine and has a course of N. 80° W. It had a length of about 2,000 feet in September, 1900—all in andesitic breccia. As a rule the rock is solid, very little timbering being required. But in the neighborhood of 1,000 feet from the mouth some zones of recent brecciation, dipping eastward, are cut through. These indicate faulting, possibly connected with the movements of landslide character revealed in the topography above the tunnel.

These tunnels are evidently designed to cut the continuation of the Tomboy lode and other less known and undeveloped lodes traversing the high ridge west of Red Creek. A very large outlay of capital will be required to complete them. Without expressing an opinion upon these particular enterprises, it may be pointed out that, in so far as the projectors of long tunnels count upon finding richer or more abundant ore than is indicated near the surface, they are playing a game of chance in which the probabilities are decidedly against them. But where the object is to cut a lode of proved value and continuity, as was the case with the Virginius lode tapped by the Revenue tunnel, the unertaking involves no more than the legitimate risks incident to most large business enterprises.

MINES OF SULTAN MOUNTAIN.

North Star mine.—Situated on the northern slope of Sultan Mountain, about a mile southwest of Silverton, this mine was at one time one of the largest and most productive in the quadrangle. It was located in 1876, but not extensively worked until about 1881, when it became the most productive mine in San Juan County, some 3,000 tons of ore having been mined. At the close of 1882 the workings comprised about 1,700 feet of drifts, and, according to the report of the Director of the Mint, over 100,000 ounces of silver and 1,500,000 pounds of lead had been extracted. The ore at that time was shipped to St. Louis, and is said to have carried from 70 to 80 ounces of silver and 35 per cent of lead. In 1883 the output is given by the Mint reports as $200,000, and in 1884 as 1,100 tons, value not stated. The same source gives the product for 1888, 1890, and 1891 as $46.522, $206,072, and $337,192, respectively, the silver being reckoned at its coinage value. The mine was finally closed about 1897, it having been practically worked out down to the tunnel level.

The main adit is a tunnel (No. 7) 7 feet in the clear and 2,100 feet in length, which has its entrance about 50 feet above Mineral Creek and penetrates the mountain in a southwesterly direction, cutting the lode nearly at right angles. The latter is a fissure vein in monzonite, striking nearly northwest and southeast, but showing considerable variation in trend. The dip is to the southwest at from 45° to 60°. The monzonite near the vein is decomposed and soft, and when the mine was visited in 1899 it was found impossible, on account of caving, to reach the ends of any of the drifts or to enter the stopes. The vein contains much quartz, and is usually wide, the average width according to one account being about 8 feet. The pay streak, generally about a foot wide, and consisting of galena and tetrahedrite, occurs very persistently along the footwall side of the vein. The pyrite occurring in the more quartzose portion of the vein is said to carry some gold, and small amounts of the latter have been found free. The average tenor of the ore is reported to have been about a half ounce of gold, 40 ounces of silver, and 50 per cent of lead.

About 100 feet southwest of the North Star vein is the Crown Point and Wheal Alfred lode, a nearly parallel quartz vein carrying some gold. The dip of this lode is steeper than that of the North Star. It shows pyrite, chalcopyrite, and tetrahedrite, with a little specularite, in a quartz gangue. Barite is probably present in both lodes.

An abundant flow of water issues from the main tunnel, and proved troublesome when the mine was being worked on account of the rapidity with which it corroded rails and pipes.

The North Star mill, erected about 1890, is equipped with 2 Blake crushers, 3 sets of rolls, 8 three-compartment jigs, 2 four-compartment jigs, 20 stamps, and 3 Woodbury tables. The power is water. This plant was running in 1899 as a custom mill.

Belcher mine.—This mine, now abandoned, is situated in the high basin opening to the northeast on the northeast side of Sultan Mountain, the principal adit being at an elevation of about 11,250 feet. It was located in 1876, yielded over 300 tons of ore in 1881, and in 1882 had become a steady producer. In 1883 there were three levels opened by tunnels, and a product of 1,700 tons of ore is recorded, averaging 38 ounces of silver and 40 per cent of lead. The outputs for the years 1883, 1884, 1890, 1891, and 1892, as given in the Mint reports, were $85,000, $13,500, $8,467, $34,995 and $63,510, respectively. The Belcher lode is in monzonite, and is said to be very similar to the North Star lode and to have nearly the same strike and dip. The workings were wholly inaccessible in 1899, and no observations could be made on the lode itself. According to Mr. W. H. Thomas, at one time part owner of the mine, the galena ore was less abundant in the lower levels, its place being largely taken by sphalerite. According to another former owner, the lode was practically worked out down to the tunnel level, and there is no reason to suppose the good ore to be any less abundant below the bottom workings.

Fragments of ore seen on the dump of the lowest tunnel showed galena, sphalerite, pyrite, chalcopyrite, and tetrahedrite in a gangue of quartz and barite with some calcite. The ore was sorted, sacked, and carried on burros down a steep trail to Silverton.

Empire group.—Through the present Empire tunnel it is intended to work a group of claims on the northeast slope of Sultan Mountain, overlooking Silverton and Bakers Park. The more important of these claims are the Empire, Little Dora, Victoria, Jennie Parker, Ajax, and Hercules. They cover portions of three important lodes, which are only roughly parallel and which have a general trend of about N. 20 W. The Jennie Parker and Hercules are supposed to form a continuation of the North Star lode.

The Empire and Jennie Parker were worked at a very early date by Melville & Summerfield, who in 1876 had a shaft 50 feet deep on the Empire lode and were erecting reduction works near the Animas River. They were apparently not successful, and the mine was idle for some years. In 1882 the Little Dora lode, which in 1881 was shipping ore from a shaft, was struck in an upper tunnel. This is a clean-cut fissure vein, usually about 2 feet wide, and, although small, has produced a fair amount of ore. The Empire at this time was opened by a shaft 125 feet deep and is said to have had about 18 inches of excellent ore, largely tetrahedrite, which averaged 100 ounces of silver and 10 per cent of copper. In 1883 the Ajax and Victoria were being developed by tunnels. None of these lodes has thus far been worked on an extensive scale.

They are all simple, nearly vertical fissure veins in monzonite. The ore is mineralogically similar to that found in the North Star lode, viz, galena, sphalerite, tetrahedrite, chalcopyrite, and pyrite, in a gangue consisting chiefly of quartz with some barite. With the prevailing price of silver, the ore is, as a whole, low grade.

The Empire tunnel enters the mountain about 150 feet above the bed of Mineral Creek, and has a straight course of S. 35° W. It cuts the Little Dora vein about 1,230 feet from the tunnel mouth. About 900 feet farther in it passes through a fissure about 10 feet wide, filled with crushed rock and gouge, but containing no ore. This is probable the Jennie Parker-Hercules lode, which it was expected to cut about 800 feet beyond the Little Dora. The tunnel was 150 feet beyond the large fissure in August, 1900, and still driving on in solid monzonite. An upper tunnel, 208 feet vertically above and about 400 feet west of the Empire tunnel, formerly known as the Montezuma and used to work the little Dora, has been relocated as the Boston tunnel and extended through to the Hercules lode, here found to be a strong vein up to 10 feet wide, dipping to the southeast at about 85°. It showed evidence of post-mineral movement along the fissure, particularly along the foot wall. The Empire-Victoria vein, as drifted on from the Boston tunnel, is fairly regular and shows a width up to 7 feet. Its dip is somewhat variable—in places as low as 65° to the southwest. Hübnerite, associated with fluorite, occurs as crystalline streaks and bunches next the hanging wall, and usually surrounded by soft gouge. It is probably of later formation than the ore.

On account of the very favorable location of the Empire tunnel it is expected that ore considerably below what is commonly regarded as the lowest grade for profitable working in this region can be here successfully handled.

Other mines and prospects.—The Fairview is a prospect a little more than a mile south of Silverton, at the contact between the intrusive monzonite stock of Sultan Mountain and the Algonkian schists and overlying limestone. The property includes a nearly north-and-south vein, carrying ore mineralogically similar to that of the Hercules lode. The Devonian limestone also, near the monzonite, has been mineralized and carries particles of native silver. Argentite is also reported. Two tunnels, one 200 and the other 600 feet in length, have been run, but there is not enough development to allow satisfactory study of the deposit.

The King lode is a quartz vein in Algonkian schists in Cataract Gulch, and outcrops just below the basal conglomerate (Cambrian?) which underlies the Devonian limestone and rests unconformably upon the edges of the schists. The strike of the vein is N. 6° W., and it dips westerly at 78°. It varies in width from 18 inches to 4 feet, and is destitute of gouge. The inclosing schists strike N. 85° E., and their planes of schistosity stand nearly vertical. The vein thus crosses the schistosity nearly at right angles. No offsetting of the schists could be detected on opposite sides of the fissure. The King lode is faulted by a small vein dipping south about 50°, and the northern portion of the lode is thrown about 12 feet to the west.

The ore of the King lode consists of galena, gray copper (probably tennantite), and chalcopyrite, with some sphalerite and pyrite, in a quartz gangue. It carries about 0.1 ounce of gold, 55 ounces of silver, and 15 per cent of lead.

Close to the King lode is a deposit of chalcopyrite, which occurs at the contact between the limestone and the underlying quartzite. This deposit is apparently directly connected with a fissure of rather flat southerly dip, which traverses both quartzite and limestone.

The Moles mine, now abandoned, is on a nearly vertical fissure vein, about 2 feet wide, in limestone, chert, and shales of Devonian age. The vein strikes N. 72° W., and occupies a fault fissure.

MINES OF CEMENT CREEK.

General.—The ore deposits of Cement Creek occur in lodes, which in Ross Basin attain great prominence. The ores are prevailingly low grade, carrying silver and lead, or, as in the case of the Gold King mine, a large proportion of gold. The occurrence of hübnerite, although it is not at present worked on a commercial scale, lends additional interest to the ore deposits of this portion of the quadrangle. A few of the mines within the Cement Creek drainage have been described in connection with adjacent portions of the region, for, as elsewhere pointed out, the geographical grouping of the ore deposits of the quadrangle has no necessary significance beyond convenience of description, and is accordingly not rigidly adhered to when other courses seem desirable.

Gold King mine.—This property, with which are included the Sampson and American mines, now under the same ownership, is by far the most important mine of the Cement Creek district. It lies on the western slope of Bonita Peak, with the present main adit tunnel at about 12,500 feet elevation. Two lodes are worked, termed, respectively, the Davis and Gold King. Both have a general northeast course, and on the main level are about 70 feet apart. The Gold King lode, which is the northwesterly one, dips southeast at about 75° on the main level. The dip of the Davis is to the northwest, but nearly vertical. In the lower level, 107 feet below the adit, these lodes are closer together. Both are strong, typical stringer lodes without any gouge. So little have they been disturbed by post-mineral movements that there is practically no water in the mine. The ore is a network of irregular stringers, inclosing much mineralized andesitic country rock. Distinct walls are lacking, and the ore is taken out to a width determined by its value. This is sometimes over 30 feet.

The Gold King lode is particularly irregular, and its character led almost inevitably to the litigation between the Gold King and Sampson companies, which was wisely settled by the purchase of the latter by the former.

It shows a marked tendency to split, inclosing large, irregular horses of the country rock, and dividing locally into two or more apparently distinct lodes of varying dips and strikes. It thus, with our present mining laws in force, opened unending possibilities of contention, so long as the two mines were under separate ownership. One of the most remarkable instances of this tendency seems to be developing on the main Gold King level, about 400 feet west of the adit crosscut. Here a large, flat body of ore, continuous with the ore of the main Gold King lode, has been followed out for 30 or 40 feet to the northwest with a very gentle northwest dip. This ore apparently rests upon the top of a large horse, as in the northern part of the stope the dip of the ore is now found to be suddenly increasing. Possibly a still more striking case is the relation between the Gold King and Davis lodes. It seems very probable that these are not distinct, but merely branches of a very large, irregular stringer lead separated by a large horse of country rock, in which case they may be expected to come together in dip and strike.

At the point where it is reached by the adit crosscut (which runs nearly north) the Gold King lode is faulted by the so-called "Red vein," which has a northwesterly course and dips northeast about 75°. The horizontal displacement is about 30 feet, the northeast portion of the Gold King being thrown to the northwest. A supposed "bend" in the Gold King lode on the level below is almost certainly the same fault. The Red vein contains rhodonite (hence its name), which is not found in the Gold King and Davis lodes. It contains no ore of value, as far as known.

The characteristic ore of the Gold King and Davis lodes consists of white quartz with abundant pyrite, often showing banding by disposition. The pyrite, when massive and nearly free from quartz, is of little value in the Gold King lode, but carries gold in the Davis lode. The best ore always contains quartz, which may occasionally show some free gold. Galena, usually fine granular, occurs in bunches, being more abundant in some portions of the lodes than in others. As a whole, the ore is low grade, and probably no mine in the quadrangle can set as small nether value for workable ore.

Extensive preparations are being made for the future working of this property. The American tunnel cuts what is thought to be the same lode at a vertical distance of several hundred feet below the present adit; while a second tunnel, about 700 feet in length in September, 1900, is being driven from the mill at Gladstone. This will eventually tap the lode about 1,500 feet below the present workings.

The ore is carried down to the mill at Gladstone on a well-built Bleichert wire-rope tramway, running from 300 to 400 buckets of 700 or 800 pounds per day. A second tramway is being built to the American tunnel. The mill is equipped with a Gates crusher, 80 stamps, and 40 Frue vanners. Power is supplied by a 250-horsepower Westinghouse engine. The free gold is amalgamated, but the bulk of the product is in the concentrates, which are shipped to the smelter. The owners of the Gold King built and control the narrow-gauge railway from Silverton to Gladstone.

Tungsten ore deposits.—The Adams claim, which has for several years furnished cabinet specimens of the tungstate hübnerite, lies on the western slope of Bonita Peak at an elevation of about 11,300 feet. The hübnerite occurs in a lode which strikes N. 10° E. and dips east about 85°. In the main it is a sheeted zone 3 or 4 feet wide, in altered andesite of the Silverton series, the fissures being filled with quartz and fluorite. The individual stringers or veinlets of the lode are rarely more than 6 inches wide and are adherent or "frozen" to the walls. The hübnerite does not occur in all portions of the lode, but in isolated and irregular bunches, streaks, or nests of bronze-brown radial crystals embedded in quartz and fluorite. The development of this deposit is very superficial, consisting of a short tunnel and some open cuts, and the tungsten ore has never been worked on a commercial scale. The prospect was evidently opened in search of other ore, as the hübnerite has been thrown with the waste on the dump.

In Dry Gulch hübnerite occurs in a strong lode, on which have been located the Dawn of Day, Sunshine, and Minnesota claims. This lode has a course of about S. 60° E. and dips southwest at 65°. Several prospecting openings have been made on it on the south side of the gulch. One of these shows a solid vein of quartz 3 or 4 feet wide in Silverton andesite. The hübnerite occurs in the quartz in two or three small streaks, from 1 to 3 inches wide, and in small isolated bunches. The crystals are smaller than those in the Adams lode.

Owing to its weight (sp. gr., 7.2) and its occurrence in otherwise nearly barren quartz and fluorite, the hübnerite of Cement Creek can be readily and cleanly concentrated. Some 5 tons of concentrates have been produced from the Dry Gulch claims; but whether the material occurs in sufficient abundance and in sufficiently continuous bodies to pay for mining is a question not yet decided.

In its occurrence the hübnerite is purely a vein mineral, occurring in a quartz or quartz and fluorite gangue, as do galena, pyrite, and other ore minerals of the region.

Red and Bonita mine.—The adit tunnel of this mine runs in an easterly direction into Bonita Mountain, from a point about 100 feet above Cement Creek. About 3,000 feet of work has been done from this tunnel, but the ore could not be made to pay and the attempt was abandoned. The workings are no longer accessible and the lode was not seen. The Red and Bonita mill is equipped with Gates crusher, 2 sets of rolls, jigs, 10 stamps, and 4 Frue vanners.

Queen Anne mine.—This is near the mouth of Ross Basin, on a lode which here consists of two or more parallel veins close together. The one worked in the Queen Anne strikes N. 50° E. and dips northwest at 85°. It consists of solid banded quartz and bismuthiferous galena ore up to 2 feet wide, "frozen" to the walls. Toward the southwest the vein pinches and carries a little gouge. It could not be traced on the surface beyond a large northwest-southeast lode which passes up into Ross Basin.

Columbia mine.—This, one of the oldest mines of the quadrangle, lies just northeast of the Queen Anne and is apparently on the same lode. It has been an intermittent producer of argentiferous galena ore on a small scale.

Ross Basin lodes.—There is at present no mine in operation in Ross Basin, and none of the developments have thus far passed beyond the prospecting stage, but the basin is traversed by several lodes of unusual size with well-exposed croppings. The most prominent one has a nearly northwest course and dips southwest at 80°. It outcrops boldly along the northeast side of the basin, especially near the head of the latter, where it passes over the ridge into Sunnyside Basin and is apparently the same as the George Washington lode. Toward the northwest it continues probably into Grey Copper Gulch. Some ore, resembling in mineralogical character that found in Sunnyside Basin, but apparently in irregular bunches, occurs in this huge lode, which is often over 50 feet wide. Several other lodes with nearly northeast courses come into the main lode from the southwest, but can not as a rule be traced on the surface to the northeast. The lodes shown on the map are only a few of those which traverse the floor of the basin in various directions, although most of them have a northeast trend. They are not all of great length, for one vein often stops short at the junction with a second vein of different course. In one case two well-exposed, strong quartz veins, one striking southwest and the other S. 80° W., come together in a massive cropping of quartz, beyond which there is no surficial sign of the continuation of either lode. Distinct evidence of one lode faulting another could not be obtained. They are apparently all of substantially the same age. The fissure fillings are usually solid white quartz with a little rhodonite and bunches of low-grade ore. In some cases the lodes contain numerous fragments of altered mineralized country rock.

Henrietta mine.—This property, in Prospect Gulch, is on a lode striking S. 40° W. and dipping northwest at 65°. Three tunnels on the south side of the gulch give access to the workings, while a fourth tunnel, destined to be the future main adit, is being driven to the westward to intersect the lode from a point about 150 feet above the qottom of the gulch. All the workings are in andesitic rocks, chiefly breccias, of the Silverton series. The ore occurs in a brecciated zone which is usually from 3 to 5 feet wide. The hanging and foot walls are well defined by planes of recent slipping, that on the hanging wall being especially regular and continuous. These slip planes usually contain a little soft gouge. Between these walls the ore occurs in stringers and bunches, partly replacing the shattered country rock included within the breccia zone. It consists of pyrite and chalcopyrite and occasionally a little galena. Pyrite is very abundant, occurring in solid masses, which, however, are not mined unless they contain chalcopyrite. Next the hanging wall there is a persistent seam, sometimes a foot in width, which consists of crushed country rock with sometimes a little white quartz. This portion of the lode is worthless, although a little sphalerite and tetrahedrite are said to occur in the quartz. No quartz was seen with the pay ore. The lode is not everywhere productive. There are frequent barren intervals, where the space between the walls is filled with broken and altered country rock impregnated with pyrite. The wall rock is impregnated with pyrite for at least 150 feet from the lode, but no ore is mined outside of the walls of the fissure.

The ore is low grade as a whole, containing from 0. 05 to 0. 1 ounce of gold and 12 to 30 ounces of silver per ton, with up to 30 per cent of copper. In 1900 the mine was reopened after some years of idleness, and it was proposed to treat the ore in a new smelter then nearing completion at the point where Cement Creek enters Bakers Park.

The lowest tunnel, which in September, 1900, had a length of about 550 feet and had not yet cut the lode, exhibits well the rather remarkable mineralization of the Silverton breccias in this vicinity. All the rock passed through is heavily impregnated with fine pyrite and frequently contains small sporadic bunches and stringers of galena and sphalerite. In some places the andesitic breccia has been locally silicified.

The Henrietta lode is of theoretic interest from the fact that the ore, in its freedom from quartz, presence of kaolin, and partial replacement of country rock, as well as in mineralogical character, is similar to the ores occurring in the stock deposits of Red Mountain. At the Henrietta, however, it is found, not in stock, but in a distinct fissure.

Other mines and prospects.—From Prospect Gulch southward to Silverton are many prospects along Cement Creek and in the side gulches. None of them are yet of much importance and not all were visited. The most ambitious of these is the Yukon tunnel of the Boston and Silverton Mining Company. No ore has yet been produced, but the tunnel is supplied with a new mill equipped with Blake crusher, rolls, jigs, 10 stamps, 2 Wilfley tables, and 4 Frue vanners. The power is steam.

INDEX.

O

PUBLICATIONS OF UNITED STATES GEOLOGICAL SURVEY.

[Bulletin No. 182.]

The serial publications of the United States Geological Survey consist of (1) Annual Reports, (2) Monographs, (3) Bulletins, (4) Mineral Resources, (5) Water-supply and Irrigation Papers, (6) Topographic Atlas of United States—fol os and separate sheets thereof, (7) Geologic Atlas of United States—folios thereof. A circular giving complete lists may be had on application.

The Bulletins treat of a variety of subjects, and the total number issued is large. They have therefore been classified into the following series: A, Economic geology; B, Descriptive geology; C, Systematic geology and paleontology; D, Petrography and mineralogy; E, Chemistry and physics; F, Geography; G, Miscellaneous. This bulletin is the twelfth in Series A, the complete list of which fo lows:

BULLETINS, SERIES A, ECONOMIC GEOLOGY.

21. Lignites of Great Sioux Reservation: Report on region between Grand and Moreau rivers, Dakota, by Bailey Willis 1885. 16 pp., 5 pls. Price, 5 cents.

46. Nature and origin of deposits of phosphate of lime, by R. A. F. Penrose, jr., with introduction by N. S. Shaler. 1888. 143 pp. Price, 15 cents.

65. Stratigraphy of the bituminous coal field of Pennsylyania, Ohio, and West Virginia, by Israel C. White. 1891. 212 pp., 11 pls. Price, 20 cents. (Exhausted.)

111. Geology of Big Stone Gap coal field of Virginia and Kentucky, by Marius R. Campbell. 1893. 106 pp., 6 pls. Price, 15 cents.

132. Disseminated lead ores of southeastern Missouri, by Arthur Winslow. 1896. 31 pp. Price, 5 cents.

138. Artesian-well prospects in Atlantic Coastal Plain region, by N. H. Darton. 1896. 228 pp., 19 pls. Price, 20 cents.

139. Geology of Castle Mountain mining district, Montana, by W. H. Weed and L. V. Pirsson. 1896. 164 pp., 17 pls. Price, 15 cents.

143. Bibliography of clays and the ceramic arts, by John C. Branner. 1896. 114 pp. Price, 15 cents.

164. Reconnaissance in the Rio Grande coal fields of Texas, by Thomas Wayland Vaughan, including a report on igneous rocks from the San Carlos coal field, by E. C. E. Lord. 1900. 100 pp., 11 pls. and maps. Price, 20 cents.

178. El Paso tin deposits, by Walter Harvey Weed. 1901. 15 pp., 1 pl. Price, 5 cents.

180. Occurrence and distribution of corundum in United States, by J. H. Pratt. 1901. 98 pp., 14 pls. Price, 20 cents.

182. Report on economic geology of Silverton quadrangle, Colorado, by F. L. Ransome. 1901. 265 pp., 16 pls. and maps. Price, 50 cents.

I

Other Publications

by

Miningbooks.com

- Placer Gold Deposits of Nevada
- Placer Gold Deposits of Utah
- Placer Gold Deposits of Arizona
- Gold Placers of California
- Browns Assaying
- Arizona Gold Placers and Placering
- Arizona Lode Gold Mines and Gold Mining
- Dredging for Gold in California
- Metallurgy
- Gold Deposits of Georgia
- Placer Examination: Principles and Practice
- Geology and Ore Deposits of the Creede District, Colorado
- Gold in Washington
- Placer Mining in Nevada
- Gold Placers and their Geologic Environment in Northwestern Park County, CO
- Placer Mining for Gold in California
- Geology and Ore Deposits of Shoshone County, Idaho
- Gold Districts of California
- Gold and Silver in Oregon
- The Porcupine Gold Placer District Alaska
- Gold Placer Deposits of the Pioneer District Montana
- Economic Geology of the Silverton Quadrangle, Colorado
- The Ore Deposits of New Mexico

- Roasting of Gold and Silver Ores, and the Extraction of their Respective Metals without Quicksilver

- Geology and Ore Deposits of the Summitville District San Juan Mountains Colorado

www.ingramcontent.com/pod-product-compliance
Lightning Source LLC
Chambersburg PA
CBHW031918190326
41519CB00007B/341